# ptoelectronics, Imaging and Sensing Series

eries editors

Dr. A. T. Augousti, Kingston University, UK
Professor K. T. V. Grattan, City University, UK
Professor G. Parry, Imperial College, London University, UK

**Optoelectronics, Imaging and Sensing** brings together the best elements of two successful series, the **Optical and Quantum Electronics Series** and the **Sensor Physics and Technology Series**. The new series will focus on exciting new developments and applications in the rapidly changing areas of optoelectronic sensing and imaging technology. The volumes cover both systems and devices, and are aimed at scientists involved in research and development as well as practising engineers. Advanced undergraduate and graduate textbooks are also included, giving tutorial introductions that are essential for those wishing to work in the challenging and multidisciplinary areas of optoelectronics, imaging and sensing. This up-to-date series will include electronic products as well as conventional books, meeting the needs of its users effectively in the most modern formats.

**Titles Available:**

**Electromagnetic Waves**
C. G. Someda

**Optical Fiber Sensor Technology Volume 2**
**Devices and Technology**
Edited by K. T. V. Grattan and
B. T. Meggitt

**Optical Fiber Sensor Technology Volume 4**
**Chemical and Environmental Sensing**
Edited by K. T. V. Grattan and
B. T. Meggitt

**Optical and Quantum Electronics Series**

1 **Optical Fiber Sensor Technology**
Edited by K. T. V. Grattan and B. T. Meggitt

2 **Vision Assistant Software**
A practical introduction to image processing and pattern classifiers
C. R. Allen and N. C. Yung

3 **Silica-based Buried Channel Waveguides and Devices**
F. Ladouceur and J. D. Love

4 **Essentials of Optoelectronics**
With applications
A. Rogers

**Sensor Physics and Technology Series**

1 **Biosensors**
T. M. Cahn

2 **Fiber Optic Fluorescence Thermometry**
K. T. V. Grattan and Z. Y. Zhang

3 **Silicon Sensors and Circuits**
F. Wolffenbuttel

4 **Ultrasonic Measurements and Technologies**
S. Kocis and Z. Figura

5 **Data Acquisition for Sensor Systems**
H. R. Taylor

# OPTICAL FIBER SENSOR TECHNOLOGY

Volume 3

# Optical Fiber Sensor Technology

## Volume 3

## Applications and Systems

Edited by

**K. T. V. Grattan**

**and**

**B. T. Meggitt**

SPRINGER-SCIENCE+BUSINESS MEDIA, B.V.

Library of Congress Cataloging-in-Publication Data

ISBN 978-1-4419-4736-9 ISBN 978-1-4757-6077-4 (eBook)
DOI 10.1007/978-1-4757-6077-4

---

*Printed on acid-free paper*

# Contents

# Contributors

**Professor K. T. V. Grattan**
Department of Electrical, Electronic & Information Engineering
City University
Northampton Square
London EC1V 0HB

**Professor B. T. Meggitt**
Department of Electrical, Electronic & Information Engineering
City University
Northampton Square
London EC1V 0HB

**Drs. C. I. Merzbacher and E. J. Friebele**
Optical Science Division
4555 Overlook Avenue
Naval Research Laboratory
Washington DC 20375
USA

**Dr. P. A. Crosby**
Department of Materials Engineering
Brunel University
Uxbridge
Middlesex UB8 3PH

**Professor A. J. Rogers**
Department of Electronic Engineering
King's College
Strand
London WC2R 2LS

**Dr. S. Chen**
Department of Mechanical Engineering
University of Maryland at College Park
Maryland, USA

**Drs. Z. Y. Zhang and B. M. A. Rahman**
Department of Electrical, Electronic & Information Engineering
City University
Northampton Square
London EC1V 0HB

**Dr. J. W. Berthold**
Research and Development Division
Babcock and Wilcox
1562 Beeson Street
Alliance, OH 44601
USA

**Dr. A. D. Kersey**
CIDRA Corporation
Dow–United Technologies Building
15 Sterling Drive
Wallingford CT 06492
USA

**Drs. G. Fernando and T. Liu**
Department of Engineering Systems
Royal Military College of Science
Cranfield University
Scrivenham
Swindon SN6 8LA

# Preface

**Systems and Applications in Optical Fiber Sensor Technology**

The essential technology which underpins developments in optical fiber sensors continues to expand, and continues to be driven to a very large extent by advances in optoelectronics which have been produced for the ever-expanding optical communications systems and networks of the world. The steps forward in the technology, often accompanied by a reduction in the price of associated components, have been, and continue to be, adapted for use in a wide variety of optical fiber sensor systems. These include, for example, the use of photoinduced gratings as fiber sensor components, coupled with the wider availability of shorter wavelength lasers, bright luminescent sources and high-sensitivity detectors which have opened up new possibilities for both novel fiber optic sensor applications and new sensing systems. This is to be welcomed at a time when, coupled with integrated optic miniaturized devices and detectors, real possibilities of systems integration, at lower cost and increased utility, can be offered. The fiber laser, and the expansions of the types and availability of the doped fiber on which it is based, offer further examples of the integration of the essential components of advanced optical sensor systems, fitted for a new range of applications. The underpinning science and engineering aspects remain the same, but the opportunities to exploit fiber optic sensors more widely can be seen to be increasing, coupled, where appropriate, with the intelligence and signal processing capabilities of the explosive growth in computer power which dominates this information age.

*Optical Fiber Sensor Technology 3 – Applications and Systems* is the third volume in the series, as a complementary text to the first, *Optical Fiber Sensor Technology*, which focused on fundamentals and principles, and the second, *Optical Fiber Sensor Technology 2 – Devices and Technology*, which contained a detailed discussion of the range of techniques and devices which have appeared in the field over recent years. A companion volume, *Optical Fiber Sensor Technology 4 – Chemical and Environmental Sensing*, adds to the comprehensive coverage of optical fiber sensor topics in the series. In total, in over 1500 pages of carefully illustrated text, this series bring together some 50 chapters written by a wide range of internationally known and recognized experts in the field from the UK, the USA, mainland Europe and the Far East, with many thousands of topical references to the essential international literature in the field. The series draws its authority from the professionalism and skills of the authors, and through careful editing blends together these authoritative contributions into a coherent series of subject discussions, clustered in terms of their underlying themes, and written as a unified and integrated series over a period of a few years to ensure the best topical coverage of the subject.

The field of photoinduced gratings in optical fibers is one which has seen explosive growth in recent years and which features strongly in the series. The second volume included three chapters on the principles, fabrication, and broad sensor and signal processing aspects of such optical fiber Bragg gratings. This volume takes the subject further, and opens with four chapters on the applications of such gratings in 'smart structure' monitoring. The first of these is by Celia Merzbacher and her colleagues at NRL Washington, a center of optical fiber sensor research since the 1970s. This chapter is on the use of fiber sensors, especially grating-based devices, in concrete structures, reflecting a range of applications of interest to the civil engineering community in this field. Gerald Fernando and his colleagues at Brunel University have produced an extensive study of the use of optical fiber sensors in a range of composite materials, discussing in some detail the underlying problems and issues involved in the use and implementation of such sensors in these varied and complex materials. A series of applications and cross-comparisons of their performance characteristics is given in these linked chapters, together with a valuable discussion of the underlying nature of the materials themselves and their compatibility with fibers installed for structural measurement and monitoring.

Modelling techniques and the potential for use in modern modelling methods are rarely considered in the context of optical fiber sensors. However, Aziz Rahman and Kenneth Grattan discuss this topic, and then give a detailed review of the implementation and potential of modern numerical methods in the design and use of a range of optical fiber sensors and associated components. In a further chapter, Alan Rogers revisits the topic of distributed optical fiber sensors, on which he has had a pioneering influence, in his review of polarimetric distributed sensors.

Optical fiber current measurement has had a long and somewhat troubled history, with early promising sensors falling foul of interference from environmental factors, and Kenneth Grattan and Ya Nong Ning discuss both the developmental background and the present technology in a comprehensive review of the field. The wide range of applied optical systems coverage continues with John Berthold's review of the field of microbend fiber optic sensors. This is a familiar, widely exploited, yet often ignored early sensor concept which has had a considerable variety of applications.

To complement earlier chapters in previous volumes on white light, and on low coherence interferometry, Stephen Chen and Beverley Meggitt have provided a short chapter on the development and implementation of an intrinsic fiber position sensor technique using white light interferometry in an innovative way.

The marketplace for optical fiber sensors, including commercial developments and a cross-comparison of available devices, is an important part of the study of the subject. This often takes its lead more from developments in the protection of valuable intellectual property through patents rather than through work reported in the scientific journals. Edward Zhang and Kenneth Grattan have produced a carefully illustrated chapter discussing recent work in the commercial patent field for the strategically important US market, showing the trends in patent protection, and concentrating on activity in the last few years.

Overall, the volume forms an integral part of the *Optical Fiber Sensor Technology* series, which can justifiably claim to be the most comprehensive and wide-ranging in the field, drawing its strength from careful editing of the most topical material from a wide range of well-known and highly respected international authors. The contribution that it makes in terms of sensor applications and systems, linking with the review of devices and technology in the second volume, and built upon the formation of principles and theory in the first, reveals the dynamic nature of the subject in a well-referenced and illustrated form, carrying the authority of its international base of contributors. Developments will undoubtedly continue, but the essential knowledge of the subject contained in these volumes will have lasting value as a series of reference texts on the exciting subject of optical fiber sensor technology. As editors, we hope you will both enjoy, and benefit in your research and scholarship from the work in this volume and this series overall.

1

# Fiber optic sensors in concrete structures: a review

C. I. Merzbacher, A. D. Kersey and E. J. Friebele

## 1.1 INTRODUCTION

The overall deterioration of the national civil infrastructure due to aging and usage beyond the anticipated loads and lifetimes for which it was designed, combined with the increasing cost of maintenance and repair, has resulted in the need for improved techniques for non-destructive evaluation of the structural health of reinforced concrete. A recent review of the available statistics reveals that almost 40% of United States bridges are 'structurally deficient' or 'functionally obsolete' [1]. New reinforced concrete constructions would also benefit from *in situ* structural monitors which could detect a decrease in performance or imminent failure, thereby optimizing lifetimes without compromising safety. Finally, although modeling the behavior of some structures made from well-characterized materials is fairly accurate, the use of new materials, unusually complex designs, or variability in strength-related factors such as void fraction or moisture content can lead to unexpected structural weakening, damage or failure. The inadequacy of the nation's highways, bridges, etc. prompted the initiation in 1993 of a National Science Foundation program, with the goal of developing new technologies aimed at 'prolonging the life and enhancing the capacity of our existing and future civil infrastructure systems' [2]. In response to the increased need, various techniques are being developed, and some of the most promising are based on the use of fiber optic sensors (FOS).

Fiber optic sensors have been the subject of considerable research for the past 20 years, with initial applications focused on military and aerospace uses. The use of fiber optic sensors in concrete was first suggested by Méndez *et al.* [3]. Subsequently, several research groups in North America and Europe have reported on a variety of fiber optic sensors embedded in, or mounted on, reinforced concrete structures. While early work focused on laboratory demonstrations of the ability to detect strain, deformation, vibration or corrosion, more recently a number of 'real-life' structures have been constructed with embedded fiber optic sensors and the associated detection instrumentation. These field projects are critical to a

*Optical Fiber Sensor Technology*, Vol. 3. Edited by K. T. V. Grattan and B. T. Meggitt.
Published in 1998 by Kluwer Academic Publishers, London. ISBN 978-1-4419-4736-9

successful transition from the technology development stage to general acceptance and use in the construction industry.

Although fiber optic sensors are ideally suited for high-resolution monitoring of the structural health of reinforced concrete, some logistical challenges to their routine installation during the construction process must be addressed. Protection of the fiber leads at the point where they exit from the concrete is critical to their survival. Detailed planning of sensor locations is required, both within the structure, in order to measure the desired properties, and at the point of exit, so that access points are convenient after construction is complete. Adequate consultation and communication with the builder in charge of the job site is necessary to ensure coordination between the workers laying the fibers and those pouring the concrete. Finally, appropriate measures must be taken to avoid damage to the fibers from the harsh environment that exists on most construction sites.

This is a review of the work that has been done to date on using fiber optic sensors to detect strain in concrete. We review all of the published work up to the present, but the reader may find two early reviews of this field by Méndez and Morse [4] and Measures [5] particularly useful.

## 1.2 FIBER OPTIC SENSORS IN CONCRETE

An optical fiber consists of a glass core surrounded by a glass cladding, with a lower index of refraction, and an outer polymer coating, which is applied during the fiber drawing process. The polymer coating protects the glass fiber surface from being flawed by abrasion, which, especially in the presence of moisture, weakens the fiber and accelerates crack growth. In concrete, the correct polymer can also protect the glass fiber from the extremely alkaline (pH 12) environment which is corrosive to standard glass fiber. In addition, the polymer buffers the fiber against small-radius bends due to the roughness of the surrounding medium. This is desirable in telecommunications, where microbending can lead to transmission losses. For strain sensing, however, the lower elastic modulus of the polymer coating affects the transfer of strain from the concrete matrix to the fiber. This effect can be accounted for in the data analysis, or it can be minimized by keeping coatings as thin as possible or using relatively stiff polymers.

Fiber cables can have an additional plastic jacket outside the thinner coating for further protection. However, pullout tests have shown bonding between the jacket and the cement to be extremely poor [6]. On the other hand, similar tests indicate that bonding between cement and standard coated fiber is sufficiently strong for good strain transfer [7].

There are many different types of fiber optic sensors (see ref. [8] for a complete review). Fiber optic sensors may be categorized according to various classification schemes. For example, they are classified as **intrinsic** if the effect of the measurand on the light being transmitted takes place in the fiber. The sensor is considered **extrinsic** if the fiber carries light from the source and to the detector, but the modulation occurs outside the fiber. Both types have been used in concrete sensing applications. Sensors for detecting chloride ions in concrete tend to be extrinsic, whereas those for measuring strain are typically intrinsic.

Fiber sensors can also be divided according to whether sensing is distributed, localized (point) or multiplexed. If sensing is distributed along the entire length of the fiber, extraneous signals can be generated in the leads, and the location of any variation in the measurand along the length of the fiber must be determined by additional probing, e.g. optical time domain reflectometry (OTDR). On the other hand, a single fiber can monitor the entire object into which it is embedded very efficiently. Localized or point sensors, as the name implies, detect measurand variation only in the vicinity of the sensor. This localization aids in determining where strain, temperature, pressure, etc. is changing, but if many individual sensors are required in order to map a large structure, it can lead to a large number of fiber leads and connections. This is a strong disadvantage at a construction site, where minimizing the number of fiber connections outside the structure is desirable in order to reduce the possibility of inadvertent damage. Some localized sensors lend themselves to multiplexing, in which multiple localized sensors are placed at intervals along the fiber length. Each sensor can be isolated by wavelength, time, or frequency discrimination, thereby allowing parameters throughout a structure to be profiled in real time, with a minimum number of fiber connections.

Finally, fiber optic strain sensors may be divided into two broad categories – those that are intensity-based and those that are not. Intensiometric sensors are simply based on the amount of light detected through the fiber, and are generally made from multimode fiber. In its simplest form, damage may be detected by a fiber breaking and transmission going from on to off. OTDR can also be used to locate where the fiber broke. A disadvantage of intensiometric sensors is their sensitivity to fluctuations in source intensity, bend loss, or detector sensitivity.

Sensors that are not intensiometric are mostly based on **interferometers** or **fiber Bragg gratings**. Among interferometric sensors, systems using Mach–Zehnder or Michelson interferometers have relatively long gauge lengths, making them particularly appropriate for measuring strain over lengths of the order of meters. A low-coherence, double Michelson interferometer system has been developed by Inaudi and co-workers especially for embedding in concrete structures [9, 10]. Several variations of Fabry–Perot interferometer (FPI) sensors have been designed, both intrinsic and extrinsic in nature. Intrinsic FPIs are formed by two partially reflective interfaces in the fiber, with the gauge length defined by the distance between the two reflectors. Extrinsic FPI (EFPI) sensors are formed by two fiber ends inserted into a tube with an inside diameter slightly greater than the fiber diameter. The gauge length defined by the points at which the two fibers are anchored can be relatively short, but the physical discontinuity at the sensor can lead to stress concentration when embedded in structural material. A drawback to EFPI sensors is that they must be calibrated and, in some designs, they measure only relative strain and therefore cannot tolerate power interruptions. In-line fiber etalons (ILFE) combine the advantages of both intrinsic and extrinsic Fabry–Perot sensors. A cavity is formed between two fiber end-faces by splicing them to a short length of hollow-core fiber of the same outer diameter as the fiber [11]. Unlike the EFPIs, this sensor has a smooth profile, well-defined gauge length and relatively high strength. Another advantage of the ILFE and EFPI sensors is their low thermal sensitivity. Due to the fact that the cavity is in air, thermal apparent strain arises only from the low thermal expansion ($10^{-7}$ per °C) of the hollow tube. A disadvantage of the FPI sensors is that it is not possible to multiplex large numbers on one fiber.

Fiber Bragg gratings (FBG) are optically induced modulations in the index of refraction of the optical fiber core which reflect light at a specific wavelength determined by the effective index of refraction and the period of the modulation [12]. The wavelength shift of the reflected light can be very precisely measured and correlates directly with strain. FBG sensors have several advantages over other sensing schemes, especially in structural applications. The foremost advantage is that the sensed information is encoded directly into wavelength, which is an absolute parameter. Therefore, the output does not depend directly on the total light levels, losses in the connecting fibers and couplers, or recalibration or re-initialization of the system. In addition, wavelength division multiplexing is readily achieved by fabricating each grating with a different period, so that it reflects at a different frequency within the broadband source spectrum.

The ideal FOS for strain measurement in civil structures would be:

1. stable;
2. localized;
3. of adequate sensitivity and dynamic range;
4. linear in its response;
5. sensitive to the direction of measurand field change;
6. single-ended, to minimize the number of leads;
7. insensitive to thermal fluctuations;
8. capable of absolute measurement;
9. nonperturbative to the structure;
10. immune to power interruption;
11. able to multiplex;
12. easy to mass produce;
13. durable for the lifetime of the structure.

No one fiber sensor fulfills all of the above requirements at this time, and therefore the type of sensor employed will depend on the parameter to be measured, the sensitivity and dynamic range required, etc.

No matter what type of fiber optic sensor is selected, the embedding of optical fibers in concrete requires both chemical and mechanical durability. The fibers must withstand the highly alkaline chemical environment as well as the physical abuse associated with placing fiber into concrete or pouring the concrete mix over the fiber. In early studies, sensors were often not localized, and the pieces of fiber in contact with the concrete were long. In these systems, the polymer coating was the only protection between the glass fiber and the surrounding environment, and

**Table 1.1** Properties of fiber optic polymer coatings

| | Polyimide | Acrylate | Silicone |
|---|---|---|---|
| Elastic modulus (MPa) | 2400 | 700 | 2 |
| Tensile strength (MPa) | 130 | 26 | 5.5 |
| Static fatigue resistant | Yes | No | No |
| Chemical durability* | Poor | Good | ? |

*Based on results from Habel and Polster [14].

it therefore had to be strong and chemically impervious, ideally with an elastic modulus that closely matched that of the glass fiber. Escobar *et al.* [13] compared two types of coating, a soft acrylate and a teftzel-silicon. The silicon-coated fiber displayed slightly higher sensitivity to strain when embedded in a concrete test article, and was somewhat more resistant to damage during handling. They concluded that the teftzel-silicon transferred strain better than the soft acrylate.

The properties of several common fiber coatings are shown in Table 1.1. The chemical durability of three polymer coatings has been studied by embedding fibers in concrete (pH 12.4–14) for 28 days, as well as by soaking coated fibers in alkaline solutions (pH 7, 11, 12.4 and 14) for periods of from 1 day to 5 months [14–16]. This is the only study in which coatings have been carefully examined **after** being embedded in, and then removed from, concrete. The results indicate that polyimide was strongly affected by a basic environment – the coating softened and deformed in both the solution and the concrete exposure tests. Acrylate was only slightly affected by the solution test, but bonding between the concrete and the fiber coating was affected in some of the test samples, and microcracks appeared in the sample embedded in concrete for 28 days. Although relatively durable, fluorine-thermoplastic showed evidence of softening after embedment in concrete for 15 days.

In recent applications, direct contact between the fiber and the concrete has been eliminated. Coated fibers are typically epoxied to reinforcing bars or to metal anchors, which transfer strain from the concrete matrix to the fiber. The sensor can then be overcoated with a protective layer of epoxy or silicone, and the fiber leads, which simply carry the signal to the detector, can be inside a plastic jacket. In these systems, it is important to characterize the entire system, not just the fiber optic, under the anticipated range of environmental conditions. Sennhauser *et al.* [17] evaluated several high-temperature glues and epoxy adhesives for use in a steam pipe application. Only one of the compounds survived the initial testing at 120°C and 2 bars pressure. This adhesive, however, failed another test when a fiber attached to a pre-strained steel bar slipped when heated to 100°C. In another test, pre-strained fibers coated with polyimide or acrylate were attached to carbon fiber rods, similar to those to be installed in a Swiss bridge, and held at 85°C and 85% relative humidity. The tension in the optical fibers decreased to 60% after 24 h and went to zero after 120 h. Furthermore, the coatings also suffered damage, particularly the acrylate, which split and exposed bare fiber.

Physical, as well as chemical, durability is critical to the successful application of fiber sensors in concrete structures. Inaudi *et al.* [18] performed a basic four-point bending fatigue test on fiber optic sensors embedded in a concrete slab. After 17 million cycles at 4 Hz, one of the reinforcement bars broke, but all of the fibers survived and exhibited no detectable creep.

Optical fibers have been shown to be capable of withstanding the physical rigors of being handled on a construction site. A group at the University of Vermont has been very active in installing fiber optic sensors in full-scale civil structures, including pedestrian, highway and railway bridges, a dam and a five-storey building [19]. Their experience has led to the development of several techniques to optimize the survival rate for installed fibers. They attached the fibers to the bottom side of the steel reinforcing bars to protect them during the concrete pour. In order to protect the leads at the point where they exit from the concrete, they fed the ends into a standard electrical box and attached the box to the inside of the forms. This becomes a fiber 'outlet' when the forms are removed after curing [20]. Researchers have repeatedly emphasized the unexpected hostility (with respect to fibers) of the job site and the need to protect exiting fibers from being driven over, soaked in water, having things tied to them or dropped on them, etc. [18, 21]. According to the University of Vermont researchers, the primary cause of failure when embedding fiber sensors in concrete structures is burying the entire fiber during the concrete pour [21]. If a fiber survives the embedding process, its prospects for long-term survival are good.

Because the study of embedded optical fibers in concrete is relatively recent, the effect of aging has yet to be determined. Escobar *et al.* [13] reported that they observed no change after 5 years in the strain-sensing capability of an interferometric sensor in a concrete test article in the laboratory. However, the lifetimes of civil structures such as bridges and dams are often 50 years or more. The uncertain effects of aging add to industry concerns, which can only be overcome by observing the performance of fiber optic sensors in the field.

## 1.3 PREVIOUS WORK

### 1.3.1 Laboratory studies

A number of groups have performed laboratory studies that demonstrate the ability of fiber optic sensors embedded in concrete to detect strain [6, 7, 13, 16, 22–34]. Hendrick *et al.* [22] also epoxied polarmetric fiber optic sensors in polyurethane samples with a range in elastic moduli (2000 to 50 000 p.s.i.) that simulated typical soils. An example of the output from a fiber Bragg grating sensor as a function of applied strain is shown in Fig. 1.1. In general, the fiber sensor was the only element embedded in the concrete, with fiber leads connecting to the light source and detection system. In one study, however, an integrated package, including a light-emitting diode (LED) source and a photodetector, was embedded with only the electrical outputs exiting from the concrete sample [23].

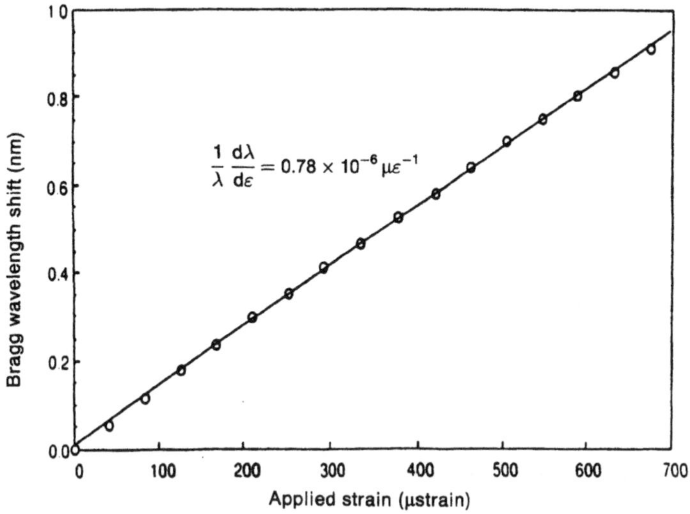


**Fig. 1.1** Wavenumber reflected from fiber Bragg grating vs. applied strain.

One of the larger laboratory tests of FOS in concrete was a full-scale single-lane bridge span outfitted with 48 FBG sensors [35–37]. Separate detection systems were used to measure static and dynamic strain information from the same sensor array. Strains associated with the concrete pouring and curing processes were continuously monitored for 38 days, and clearly showed initial thermal expansion, followed by shrinkage and finally an increase in strain upon the removal of the forms. Dynamic strain measurements made before and after the bridge was damaged by cutting one of the girders, as well as after the damage was repaired, detected distinct changes in the modal behavior of the structure.

Several studies have compared fiber optic strain sensors with traditional foil, or resistance, strain gauges [16, 24, 25, 27, 28, 30, 31, 33, 38–40]. When the two sensors are co-located, the results are in good agreement (Fig. 1.2); in fact, the fiber sensor can have a significantly larger signal-to-noise ratio (Fig. 1.3) due to its inherent immunity to electromagnetic interference.

Nanni *et al.* [7] considered the effect of the polymer coating on the strain transfer from the concrete to the glass fiber. They assumed typical values for the physical properties of the glass and polymer (elastic modulus of the polymer, $E_p = 1/200$ that of glass; shear modulus of the polymer $= 0.4E_p$), with a glass diameter of 120 microns and a coating thickness of 60 microns, and a 'perfect bond' between the concrete and the fiber. For fibers parallel to the applied load, a gauge length of 3 mm is required to achieve strain in the fiber core equal to that in the matrix. The strain in the fiber when perpendicular to the applied load is about 100 times less than in the matrix, irrespective of the probe length.

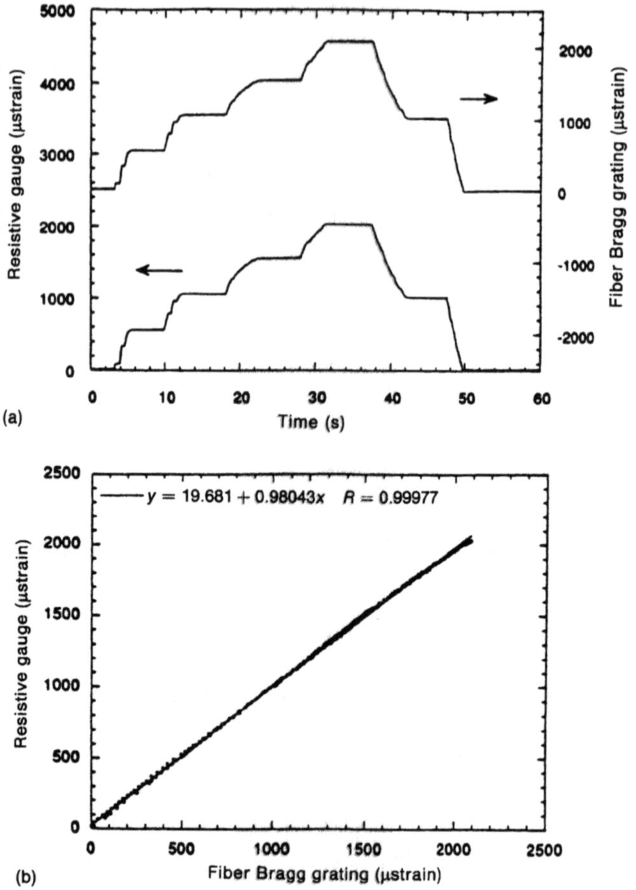


**Fig. 1.2** (a) Strain response of FBG sensor and resistive strain gauge. (b) Correlation between FBG sensor and resistive strain gauge results (reproduced with permission from [39]).

Polarimetric, interferometric and Bragg grating sensors have all been used in the laboratory studies mentioned above. In general, the leads of the fibers in and out of the concrete were found to be especially vulnerable, and were protected by plastic or metal tubing. When reinforcing bars were also emplaced, fiber sensors were epoxied to the bar after filing a flat side and, in some cases, machining a groove into which the fiber was glued (e.g. ref. [27]). For added protection, Davis *et al.* [35] put an overcoat of epoxy over the sensor, whereas Masri *et al.* [30] put a layer of waterproofing silicone gel.

Prohaska *et al.* [24] and Guerin *et al.* [31] epoxied the fiber optic sensor in a stainless steel tube for added protection prior to embedding in concrete. In these cases, the strain detected by the sensor must be corrected for the relatively

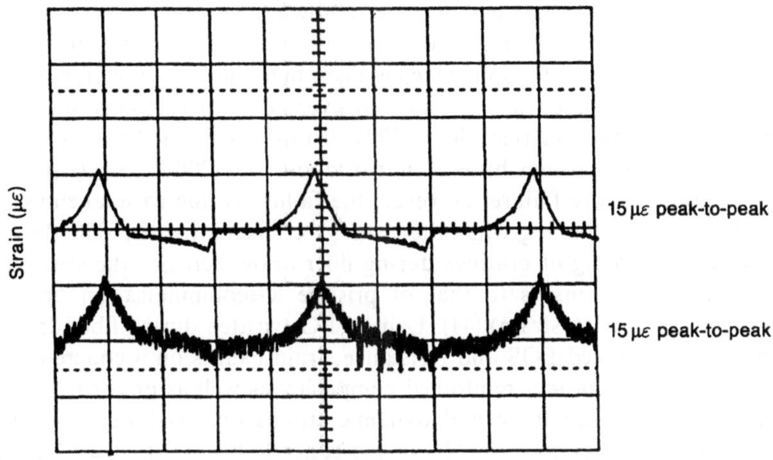


**Fig. 1.3** Comparison of frequency response of extrinsic Fabry–Perot interferometer (EFPI) (top) and foil strain gauge (bottom), both surface-mounted on a vibrating test article (reproduced with permission from [28]).

low modulus of the epoxy in order to get an accurate measure of the absolute values of strain in the concrete.

Fiber optic strain sensors can be extremely sensitive, with detection limits as low as one part in $10^7$, or 0.1 microstrain, for static strain, and <1 nanostrain/$\sqrt{Hz}$ for dynamic strain. However, the strains that occur in concrete structures can be several thousands of microstrains, and therefore dynamic range is a more important issue. Interferometric sensors which are based on an air cavity, such as EFPI sensors, can be mounted so that the two fiber ends which define the cavity are free to move and are bonded to the concrete matrix at some distance from the sensor head. Although this type of sensor can withstand virtually unlimited strain, the maximum strain that can be detected is limited by the decreased coupling efficiency between the fibers, causing decreasing light output with increasing cavity size.

Strains as large as several percent also occur in reinforcement bars when they are pre-stressed. By placing the reinforcing member under tension before concrete is poured and cured, the resulting structure is under compressive stress when the tension is released. This adds strength to the structure, since concrete is much stronger in compression than in tension. It would be of interest to know the strains at the interface between the concrete and the reinforcing member; however, most fibers cannot withstand the extension of several percent during pre-stressing. De Vries *et al.* [33] demonstrated the ability of absolute EFPI sensors attached to pre-stressing tendons (not embedded in concrete) to withstand, and measure, over 10 000 microstrains in tension. They noted that at the maximum strain levels, several of the co-located resistance strain gauges suffered permanent damage.

Fiber Bragg grating sensors have also been shown to be robust at relatively high strains. Alavie *et al.* [32] attached a FBG sensor as well as a resistance strain gauge to a polymer/carbon fiber tendon, which is a new material being used for concrete reinforcement. The sensor was strained to over 8000 microstrains and cycled between 0 and −2000 microstrains >320 000 times without detriment to the sensor. Their Bragg gratings have a limit of about 100–200 k.p.s.i. (equivalent to 1–2% strain) before failure; however, the failure is due to weakening that occurs during fiber handling when the gratings are being written. The development of on-line writing of gratings during fiber draw increases the strength of fiber Bragg grating sensors to that of pristine telecommunication fiber, i.e. about 700 k.p.s.i. (~7% strain) [41]. Using FBGs written during fiber drawing, Nellen *et al.* [42] placed FOS and resistance strain gauges on a concrete beam laminated with carbon-fiber-reinforced composite, as well as on a carbon-fiber pre-stressing cable. The static and dynamic strains observed were similar to those recorded by Alavie *et al.* [32]. In the concrete beam tested by Nellen *et al.* [42], FBGs were installed both as local strain sensors and as longer gauge (0.1–1 m) extensiometers. Their data show that inhomogeneities in the concrete larger than one-fifth of the sensor gauge length cause broadening of the peak reflected from the localized sensor upon loading. However, the signal from the extensiometric grating sensor was not affected due to its longer length.

Nellen *et al.* [43] have also developed a sensor package within a 2-m-long steel rod for embedding in large structures such as dams. The rod contains three 0.5-m gauge length FBG sensors, each of which consists of a ~1-cm-long grating attached to the strained steel rod and a grating attached to an unstrained 'dummy' piece of steel. The latter undergoes only thermal strain and allows for temperature compensation. Resistance strain gauges were also co-located near the sensing FBGs in the rod. After embedding the rod in a concrete test article, strain transfer to the internal sensors was verified by comparing the FBGs and resistance strain gauge outputs to deformation measurements made externally; all three readings agreed.

In two studies, fiber sensors have been embedded in carbon-fiber-reinforced epoxy wrapped around concrete cylinders [44, 45]. Wrapping concrete columns with composite adds considerable strength, and is a method for retrofitting structural supports in seismically active areas. Lou *et al.* [44] used an OTDR-type FOS in which the sensitivity of the sensor is enhanced by overlapping several loops of fiber in a patch that can be surface-mounted or embedded between layers in a composite. Strain in the composite layer was measured during flexural testing in which the base of the column was held fixed and the top was displaced horizontally. Davis *et al.* [45] prepared a filament-wound, composite-wrapped cylinder with axial and hoop fiber Bragg grating sensors embedded under the outermost ply of reinforcing fibers. The cylinder was filled with concrete and cured. Compressive strains of up to 3.8% were measured during axial loading to 3000 kips of the 30.48 cm (12 in.) diameter columns.

Finally, although not directly in concrete, a simple intensiometric fiber optic sensor was installed in an elastomer layer of a model bridge bearing [46]. The bearing consists of several alternating layers of steel reinforcement plates and elastomer material. Light intensity was directly related to compressive and shear force.

### 1.3.2 Field applications

In addition to the laboratory studies described above, fiber optic sensors have been used to measure strain and vibration in a number of newly constructed structures. They have also been attached to existing structures for nondestructive evaluation of structural health, such as crack growth, and monitoring use, such as traffic load on roadways.

Habel and Hofmann [47] placed interferometric sensors on a bridge in Berlin which had visible cracks. They attached EFPI sensors directly to exposed steel reinforcement bars near the crack and measured both deformation and vibration due to a test load as well as normal traffic. They detected strains in the order of tens of microstrains with uncertainty of $\sim$1 microstrain. The data collected with normal traffic suggested that the weight of individual trucks could be detected by this system, following a calibration.

Habel and co-workers also embedded EFPI sensors in a reinforced concrete wall and measured the internal strain during initial curing [16, 47]. In order to protect the sensor during emplacement, metal tabs were attached to the sensor head to transfer strain from the concrete. The sensor head was then covered in silicon rubber, with the tabs protruding, and the coated sensor was embedded in a concrete block with flexible metal tubing to protect the leads. This block was subsequently embedded in the wall during a routine concrete pour. In order to detect the slow changes during the curing process, self-calibration was achieved by imposing a known displacement. A disadvantage of this technique was the concentration of stress at the interface between the block and the surrounding concrete. An alternative set-up for orienting and protecting sensors during emplacement involves jacketed fibers held in a rigid frame [16]. While preliminary results indicate successful embedment using this rig, no strain data have yet been published.

Two types of fiber optic sensors were installed on a concrete dam in Germany [48]. One is a rather low-sensitivity fiber strain sensor embedded in the bulk concrete. The other is an extrinsic fiber sensor with a mechanical sensor head to detect displacement across joints between concrete panels. If movement across the joint is greater than some specified limit, the fiber is severed, and the loss of transmission triggers an alarm. The latter were set up in triplicate to ensure reliability and had been in operation for 2 years at the time of the report.

A bridge in Keverkusen, Germany, was constructed with fiber optic sensors embedded in the upper and lower surfaces to monitor corrosion and crack formation [49]. Copper wire sensors were embedded during fabrication of the pre-stressed fiberglass tendons which were used to reinforce the concrete, and the potential for using fiber sensors in this role was discussed in the article. Elongation detected interferometrically over 2 months (prior to the bridge

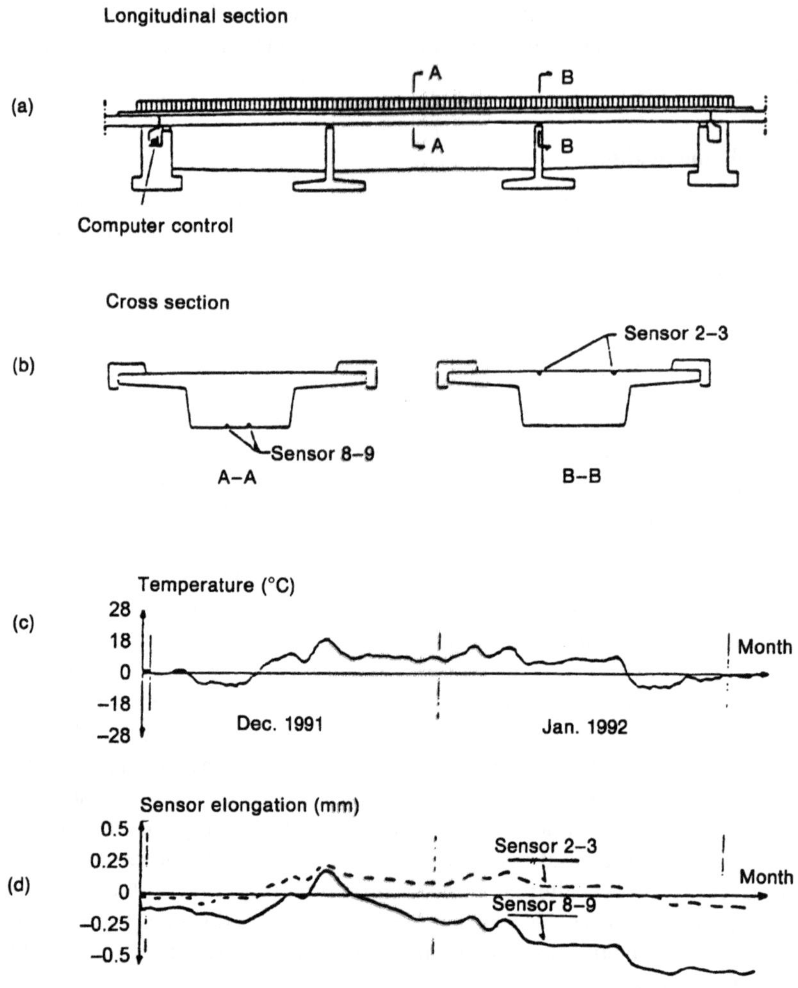


**Fig. 1.4** (a) Longitudinal and (b) cross-sectional view of a bridge in Keverkusen, Germany, showing sensor locations. (c) Temperature and (d) sensor elongation as a function of time (reproduced with permission from [49]).

being open to traffic) was due to temperature variation and creep (Fig. 1.4). The dominant thermal effects can be corrected if temperature is measured independently. Extrinsic fiber corrosion cells were also installed on this bridge in order to monitor weathering of the cement and penetration of chlorides from road salt.

An interstate highway bridge in New Mexico was instrumented with FBG sensors [50]. In the first phase of the study, two FBG sensors were installed. Using a broadband source with a tunable filter, one sensing element at a time could be interrogated; however, the system permits any number of elements to be interrogated sequentially. The goal of the project was to determine the strain levels and frequency response of the bridge with traffic loading. Therefore a system was designed with high sensitivity (5 nanostrains), sufficient dynamic range (750 microstrains) and a ~1 kHz bandwidth to cover a wide range of potential strain response characteristics. Both of the sensors, which were installed on the lower flange and web of a girder, were able to detect the number and relative weight of vehicles, as well as to determine the overall frequency response of the bridge.

Surface-mounted intrinsic Fabry–Perot interferometers (IFPIs) on the Sungsan Bridge over the Han River in Seoul are the first reported use of FOS in Korea [51]. Dynamic tests were performed by driving a truck of known weight across the bridge at various speeds. The strain increased substantially at speeds greater than 20 km $h^{-1}$, but seemed to decrease slightly at speeds greater than 40 km $h^{-1}$.

A bridge in Calgary was the first bridge to use carbon fiber composite pre-stressed tendons in some of the concrete girders. Researchers from the University of Toronto embedded fiber Bragg grating sensors in both the composite- and steel-reinforced girders [52–54]. In order to circumvent the limited extensional strain that glass fibers can withstand, the fiber sensors were attached to pre-stressed tendons after they were placed under tension. After the concrete was cured and the tension released, the Bragg gratings measured only compressive strain (Fig. 1.5). The object of the sensors was to monitor the performance of the new material (carbon fiber composite) compared with traditional steel, over time. Data were collected periodically, not continuously, by taking instrumentation hardware to the bridge and plugging it into a connection port. Results from

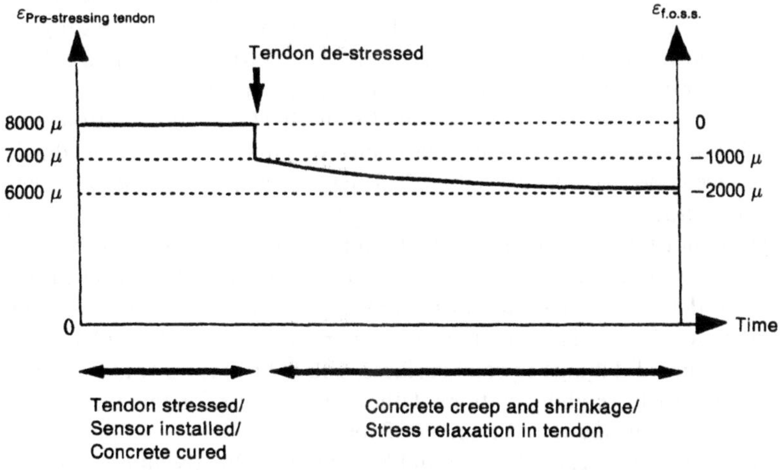


**Fig. 1.5** Strain measured by fiber Bragg grating fiber optic strain sensors (f.o.s.s.) attached to pre-stressed tendons and then embedded in concrete as a function of time (reproduced with permission from [52]).

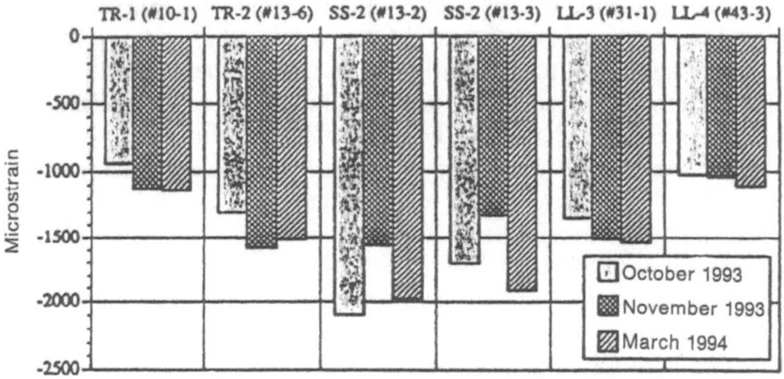


**Fig. 1.6** Internal strains measured in girders reinforced with both carbon/polymer (TR and LL) and steel (SS) pre-stressed tendons over a 6-month period (reproduced with permission from [53]).

the first 6 months are shown in Fig. 1.6. The ability to collect intermittent data without any recalibration is a significant advantage of fiber Bragg grating sensors. Strain information is directly related to the wavelength of the output signal as long as no creep occurs between the sensor and the strain-transmitting component to which it is attached.

An all-fiber corrosion sensor was planned for three bridges in Vermont where heavy use of de-icing compounds is associated with substantial infrastructure degradation [55]. The sensor is based on absorption by the steel reinforcement bar of light propagating in the evanescent wave, which is a function of the degree of corrosion. Alternatively, colorimetric sensing of the presence of chloride ions can be performed *in situ* by a fiber sensor in contact with a sol-gel containing a fluorescent agent [56]. When chloride reacts with the agent, a color shift in the light transmitted by the fiber is measured spectroscopically.

The potential use of fiber optic sensors for roadbed sensors has attracted considerable attention. For example, polarimetric sensors have been embedded in a road surface to detect vehicle weight [57]. The number and frequency of fringes in the sensor output increased with vehicle weight and speed, respectively (Fig. 1.7). The system must be calibrated in order to obtain absolute weights, and to obtain total vehicle weight, it is necessary to sum the force from all the wheels. It is also possible to detect the dimension of the area of contact between the tire and the pavement, which is related to the inflation pressure (Fig. 1.8). Overinflation leads to increased damage to road surfaces, and fiber sensors could expedite its detection at truck weighing stations. Other applications for this type of sensor include monitoring speed and number of vehicles.

Inaudi and co-workers in Switzerland have taken a slightly different approach to the problem of monitoring civil structures [18, 58]. Rather than measure local strain at a large number of positions in order to deduce changes in the behavior of the entire structure, they have focused on the development of longer gauge-length

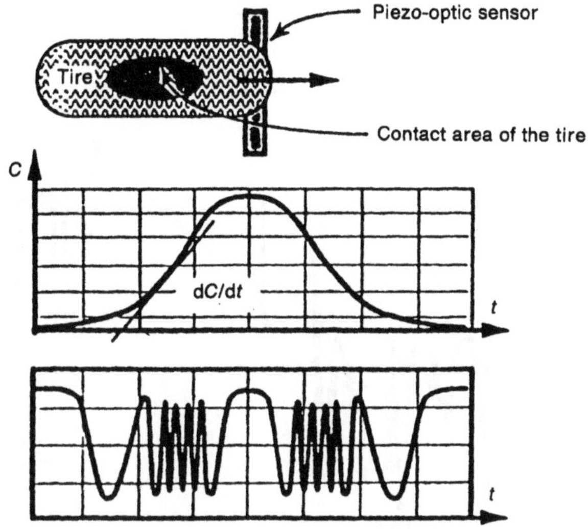


**Fig. 1.7** Fiber optic sensor response to a vehicle in motion. The bottom plot shows the 'raw output' from a polarimetric sensor as a function of time. The top plot is the transformed data showing the weight on the sensor as a function of time (reproduced with permission from [57]).

(up to tens of meters) deformation sensors that detect variation in the overall geometry. This approach will detect cracking, creep, etc. only if it impacts on the shape of the structure on a larger scale. The sensor is based on a low-coherence double interferometer [9], although the technique can be used with FBGs and other short gauge-length sensors as well. These researchers have developed a self-contained portable reading unit that can be marketed commercially. The

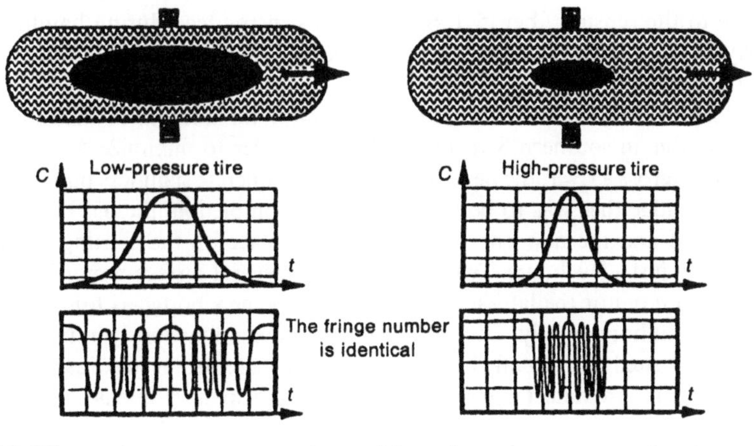


**Fig. 1.8** Fiber optic sensor response when weight and speed are constant, but tire pressure is varied (reproduced with permission from [57]).

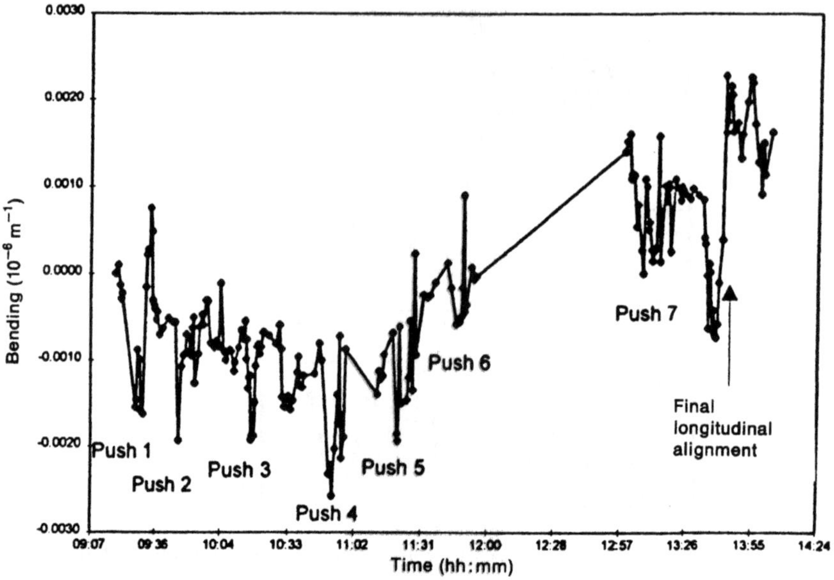


**Fig. 1.9** Horizontal bending of railway bridge while being pushed (reproduced with permission from [62]).

briefcase-sized unit is connected to the sensing fibers at a central connection box located on the structure.

Over 200 deformation sensors, with lengths from 0.1 to 50 m, have been placed in various structures by Swiss researchers, including several bridges [18, 59, 60] and dams [61]. Deformation of a structure during concrete curing is readily detected. In early applications, the fibers were stripped of their polymeric coating before they were attached to the metal anchors which transfer the strain from the concrete to the sensing fiber [59]. Surface flaws introduced during handling led to early failure of these sensors due to cracking. Polyimide coating was used in subsequent installations.

Fiber optic sensors were also used to monitor an unusual installation of a railroad bridge in southern Switzerland [62]. In order to minimize disruption of rail services across a river, the new steel-reinforced concrete span was built adjacent to the old bridge that was to be replaced. Upon completion, the old bridge deck was removed and the new one was pushed into place on the existing piers and abutments in just a few hours. Two embedded and several surface-mounted sensors were used to monitor the lateral deformation of the new bridge as four hydraulic jacks pushed it horizontally. The jacks were moved simultaneously in incremental steps, and adjusted individually between steps. The incremental steps, as well as the 1-h lunch break, are detectable in the bending of the span* (Fig. 1.9).

*Bending is defined as the difference in deformation between sensors embedded on either side of the structure's neutral axis.

The Swiss group have also been active in the application of FOS to underground civil projects [10]. They point out that the behavior of geological structures in the vicinity of pilings and tunnels is often poorly understood, and monitoring deformation in these systems could provide useful information to ensure safety during construction and use. They have instrumented anchoring cables used to stabilize a rock-slide area, and are in the process of installing their system in a precast concrete tunnel lining in Fribourg, Switzerland.

Another active group in terms of embedding fiber optic sensors in concrete structures has been that of Fuhr and co-workers at the University of Vermont. They have installed fiber optic sensors to monitor vibration in several bridges, a 5-storey building and a hydroelectric dam. A brief review of their work is given here; for a more detailed review see ref. [19].

An early project (1989) involved surface-attaching a multimode fiber to the median of a multi-lane interstate highway bridge during resurfacing and expansion. During a 3-h test, vibrations due to truck traffic and jackhammers were readily detected. Motion in the fiber leads due to wind was also detected. In this short-term test, instrumentation at the site was operated from the back of a truck. More permanent sensors were embedded in the deck of a replacement railway bridge, which allowed the bridge performance to be monitored [63].

In addition to these roadway applications, Fuhr and co-workers have reported on the instrumentation of a new 65 000 square-foot building on the university campus [20]. Over 2.5 miles of single-mode and multi-mode fiber optic sensors were embedded throughout the building. The sensor network allowed measurement of vibration, wind pressures, loading, creep and parameters relating to building health, such as crack development. Temperature and strain sensors were also installed in key locations. Time-of-flight measurements of instrumented impacts were correlated with concrete curing and could be very useful in the future for monitoring the rate of curing in large structures. Typically, at present, an ample period is given to ensure sufficient curing, but a direct fiber optic measurement could significantly reduce the waiting period without compromising safety.

The most massive structure outfitted by the Vermont group is the Winooski One hydroelectric dam [26, 64]. The builder and the owner of the dam were interested in vibration levels at key structural points, as well as the water flow and water pressure on the upstream face. The limited access points led to the development and use of multiplexed fiber optic sensors. In addition, multifunctional fibers capable of simultaneously sensing vibration and pressure [64], or simultaneously sensing vibrations and transmitting optical image signals were developed and embedded [66]. Results from an accelerometer and an embedded statistical mode sensor* show good agreement (Fig. 1.10). The fiber sensors helped resolve a problem during the start-up of the hydroelectric plant when

*A statistical mode sensor is an intensiometric sensor that measures variation in the speckle pattern of a multimode fiber. This pattern in the output light intensity arises from the coherent interaction of the many modes carried by the fiber.

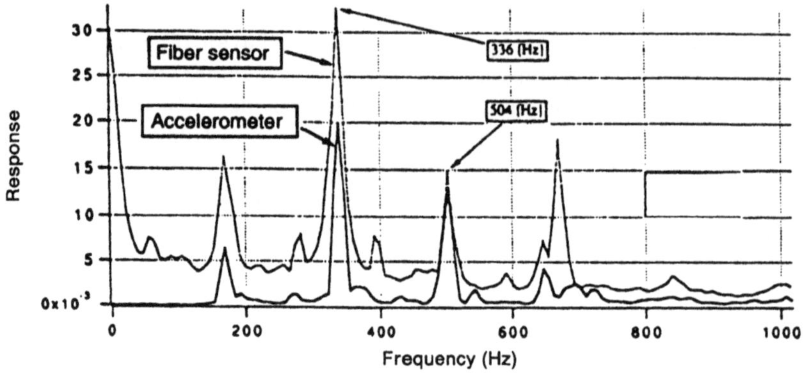


**Fig. 1.10** Comparison of outputs from a fiber optic sensor and an accelerometer embedded in a hydroelectric dam (reproduced with permission from [64]).

the expected generator efficiency was not achieved. The resonant vibrational frequency of one turbine was slightly less than expected, which pointed to a main gear which was out-of-round. After the gear was replaced, the frequency matched the anticipated value and the efficiency met specifications.

In addition, following construction, a magnitude 4.7 earthquake occurred in the area of the dam. The ability to demonstrate that the vibrational signature and generator efficiencies had not changed precluded having to do an underwater visual inspection of the structure (D. R. Huston, personal communication, 1995).

In the process of stringing miles of optical fiber in various structures, the researchers at the University of Vermont conceived a 'sensor panel' which could be fabricated in advance and then installed more efficiently at the construction site [67]. Although some panel designs were presented, none were installed. While having some advantages in terms of ease of emplacement, a bulkier panel is more likely to have an impact on the properties of the concrete, and would have to be proved not to interfere with the function of the structure.

In the course of their work, Fuhr and colleagues have had to address many of the logistical problems associated with the task of planning a large installation, working on a construction site and subsequently gathering the wealth of data that are produced. In the case of intensiometric sensing, long fiber leads and awkward access can be avoided by the use of radio telemetry (i.e. cordless phone technology) [68]. Using off-the-shelf equipment without additional amplification, signals from a statistical mode sensor were transmitted over a distance of >300 feet with an average error of <0.15% (maximum error of 0.4%).

Once a fiber optic network is operational, a steady stream of data is available and is typically stored by a computer on site. In some cases, the analysis of those data may be more conveniently done remotely. The University of Vermont group has connected the on-site computer at the Winooski One dam to the Internet global computer network [69, 70]. The computer at the dam is set up as an anonymous file transfer protocol (FTP) server, which allows any Internet user to log on for the purpose of downloading data files. The time-history data from each sensor are archived, as well as digitally formatted data on the power generation and water flow for each turbine. Downloaded data may subsequently be analyzed, for example to monitor long-term structural health.

In addition to FTP access, the Winooski One dam information is available on the World Wide Web, a distributed hypermedia environment originally developed at CERN [69, 70]. Using a hypermedia browser such as Netscape or Mosaic, information, including text, graphics and audio/video, can be viewed and downloaded from anywhere on the Internet. Video images of the water flow from a camera mounted on the dam and of a display monitor in the dam control room are currently available on the Web.

## 1.4 CONCLUSIONS

Fiber optic sensors, in particular distributed strain sensors, represent a powerful new sensing technology which may prove to be well suited to monitoring strain in concrete structures. They are small, yet robust, and capable of measuring properties in the interior of a structure in which they are embedded without perturbing the concrete itself. Once emplaced, fiber sensors can continuously or periodically monitor the health of the structure, and the output may be stored on a computer or connected to an alarm to signal deviation beyond preset limits.

Several types of sensors have been developed, including intensiometric, polarimetric, modalmetric and Fabry–Perot interferometric sensors. Fiber Bragg grating sensors have many advantages over these other types. They are intrinsic, have a linear response, require no calibration, and can be mass-produced during fiber drawing. Perhaps the biggest advantage of the FBG sensors is that they detect a change in the measurand field as a wavelength shift, which is an absolute measurement and is not susceptible to instrumental drift. In addition, the FBG sensors are by far the most readily multiplexed of all fiber optic sensors.

The use of fiber optic sensors for detecting strain in concrete structures has been demonstrated by a number of researchers using a variety of sensing schemes. Fiber optic sensors have been attached to existing structures to monitor performance and to evaluate crack growth. However, more importantly, they have been embedded within concrete, both in the laboratory and in real-world structures. They have yielded good agreement and superior signal-to-noise when compared with traditional foil strain gauges. While early applications utilized simple intensity-based sensors, recent implementations have focused on interferometric- and FBG-based systems. The EFPI and ILFE interferometric sensors are currently less temperature-sensitive, but the FGB sensors yield absolute strain, do not require quadrature detection, and temperature effects can be corrected. Some of the types of structures that have been built with embedded fiber optic sensors include railway and highway bridges, roadways, multistorey buildings and dams. Practical and logistical challenges to routine implementation of fiber optic sensor networks in commercial civil structures still remain, perhaps the largest of which is the resistance of the construction industry to change. These obstacles will be overcome by additional demonstrations that fiber optic sensors can save money, decrease risks and extend, rather than decrease, the useful life of civil structures.

## ACKNOWLEDGEMENTS

This work was supported by the Federal Highways Administration under agreement No. DTFH61-94-Y-00084.

## REFERENCES

1. Dunker, K. F. and Rabbat, B. G. (1993) Why America's bridges are crumbling. *Sci. Am.*, **March**, 66–72.
2. Liu, S. C., Chong, K. P. and Singh, M. P. (1994) Civil infrastructure systems research: hazard mitigation and intelligent material systems. *Smart Mater. Struct.*, **3**, A169–A174.
3. Méndez, A., Morse, T. F. and Méndez, F. (1989) Applications of embedded optical fiber sensors in reinforced concrete buildings and structures. *Fiber Optic Smart Structures and Skins II, SPIE*, **1170** (SPIE, Bellingham, WA), pp. 60–69.
4. Méndez, A. and Morse, T. F. (1992) Overview of optical fiber sensors embedded in concrete. *Fiber Optic Smart Structures and Skins V (Boston, MA, 1992), SPIE*, **1798** (SPIE, Bellingham, WA), pp. 205–216.
5. Measures, R. (1992) Smart structures – a revolution in civil engineering. *Advanced Composite Materials in Bridges and Structures* (K. W. Neale and P. Labossiere, eds), Canadian Society for Civil Engineering, Montreal, pp. 31–59.
6. Méndez, A., Morse, T. F. and Reinhart, L. J. (1993) Experimental results on embedded optical fiber sensors in concrete. *Smart Structures and Materials Conference (Albuquerque, NM, 1993), SPIE*, **1918** (SPIE, Bellingham, WA), pp. 420–427.
7. Nanni, A., Yang, C. C., Pan, K., Wang, J.-S. and Michael, R. R. Jr. (1991) Fiber-optic sensors for concrete strain/stress measurement. *ACI Mater. J.*, **88**, 257–264.
8. Grattan, K. T. V. and Meggitt, B. T. (1995) *Optical Fiber Sensor Technology*. Chapman & Hall, London, 499pp.
9. Inaudi, D., Elamari, A. and Vurpillot, S. (1994) Low coherence interferometry for the monitoring of civil engineering structures. *Proceedings 2nd European Conference on Smart Structures and Materials (Glasgow, 1994)*, pp. 216–219.
10. Inaudi, D., Vulliet, L., Pflug, L., Vurpillot, S. and Wyser, A. (1995) Low-coherence sensors for the monitoring of underground works. *Smart Sensing, Processing, and Instrumentation (San Diego, CA, 1995) SPIE*, **2444** (SPIE, Bellingham, WA), pp. 170–178.
11. Sirkis, J. S., Brennan, D. D., Putnam, M. A., Berkoff, T. A., Kersey, A. D. and Friebele, E. J. (1993) In-line fiber etalon for strain measurement. *Opt. Lett.*, **18**, 1973–1975.
12. Meltz, G., Morey, W. W. and Glenn, W. H. (1989) Formation of Bragg gratings in optical fibers by a transverse holographic method. *Opt. Lett.*, **14**, 823–825.
13. Escobar, P., Gusmeroli, V., Martinelli, M., Lanciani, I. and Morabito, P. (1992) Fiber-optic interferometric sensors for concrete structures. *Proceedings 1st European Conference on Smart Structures and Materials (Glasgow, 1992)*, pp. 215–218.
14. Habel, W. R. and Polster, H. (1995) The influence of cementitious building materials on polymeric surfaces of embedded optical fibers for sensors. *J. Lightwave Technol.*, **13**, 1324–1330.

15. Habel, W. R., Höpcke, M., Basedau, F. and Polster, H. (1994) The influence of concrete and alkaline solutions on different surfaces of optical fibres for sensors. *Proceedings 2nd European Conference on Smart Structures and Materials (Glasgow, 1994)*, pp. 168–171.
16. Habel, W. R. and Hillemeier, B. (1995) Results in monitoring and assessment of damage in large steel and concrete structures by means of fiber optic sensors. *Smart Structures and Materials (San Diego, CA, 1995), SPIE*, **2446**, (SPIE, Bellingham, WA), pp. 25–36.
17. Sennhauser, U., Bronnimann, R. and Nellen, P. M. (1996) Reliability modelling and testing of optical fiber Bragg sensors for strain measurements. *Fiber Optic and Laser Sensors XIV, (Denver, CO, 1996) SPIE*, **2839** (SPIE, Bellingham, WA), pp. 64–75.
18. Inaudi, D., Casanova, N., Kronenberg, P., Marazzi, S. and Vurpillot, S. (1997) Embedded and surface-mounted fiber optic sensors for civil structural monitoring. *Industrial and Commercial Applications of Smart Structures Technologies, SPIE*, **3044** (SPIE, Bellingham, WA), pp. 236–243.
19. Huston, D. R., Fuhr, P. L., Ambrose, T. P. and Barker, D. A. (1994) Intelligent civil structures – activities in Vermont. *Smart Mater. Struct.*, **3**, 129–139.
20. Fuhr, P. L., Huston, D. R., Kajenski, P. J. and Ambrose, T. P. (1992) Performance and health monitoring of the Stafford Medical Building using embedded sensors. *Smart Mater. Struct.*, **1**, 63–68.
21. Ambrose, T. P., Huston, D. R. and Fuhr, P. L. (1992) Lessons learned in embedding fiber sensors into large civil structures. *Fiber Optic Smart Structures and Skins V (Boston, MA, 1992), SPIE*, **1798** (SPIE, Bellingham, WA), pp. 194–199.
22. Hendrick, R. O., Shadaram, M., Nazarian, S. and Picornell, M. (1992) Measuring stress distribution in pavements using single-mode fiber. *Fiber Optic Smart Structures and Skins V (Boston, MA, 1992), SPIE*, **1798**, (SPIE, Bellingham, WA), pp. 200–204.
23. Pope, C. E., Wu, S.-P., Chuang, S. L., Calero, J. and Murtha, J. (1992) An integrated fiber optic strain sensor. *Optical Design and Processing Technologies and Applications, SPIE*, **1779** (SPIE, Bellingham, WA), pp. 113–121.
24. Prohaska, J. D., Snitzer, E., Chen, B., Maher, M. H., Nawy, E. G. and Morey, W. W. (1992) Fiber optic Bragg grating strain sensor in large-scale concrete structures. *Fiber Optic Smart Structures and Skins V (Boston, MA, 1992), SPIE*, **1798** (SPIE, Bellingham, WA), pp. 286–294.
25. Claus, R. O., de Vries, M., Nasta, M., Murphy, K. A. and Masri, S. F. (1993) Optical fiber sensors for the quantitative measurement of reinforced concrete structures. *Experiments in Smart Materials and Structures*, AMD, **181** (ASME, New York), pp. 97–101.
26. Fuhr, P. L., Huston, D. R., Ambrose, T. P. and Snyder, D. M. (1993) Stress monitoring of concrete using embedded optical fiber sensors. *J. Struct. Eng.*, **119**, 2263–2269.
27. Maher, M. H. and Nawy, E. G. (1993) Evaluation of fiber optic Bragg grating strain sensor in high strength concrete beams. *Application of Fiber Optic Sensors in Engineering Mechanics* (F. Ansari, ed.), ASCE, New York, pp. 120–133.
28. Masri, S. F., Mustafa, M., de Vries, M. J. and Claus, R. O. (1993) Experimental study of embedded fiber-optic strain gauges in concrete structures. *Smart Structures and Materials Conference (Albuquerque, NM, 1993), SPIE*, **1918** (SPIE, Bellingham, WA), pp. 428–434.
29. Calero, J., Wu, S.-P., Pope, C., Chuang, S. L. and Murtha, J. P. (1994) Theory and experiments on birefringent optical fibers embedded in concrete. *J. Lightwave Technol.*, **12**, 1081–1090.
30. Masri, S. F., Agbabian, M. S., Abdel-Ghaffar, A. M., Higazy, M., Claus, R. O. and de Vries, M. J. (1994) Experimental study of embedded fiber-optic strain gauges in concrete structures. *J. Eng. Mech.*, **120**, 1696–1717.

31. Guerin, J. J., Lequime, M., Toppani, E., Leygonie, M. and Chauvel, D. (1994) Embedded optical fibre strain gages for civil engineer: Application to concrete monitoring. *Proceedings 2nd European Conference on Smart Structures and Materials (Glasgow, 1994)*, pp. 224–227.

32. Alavie, A. T., Maaskant, R., Ohn, M. M., Rizkalla, S. and Measures, R. M. (1994) Application and characterization of intracore grating sensors in a CFRP prestressed concrete girder. *Smart Structures and Materials Conference (Orlando, FL, 1994), SPIE*, **2191** (SPIE, Bellingham, WA), pp. 103–110.

33. de Vries, M., Nasta, M., Bhatia, V., Tran, T., Greene, J., Claus, R. O. and Masri, S. (1995) Performance of embedded short-gage-length optical fiber sensors in a fatigue-loaded reinforced concrete specimen. *Smart Mater. Struct.*, **4**, A107–A113.

34. Kim, K. S., Yoo, J. W., Kim, S. K. and Kim, B. Y. (1996) Embedded intrinsic Fabry–Perot optical fiber sensors in the cement concrete structure. *Smart Sensing, Processing, and Instrumentation (San Diego, CA, 1996), SPIE*, **2718** (SPIE, Bellingham, WA), pp. 218–231.

35. Davis, M. A., Bellemore, D. G., Kersey, A. D., Putnam, M. A., Idriss, R. L. and Kodindouma, M. (1996) Fiber optic sensor system for bridge monitoring with both static load and dynamic modal sensing capabilities. *Proceedings of Nondestructive Evaluation Techniques for Aging Infrastructure and Manufacturing (Scottsdale, AZ), SPIE*, **2946** (SPIE, Bellingham, WA), pp. 219–232.

36. Idriss, R. L. and Kodindouma, M. B. (1996) Real-time monitoring of the after fracture response of a multi-girder steel bridge. *Proceedings of the 3rd Conference on Nondestructive Evaluation of Civil Structures and Materials (Boulder, CO, 1996)*, pp. 335–348.

37. Idriss, R. L. and Kodindouma, M. B. (1997) *Design and Implementation of an Optical Fiber Monitoring System in a Full-Scale Laboratory Bridge*. New Mexico State University, Department of Civil Engineering Report No. 97-02, 78pp.

38. Gusmeroli, V., Martinelli, M., Barberis, A., Escobar, P. and Morabito, P. (1994) Thermal expansion measurements of a concrete structure by embedded fiber-optic: an effective example of simultaneous strain–temperature detection. *Proceedings 2nd European Conference on Smart Structures and Materials (Glasgow, 1994)*, pp. 220–223.

39. Kersey, A. D., Davis, M. A. and Bellemore, D. G. (1995) Development of fiber sensors for structural monitoring. *Non-Destructive Evaluation of Aging Bridges and Highways (Oakland, CA, 1995), SPIE*, **2456** (SPIE, Bellingham, WA), pp. 262–268.

40. Kodindouma, M. B. and Idriss, R. L. (1996) An integrated sensing system for highway bridge monitoring. *Smart Structures and Materials (San Diego, CA, 1996), SPIE*, **2719** (SPIE, Bellingham, WA), pp. 132–140.

41. Askins, C. G., Putnam, M. A., Patrick, H. J. and Friebele, E. J. (1997) Fiber strength unaffected by on-line writing of single-pulse Bragg gratings. *Electron. Lett.*, **33**, 1333–1334.

42. Nellen, P. M., Bronnimann, R., Sennhauser, U., Askins, C. G. and Putnam, M. A. (1996) Strain measurements on concrete cantilever beam and carbon fiber cable with distributed optical fiber Bragg grating sensors. *Opt. Eng.*, **35**, 2570–2577.

43. Nellen, P. M., Anderegg, P., Bronnimann, R. and Sennhauser, U. (1997) Application of fiber optical and resistance strain gauges for long-term surveillance of civil engineering structures. *Smart Systems for Bridges, Structures, and Highways, SPIE*, **3043** (SPIE, Bellingham, WA), pp. 77–86.

44. Lou, K.-A., Yaniv, G., Hardtmann, D., Ma, G. and Zimmermann, B. (1995) Fiber optic strain monitoring of bridge column retrofitted with composite jacket under flexural loads. *Smart Systems for Bridges, Structures, and Highways, SPIE*, **2446** (SPIE, Bellingham, WA), pp. 16–24.

45. Davis, M. A., Bellemore, D. G., Putnam, M. A., Kersey, A. D., Slattery, K. T., Corona, K. and Schowengerdt, M. (1996) High strain monitoring in composite-wrapped concrete cylinders using embedded fiber Bragg grating arrays. *Proceedings of the North American Smart Structures Meeting (San Diego, CA), SPIE*, **2721**, (SPIE, Bellingham, WA), pp. 149–154.

46. Caussignac, J. M., Chabert, A., Morel, G., Rogez, P. and Seantier, J. (1992) Bearings of a bridge fitted with load measuring devices based on an optical fiber technology. *Proceedings 1st European Conference on Small Structures and Materials (Glasgow, 1992)*, pp. 207–210.

47. Habel, W. R. and Hofmann, D. (1994) Determination of structural parameters concerning load capacity based on Fibre-Fabry–Perot-Interferometers. *Proceedings 2nd European Conference on Smart Structures and Materials (Glasgow, 1994)*, pp. 176–179.

48. Holst, A., Habel, W. and Lessing, R. (1992) Fiber-optic intensity-modulated sensors for continuous observation of concrete and rock-fill dams. *Proceedings 1st European Conference on Smart Structures and Materials (Glasgow, 1992)*, pp. 223–226.

49. Wolff, R. and Miesseler, H.-J. (1992) Monitoring of prestressed concrete structures with optical fiber sensors. *Proceedings 1st European Conference on Smart Structures and Materials (Glasgow, 1992)*, pp. 23–29.

50. Davis, M. A., Kersey, A. D., Berkoff, T. A., Jones, R. T., Idriss, R. L. and Kodinduma, M. (1997) Dynamic strain monitoring of an in-use interstate bridge using fiber Bragg grating sensors. *Smart Systems for Bridges, Structures, and Highways (San Diego, 1995), SPIE*, **3043** (SPIE, Bellingham, WA), pp. 87–95.

51. Kim, K. S., Ryu, J. U., Lee, S. J. and Choi, L. (1997) In-situ monitoring of Sungsan Bridge in Han River with an optical fiber sensor system. *Smart Systems for Bridges, Structures, and Highways, SPIE*, **3043** (SPIE, Bellingham, WA), pp. 72–76.

52. Maaskant, R., Alavie, T., Measures, R. M., Ohn, M., Karr, S. and Glennie, D. (1994) Fiber optic Bragg grating sensor network installed in a concrete road bridge. *Smart Structures and Materials Conference (Orlando, FL, 1994), SPIE*, **2191** (SPIE, Bellingham, WA), pp. 457–465.

53. Measures, R. M., Alavie, T., Maaskant, R., Huang, S. and LeBlanc, M. (1994) Bragg grating fiber optic sensing for bridges and other structures. *Proceedings 2nd European Conference on Smart Structures and Materials (Glasgow, 1994)*, pp. 162–167.

54. Measures, R. M., Alavie, A. T., Maaskant, R., Ohn, M., Karr, S. and Huang, S. (1995) A structurally integrated Bragg grating laser sensing system for a carbon fiber prestressed concrete highway bridge. *Smart Mater. Struct.*, **4**, 20–30.

55. Fuhr, P. L., Ambrose, T. P., Huston, D. R. and McPadden, A. P. (1995) Fiber optic corrosion sensing for bridges and roadway surfaces. *Smart Structures and Materials Conference (San Diego, CA, 1995), SPIE*, **2446** (SPIE, Bellingham, WA), pp. 2–8.

56. Fuhr, P. L., Huston, D. R. McPadden, A. J. and Cauley, R. F. (1996) Embedded chloride detectors for roadways and bridges. *Smart Systems for Bridges, Structures, and Highways (San Diego, 1995), SPIE*, **2719** (SPIE, Bellingham, WA), pp. 229–237.

57. Teral, S. (1992) Vehicle weighing in motion with fiber optic sensors. *Proceedings 1st European Conference on Smart Structures and Materials (Glasgow, 1992)*, pp. 139–142.

58. Inaudi, D., Vurpillot, S. and Casanova, N. (1996) Bridge monitoring by interferometric deformation sensors. *Laser Optoelectronics and Microphotonics: Fiber Optic Sensors (Beijing, China, 1996), SPIE*, **2895** (SPIE, Bellingham, WA), pp. 34–45.

59. Vurpillot, S., Inaudi, D. and Ducret, J.-M. (1996) Bridge monitoring by fiber optic deformation sensors: design, emplacement and results. *Smart Systems for Bridges, Structures, and Highways (San Diego, 1995), SPIE*, **2719** (SPIE, Bellingham, WA), pp. 141–149.

60. Vurpillot, S., Casanova, N., Inaudi, D. and Kronenberg, P. (1997) Bridge spatial deformation monitoring with 100 fiber optic deformation sensors. *Smart Structures and Materials (San Diego, CA, 1997), SPIE,* **3043** (SPIE, Bellingham, WA), pp. 51–57.

61. Kronenberg, P., Casanova, N., Inaudi, D. and Vurpillot, S. (1997) Dam monitoring with fiber optic sensors. *Smart Structures and Materials, SPIE,* **3043** (SPIE, Bellingham, WA), pp. 2–11.

62. Inaudi, D., Casanova, N., Kronenberg, P. and Vurpillot, S. (1997) Railway bridge monitoring during construction and bridge sliding. *Smart Structures and Materials 1997: Smart Systems for Bridges, Structures, and Highways (San Diego, CA, 1997), SPIE,* **3043** (SPIE, Bellingham, WA), pp. 58–64.

63. Huston, D. R., Fuhr, P. L. and Ambrose, T. P. (1993) Dynamic testing of concrete with fiber optic sensors, in *Application of Fiber Optic Sensors in Engineering Mechanics* (F. Ansari, ed.), ASCE, New York, pp. 134–143.

64. Fuhr, P. L., Huston, D. R., Ambrose, T. P. and Barker, D. A. (1994) Embedded sensors results from the Winooski One hydroelectric dam. *Smart Structures and Materials Conference. (Orlando, FL, 1994), SPIE,* **2191** (SPIE, Bellingham, WA), pp. 446–456.

65. Fuhr, P. L. and Huston, D. R. (1993) Multiplexed fiber optic pressure and vibration sensors for hydroelectric dam monitoring. *Smart Mater. Struct.,* **2**, 260–263.

66. Fuhr, P. L., Kajenski, P. J. and Huston, D. R. (1992) Simultaneous single fiber optical communications and sensoring for intelligent structures. *Smart Mater. Struct.,* **1**, 128–133.

67. Fuhr, P. L., Huston, D. R. and Ambrose, T. P. (1993) Prefabricated sensor panels for smart civil structures instrumentation. *Smart Structures and Materials Conference (Albuquerque, NM, 1993), SPIE,* **1918** (SPIE, Bellingham, WA), pp. 435–439.

68. Fuhr, P. L., Huston, D. R. and Ambrose, T. P. (1993) Interrogation of multiple embedded fiber sensors in civil structures using radio telemetry. *Smart Mater. Struct.,* **2**, 264–269.

69. Fuhr, P. L., Huston, D. R. and Ambrose, T. P. (1994) Remote monitoring of instrumented structures using the INTERNET information superhighway. *Proceedings 2nd European Conference on Smart Structures and Materials (Glasgow, 1994)*, pp. 148–151.

70. Fuhr, P. L., Huston, D. R., Ambrose, T. P. and Mowat, E. F. (1995) An Internet observatory: remote monitoring of instrumented civil structures using the information superhighway. *Smart Mater. Struct.,* **4**, 14–19.

# 2

# The application of optical fiber sensors in advanced fiber reinforced composites. Part 1: Introduction and issues

G. F. Fernando, P. A. Crosby and T. Liu

## 2.1 INTRODUCTION

Advanced fiber reinforced composites (AFRCs) are a class of materials which are made up of a reinforcing phase and a matrix phase. The reinforcing phase can be short fibers or continuous fibers. Typical examples of fibers which are used include carbon, glass, silicon carbide and polyaramid. The matrix phase can be a thermoplastic, thermoset, ceramic or metal. A summary of selected properties for engineering materials is presented in Table 2.1. With reference to Table 2.1, it is readily apparent that the specific properties (property of interest divided by the density) of AFRCs are superior to those of other engineering materials. This makes AFRCs ideal materials for primary and secondary load-bearing applications where weight is at a premium. Hence there is extensive utilization for aerospace and other transport-based applications. The drive to reduce the overall weight of AFRCs has resulted in the development of hollow glass and carbon fibers.

Other advantages of AFRCs include their ease of manufacture and flexibility of design to yield the desired mechanical and thermal properties. For example, unique mechanical properties can be obtained by hybridizing (mixing) two different fiber types: the impact resistance of a carbon fiber composite can be enhanced by hybridizing it with polyaramid fibers. The strength and stiffness of the hybrid composite will then be dictated by the relative volume fractions of the two fibers and their relative positions within the laminated composite.

The function of the fibers is to carry the majority of the load which is imposed on the composite. The matrix, on the other hand, transfers the applied load to the fibers and also secures the position of the individual fibers within the composite. The efficiency of this load-transfer process from the matrix to the fibers is dictated by a number of factors, including the nature and the strength of the interfacial bond. As will be seen later, this is an importance issue with the use of embedded optical fiber sensors in AFRCs.

*Optical Fiber Sensor Technology*, Vol. 3. Edited by K. T. V. Grattan and B. T. Meggitt.
Published in 1998 by Kluwer Academic Publishers, London. ISBN 978-1-4419-4736-9

**Table 2.1** Selected properties for common reinforcing fibers and matrix systems [1]

| Material | Density ($kg\,m^{-3}$) | Young's modulus (GPa) | Tensile strength (GPa) | Failure strain (%) | Thermal expansion ($10^{-6}\,K^{-1}$) | Diameter (μm) | Maximum service temperature (°C) |
|---|---|---|---|---|---|---|---|
| E-glass fibers | 2560 | 76 | 2.0 | 2–2.6 | 5.0 | 10–16 | 650 |
| S-glass fibers | 2490 | 86 | 4.4 | 2–2.5 | 2.3 | 7–10 | 700 |
| Quartz fibers | 2200 | 73 | 3.6 | 2–2.5 | 0.54 | 9–14 | 1050 |
| High modulus carbon fibers | 1950 | 330–350 | 2.3–2.6 | 0.5–0.7 | −0.7 | 7–10 | * |
| High strength carbon fibers | 1750 | 220–240 | 3.0–3.3 | 1.3–1.4 | −0.4 | 7–10 | * |
| Polyaramid (Kevlar™-49) | 1450 | 130–140 | 2.8–3.0 | 2.3–2.4 | −2 | 12–14 | 400 |
| Polyethylene | 970 | 85–99 | 2.7–3.2 | 3.5–3.7 | −10 | 27–38 | Melts at 130°C |
| Silicon carbide fibers | 3000 | 400 | 2.4 | 0.6 | 4.0 | 7–10 | |
| Aluminum | 2700 | 70 | 0.2–0.6 | 6–20 | 24 | – | – |
| Titanium | 4500 | 110 | 0.3–1.0 | 4–12 | 9 | – | – |
| Mild steel | 7800 | 208 | 0.4 | >20 | 17 | – | – |
| Aluminum oxide ($Al_2O_3$) | 3800 | 380 | 0.5 | 0.1 | 8 | – | 1200 |
| Epoxy | 1200 | 3–7 | 0.035–0.1 | 1–3 | 60 | – | 80–150 |
| Polyetheretherketone | 1250–1300 | 3.6 | 0.15–0.18 | 50 | 47 | – | 120 |

*The upper service temperature for carbon fiber would depend on the actual matrix system and the test environment.

Optical fiber sensors are a very attractive proposition for deployment in AFRCs because: (i) they can be used in aggressive and explosive environments where electrical-based sensors may not survive or may be too dangerous to be deployed; (ii) their circular cross-section and relatively small size makes them easily integrated into composite preforms at the time of manufacture, or they can be surface-mounted onto existing structures and components; (iii) similar fiber surface treatment procedures which are used for the reinforcing fibers can be applied to silica or polymer-based optical fibers; (iv) the ability to multiplex a series of sensors along a single optical fiber to obtain distributed information on chemical composition, strain and temperature at specific locations within the AFRC is very advantageous for process and condition monitoring; (v) the deployment of a single optical fiber sensor design for simultaneous and multiple measurands is equally important.

With reference to process or cure monitoring, the parameters of interest are: (i) the rate of heating and cooling through the laminated prepreg; (ii) the viscosity of the resin as a function of time and temperature; (iii) the rate of depletion of the active functional groups in the resin system; (iv) the volume of volatiles released during processing; (v) the void content; (vi) the nature and magnitude of the residual stress in the composite. Optical fiber sensors can be used to monitor all these parameters.

The requirements for damage detection in AFRCs are more demanding as they depend on the actual remit in terms of function and performance. For example, a number of papers have been published on the use of optical fiber sensors for static strain, temperature, vibration and displacement measurements. On the other hand, significant work still needs to be done to demonstrate that these sensors can be used to study the mechanisms of damage initiation and propagation in composites. In addition to this, a number of practical issues still need to be addressed. These include: (i) the effect of the diameter mismatch between the optical fiber and the reinforcing fibers; (ii) the influence of the embedded optical fiber on the long-term performance of the AFRC; (iii) the reliability and long-term durability of the embedded optical fiber sensor; (iv) issues relating to inter-connection between the embedded sensor and the associated instrumentation. The opportunities for cost-effective, simple, robust and reliable optical fiber sensors for use in the composites field are considerable.

A range of selected properties for different optical fiber systems is summarized in Table 2.2. It should be apparent from the data presented in Table 2.2 that commercially available optical fibers can be used over a wide range of temperatures and the electromagnetic spectrum. No attempt has been made to compile the current costs of these optical fibers, but they range from a few cents per metre to over hundreds of dollars per metre. Issues which have to be considered with respect to embedding these fibers are: (i) the maximum service temperature of the coating and the optical fiber, including the processing pressure for the prepregs; (ii) the chemical reactivity or inertness of the optical fiber surface to the matrix; (iii) the coefficient of thermal expansion of the optical fiber system and the composite; (iv) the Poisson's ratio; (v) the maximum permissible bend radius for the optical fibers.

**Table 2.2** Selected properties for optical fiber sensors

| Optical fiber type | Density ($kg\,m^{-3}$) | Young's modulus (GPa) | Wavelength range (μm) | Thermal expansion ($10^{-6}\,K^{-1}$) | Refractive index | Poisson's ratio | Maximum service temperature (°C) |
|---|---|---|---|---|---|---|---|
| Fused silica | 2200 | 73 | 0.3–1.5 | 0.54 | 1.45–1.47 | 0.16 | 1000 |
| Chalcogenide | 4400 | 21 | 3–10 | 14 | 2.5–2.65 | 0.24 | 300 |
| Fluoride | 4610 | 56 | 0.5–4.3 | 18.7 | 1.52 | 0.3 | 250 |
| Sapphire | 3970 | 414 | 0.2–4.5 | 8.8 | 1.768 | 0.27–0.3 | 2000 |
| PMMA | 1170 | 3.3 | 0.4–0.8 | 260 | 1.492 | 0.4 | 80 |

PMMA, polymethylmethacrylate optical fibers [2].

Prior to considering aspects of process and health monitoring of AFRCs, a brief summary of the common manufacturing methods used in the production of AFRCs is presented in the next section. If optical fiber sensors are to be embedded in AFRCs at the time of manufacture, it is important to appreciate the key features of these processes. Where appropriate, the practical issues relating to each manufacturing process will also be addressed.

## 2.2 A BRIEF SUMMARY OF SELECTED MANUFACTURING TECHNIQUES FOR AFRCs

### 2.2.1 Filament winding

Filament winding is used extensively to produce structures such as rocket motor cases, pressure vessels, pipes, power transmission shafts, etc. Figure 2.1 shows the basic elements of a typical filament winding machine. The reinforcing fiber tows (a) are fed through a tensioning device and guides (b) to the resin bath (c), and through the traversing carriage (d) onto a rotating mandrel (e). The carriage then traverses horizontally along the length of the rotating mandrel laying resin-impregnated fibers in a predetermined fashion. The wind angle is controlled as a function of the traverse speed of the resin bath and the rotation speed of the mandrel.

The optical fiber sensors can be introduced at any stage during the filament winding process. However, this will require the winding operation to be temporarily stopped in order to locate the prefabricated optical fiber sensor system. Once the required number of layers have been wound onto the mandrel, they are transferred to an oven or autoclave to cure the resin. The cured composite tube is then removed from the mandrel. The actual method of separating the composite from the mandrel will depend on the type of mandrel used. In general, with the use of metal mandrels, the composite is either pushed or pulled off the mandrel. This is a vulnerable time for the embedded optical fiber sensors as they can easily be damaged. The areas where the optical fibers enter and exit the filament-wound tube can be reinforced by winding additional layers of impregnated fibers at

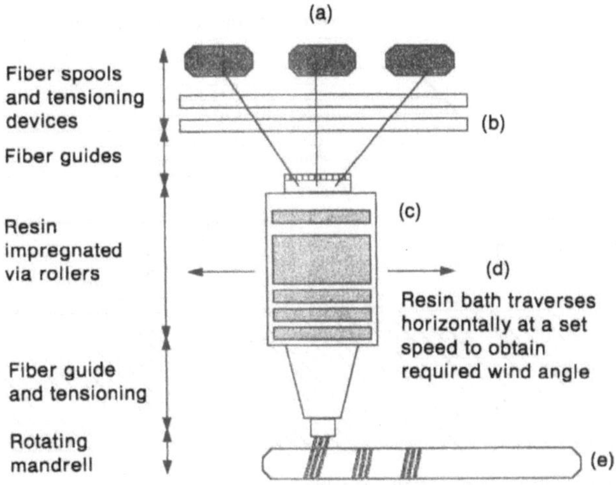


**Fig. 2.1** Schematic illustration of the wet-filament winding process [3].

the ends of the tube. Issues relating to the protection of the optical fibers are discussed in the next section.

There are a number of variations on the filament winding process, including dry-winding, prepreg and tape-based winding etc. The design of the mandrel also tends to vary depending on the actual design of the composite component. Examples of mandrel designs include low-melting-point metal alloys, collapsible mandrels, and plaster-based and water-soluble mandrels.

The market value for filament-wound products in Europe for 1990 was estimated at approximately $2 billion per annum. As an example, the UK market for reinforced composites is believed to consume 86 000 tonnes of thermosetting resin per annum [4].

### 2.2.2 Prepreg-based composites

AFRCs are generally made by stacking and consolidating together a specified number of prepregs in the required directions. A prepreg consists of an array of unidirectional reinforcing fibers which are secured in position by a resin (matrix) system. The resin consists of chemical groups which can react under the application of heat to produce a rigid cross-linked structure; this is commonly referred to as the curing process. The stacked prepregs are then subjected to a heating regime to cure the resin. A vacuum is usually applied during the cure process to assist in the removal of residual solvents and entrapped air. Pressure is also applied to consolidate the prepregs.

A typical manufacturing process using prepregs first involves the thawing-out of the prepregs from their typical storage temperature of −18°C to ambient. The

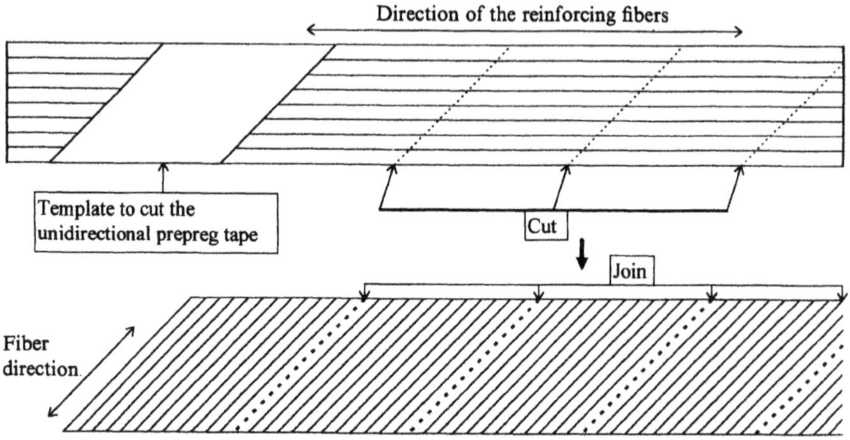


**Fig. 2.2** Schematic illustration of the production of 45° continuous prepreg from a unidirectional prepreg roll.

prepreg is usually supplied in the form of a roll, and the required shape is either cut or stamped out. Since the fibers in the prepreg are aligned in one principal direction, it is necessary to produce the required length of the prepreg with the fibers in the desired direction. The following description refers to the fabrication of a panel of unidirectional (UD), cross-ply and quasi-isotropic composite. In theory, the procedure for the UD stacking sequence is relatively simple (Fig. 2.2). However, from a practical point of view, each prepreg layer has to be consolidated to ensure that the possibility of entrapped air is minimized. In the case of the cross-ply stacking sequence, the UD prepreg is first cut to the required length and oriented perpendicular to the original direction. A series of these are produced and secured together with tape to produce a continuous length of 90° prepregs. As shown in Fig. 2.2, the production of 45° prepregs is more tedious as it requires a template to be used to produce the desired angle and then the subsequent joining of these sections to produce a continuous length.

Stacking or laminating the prepregs is usually carried out manually for small components/structures and via automated prepreg laying machines for the production of large structures such as helicopter rotor blades. Figure 2.3 illustrates three common types of lay-up sequences used in the manufacture of AFRCs. Since the relative orientation of the prepregs relative to each other is an important parameter with respect to the properties of the composite, a vacuum-assisted manual lamination process is sometimes used. A schematic illustration of such a device is shown in Fig. 2.4.

It can be seen from Fig. 2.4 that the application of a vacuum through the perforated bottom plate will result in the prepreg (with the release-paper) being attached to it. The prepreg assembly is then transferred and accurately positioned in place with the aid of a locator pin or similar device on the mould surface. The

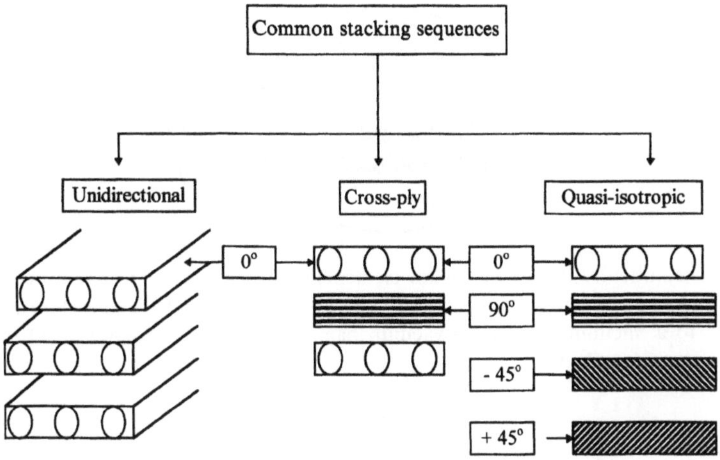


**Fig. 2.3** Illustration of three common ply stacking sequences: unidirectional, cross-ply and quasi-isotropic.

vacuum is turned off and the prepreg drops into position. The prepreg is then consolidated onto the previous layer using a roller. The release paper is removed, and the whole process is repeated until the required number of prepreg layers have been laid down.

The optical fiber sensor can be introduced after the desired number of prepregs have been laid. Due care and attention has to be paid during the consolidation phase to ensure that (i) the layers containing the prepregs are consolidated, and (ii) the optical fiber sensor is not damaged. Techniques to protect the embedded optical fibers are discussed in the next section. The laminated prepreg

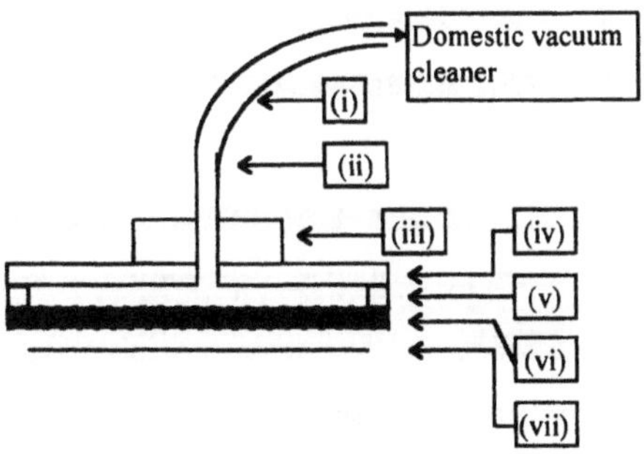


**Fig. 2.4** Schematic illustration of a vacuum-assisted prepreg laminating jig: (i) flexible hose attached to a domestic vacuum cleaner; (ii) vacuum inlet; (iii) handle; (iv) polymethylmethacrylate top plate; (v) seal between the top plate and (vi) the perforated bottom plate; (vii) the prepreg.

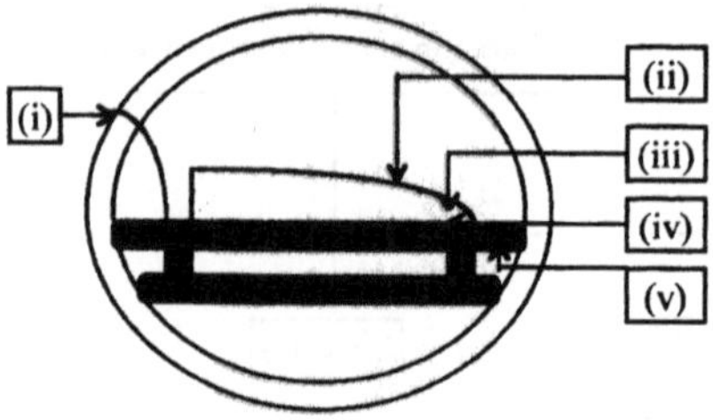


**Fig. 2.5** Schematic illustration of the key elements of an autoclave. Details of the heating and cooling elements and pressure ports have been excluded. (i) Flexible hose which is connected to a vacuum source; (ii) vacuum bag; (iii) laminated prepregs; (iv) thermocouple; (v) autoclave bed.

is usually processed in a heated press or an autoclave under pressure and vacuum. The basic elements of an autoclave are illustrated in Fig. 2.5. The application of a vacuum is achieved via the use of a vacuum-bag. This basically involves the encapsulation of the laminated prepregs in a sealed bag Fig. 2.6). In Fig. 2.6, (i) is the base of the autoclave, and (ii) is a release-spray and/or release cloth to protect the processing equipment surfaces from the curing resin. The former is usually a silicone or polytetrafluoroethylene (PTFE)-based compound and the latter is a fabric (glass) which is coated with a nonstick chemical agent. An optional, bondable surface is provided by (iii) and (v), (iv) represents the laminated prepregs, (vi) is a perforated release cloth which separates the cured composite from the other material used in the vacuum-bagging process, (vii) is

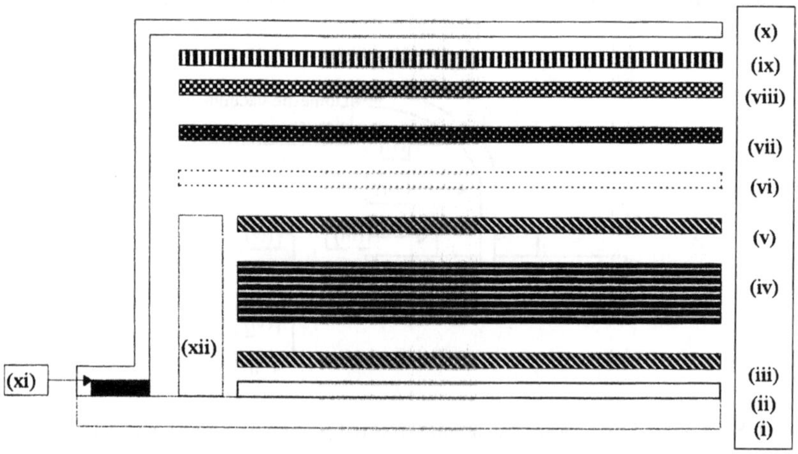


**Fig. 2.6** Schematic illustration of a simple vacuum-bagging system [1]. The illustration is not to scale and the various components have been isolated for clarity. See text for details.

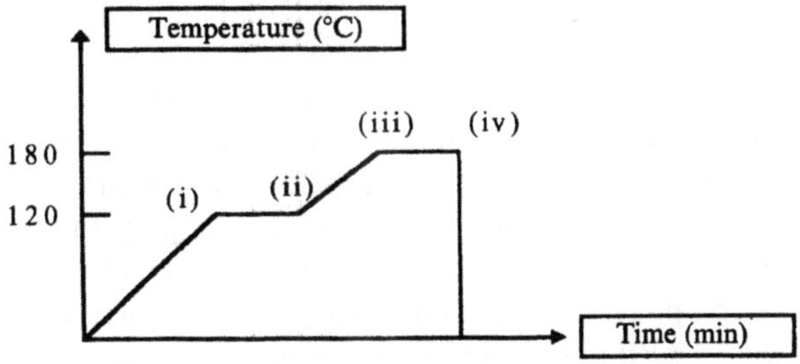


**Fig. 2.7** Illustration of a typical cure schedule used in the processing of epoxy/amine-based prepreg systems [5].

known as the bleeder-cloth and its function is to absorb the excess resin from the prepreg, (viii) is a barrier-type fabric and its main function is to limit the migration of the resin to the 'bleeder cloth', (ix) is known as the breather cloth and its function is to evacuate the vacuum bag and to enable the transfer of the required

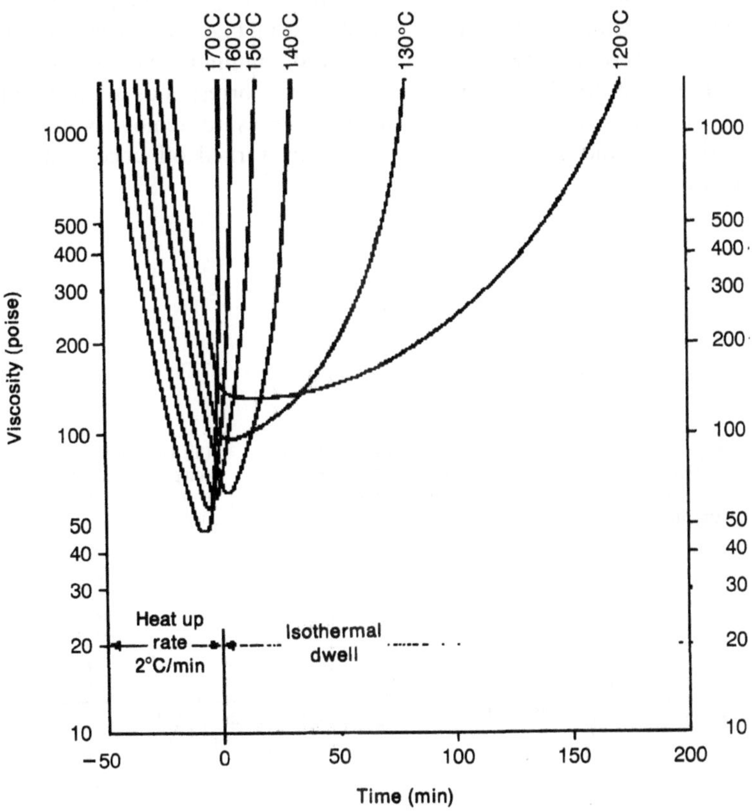


**Fig. 2.8** A typical viscosity/temperature profile for Fibredux 914 [5].

autoclave pressure, (x) corresponds to the vacuum bag itself, (xi) is the vacuum sealing tape, and (xii) corresponds to the dam.

Zero-bleed prepreg systems are also available and these do not require the bleeder-cloth. The viscosity characteristics are generally obtained by careful chemical formulation and not by advancement of the cure process (usually referred to as B-staging). Viscosity modifiers such as polyethersulphone (PES) are used in some prepreg systems to control the viscosity of the resin. The PES is soluble in one component of the resin system. However, when the cross-linking reaction proceeds, the PES precipitates out and appears as fine spherical particles (known as the second phase). This second phase is known to enhance the toughness of the resin system.

Figure 2.7 shows a typical cure schedule for an epoxy/amine resin system: (i) heat under vacuum to 120°C at 1–20°C per minute; (ii) when the component reaches 120°C apply 700 kPa pressure and (iii) carry out an isothermal dwell for a specified period (optional); (iv) a post-cure treatment then follows where the composite is heated at a specified rate to a higher temperature and a second isothermal dwell is carried out. Finally, the sample is cooled to approximately 60°C prior to releasing the pressure and ejecting the component.

The viscosity/temperature/time profile shown in Fig. 2.8 illustrates that the viscosity of the resin is lowered significantly and a minimum viscosity is reached before gelation occurs. The application of pressure corresponds to this minimum viscosity. The primary function of the vacuum is to remove volatiles and entrapped air in the prepregs, and it is usually turned off once gelation has been achieved.

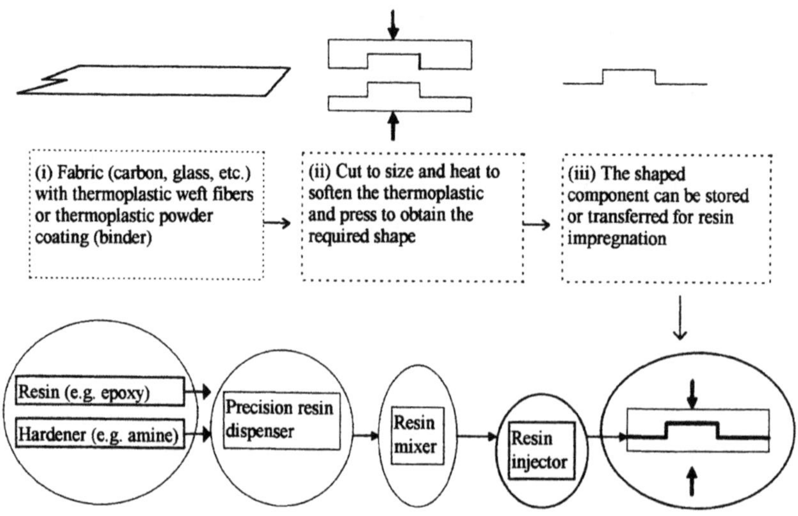


**Fig. 2.9** Schematic illustration of the resin transfer moulding process (RTM).

Apart from unidirectional prepreg material, other preforms are available, including: (i) woven fabrics; (ii) knitted fabrics; (iii) chop-strand-mats; (iv) braided fabrics; (v) three-dimensional fabrics. Obviously the integration of optical fiber sensors into these preforms is not as simple as that previously encountered for prepregs as the undulation of the reinforcing fibers may influence the performance of the embedded optical fiber sensor.

### 2.2.3 Resin transfer moulding

A schematic illustration of the resin transfer moulding manufacturing process is presented in Fig. 2.9. This manufacturing process is used extensively in the production of fiber-reinforced components for the automotive industry and is being used increasingly by the aerospace sector. It involves the use of a fabric which contains a thermoplastic binder which is pressed in a heated mould to give the desired shape. The shaped preform is then transferred to a low-pressure press where the thermosetting resin is injected into the heated mould. Once the resin has cured, the component is released from the mould.

### 2.2.4 Wet lay-up

There are a number of variations on this process, but in general it involves the use of a dry fabric which is draped over a mould. The fabric is then impregnated manually with the resin. In cases where short fibers are used, they tend to be sprayed onto a mould along with a pre-mixed resin system. These fabrications require very good ventilation and extraction facilities.

### 2.2.5 Pultrusion

Pultrusion is a manufacturing process which is used to manufacture fiber-reinforced composites which are continuous with a constant cross-section. A schematic illustration of this process is shown in Fig. 2.10. In this manufacturing

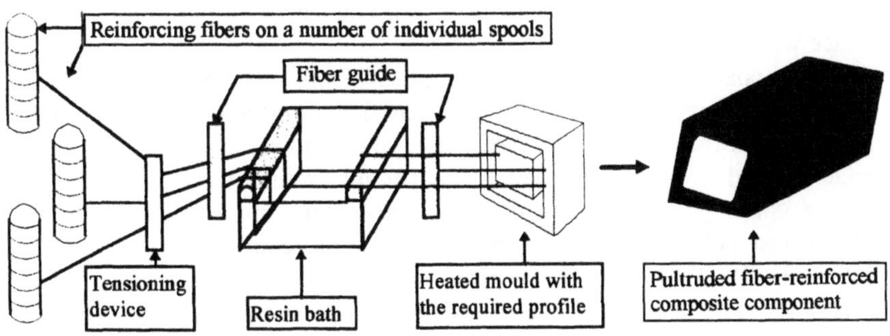


**Fig. 2.10** Schematic illustration of the pultrusion process [6].

process, the fibers are first impregnated using a resin bath and then fed through a heated die with the appropriate profile.

## 2.3 EMBEDDED OPTICAL FIBER SENSORS AND AFRCs – THE ISSUES

### 2.3.1 Protecting the optical fibers

With reference to the previous section on the various techniques which can be used to manufacture AFRCs, it is readily apparent that the integration of optical fiber sensors into the component at the time of manufacture is not a simple matter. A great deal of care and attention has to be paid to ensure that the leading and trailing optical fibers are not damaged during the manufacturing operation. A number of measures can be taken to enhance the probability of survival of the embedded optical-fiber-based sensor during the processing of the composite. For example, PTFE tubing can be used to protect the leading and trailing optical fibers [7]. However, this will entail a small section of the PTFE tube being inserted into the composite. The inserted end will also need to be sealed to stop the resin from entering the tube. For prepregs which are processed in an autoclave, it is usually necessary to build a 'moat' around the lead-in and lead-out optical fibers to prevent the resin from wicking-up. Vacuum-bagging tape is an effective material for the construction of the moat and also as the sealant for the PTFE tube. These precautions are necessary because the resin, once cross-linked, is extremely brittle and any fracture in the resin will tend to propagate through the optical fiber.

### 2.3.2 Coatings

The nature of the coating [8–10] on the optical fiber is an important aspect with respect to both enhancing the survivability during processing, and also the efficiency with which the load is transferred from the matrix to the sensor. The chemical requirements for the coating are that (i) it must survive the processing temperature of the composite, and (ii) it must be capable of forming a strong interface with the matrix.

Common coating materials used on optical fibers include UV-cured epoxy, acrylate, silicone and polyimide. Acrylate-based coatings may be affected by solvents present in the resin system. Silicone-based coatings may not offer a strong interfacial bond strength with the matrix. The maximum operating temperature for these coatings will depend on their particular composition. In general, acrylates and polyimides have a service temperature in the range −40°C to +85°C and −190°C to +350°C, respectively [11].

Excellent analytical analyses on issues relating to coatings have been presented by the research group led by Sirkis [12–15]. King and Aloisio [16] investigated the effects of single- and two-layer polymer coating systems for optical fibers. They stated that for a single-layer polymer-coated fiber, a state of radial compressive stress is said to be present as the coating consolidates over the stiffer fiber as the AFRC cools to ambient from the curing temperature. The two-layered system consisted of a relatively low-modulus polymer as the primary coating, and a higher-modulus polymer as the secondary coating. Since the coefficients of thermal expansion for the two layers were different, a state of radial tension was said to develop at the optical fiber primary coating interface. Their analysis showed that this stress was sufficient to induce debonding.

Optical fiber sensors such as fiber Bragg gratings, evanescent, refractive index-based sensors and most interferometric sensors require a section of the coating to be removed in order to fabricate the sensor. The method used to remove the coating has been reported to affect the tensile properties of the optical fiber, for example, the authors have found certain types of mechanical strippers to lower the tensile strength of multimode optical fibers. Matthewson *et al.* [17] found that removing the cladding with concentrated sulphuric acid did not have any significant effect on the tensile and flexural strength.

The technology behind the use of specified surface treatments for reinforcing fibers is well established. Silane-based surface treatments are extensively used with glass fibers, and obviously this technology is directly transferable to silica-based optical fibers. The main principle with these surface treatments is the use of an organo-silane which is capable of forming a covalent bond with the glass fiber surface, and a second active functional group which can react with the matrix. For example, an amino-silane can be used in conjunction with an epoxy/amine-based resin system. A generalized reaction scheme to illustrate the mode of operation of such an organo-silane is shown below.

$$NH_2R{-}Si{-}(OC_2H_5)_3 + 3H_2O \rightarrow NH_2R{-}Si{-}(OH)_3 + 3C_2H_5OH \qquad (2.1)$$

The organo-silane is hydrolyzed with water with the formation of hydroxy (OH) groups (silanol):

$$NH_2R{-}Si{-}(OH)_3 + HO{-}\text{glass surface} \rightarrow NH_2R{-}\overset{|}{\underset{|}{Si}}{-}O{-}\text{glass surface} \qquad (2.2)$$

The silanol then reacts with the hydroxy groups on the surface of the glass and neighboring silanols to form polysiloxane.

$$NH_2R{-}\overset{|}{\underset{|}{Si}}{-}O{-}\text{glass surface} + \text{epoxy resin} \rightarrow$$

$$\text{formation of covalent bonds between } NH_2 \text{ and epoxy functional groups} \qquad (2.3)$$

The amino group ($NH_2$) can now react with the epoxy resin matrix to form covalent bonds and is thus integrated into the cross-linked resin.

It is worth noting that the chemicals which are used as coupling agents are similar to those used in the manufacture of sol-gel-based glasses. Sol-gel-based coatings are increasingly being used for cladding optical fibers for specialist applications such as chemical, environmental and biological sensing. Matejec *et al.* [18] found the tensile strength of the sol-gel-coated optical fibers to be degraded. They proposed hydrolytic degradation as the primary cause for the observed degradation. There is a wealth of information in the literature on the hydrolytic degradation of reinforcing glass fibers [19–21] and similar degradation mechanisms will probably apply to the hydrolytic degradation of optical fiber sensors.

Aluminum and gold-coated optical fibers are commercially available, and the maximum service temperature for these fibers has been reported as 400°C and 750°C, respectively [11]. Evans [22] studied the reaction kinetics of producing *in situ* a coating of aluminum oxide in an aluminum matrix using conventional silica-based multimode optical fibers. The interfacial reaction was said to proceed according to the following reaction scheme:

$$4Al_{(l)} + 3SiO_{2(s)} \rightarrow 3Si_{(s)} + 2Al_2O_{3(s)} \qquad (2.4)$$

where (l) and (s) denote liquid and solid phase, respectively. The concept of producing *in situ* protective coatings is interesting, and it could be used to design sensor systems for high-temperature applications.

### 2.3.3 The survivability of embedded optical fiber sensors

Although numerous reports have been published on the utilization of optical fiber sensors for process and health monitoring, the majority of these studies have been carried out under laboratory-based conditions. Obviously, there is a need to evaluate the relative performance of the various sensor designs under typical in-service operating conditions. Levin and Nilsson [23] carried out a study to investigate the survivability of embedded single and multimode fibers in cross-ply carbon fiber composites during embedment, autoclaving and subsequent handling. They investigated issues such as the radius of curvature of the embedded optical fibers, protection methods from the lead-in and lead-out fibers, and also the mode of entry and exit from the laminated prepreg. The radius of curvature was found not to be an issue with respect to survivability during the autoclave-based processing of the composite. However, they reported that five out of the 15 acrylic coated fibers were damaged post-processing. This may have been due to some degradation of the coating at the processing temperature of the resin. Teflon tubing was found to be an effective means of protecting the optical fibers in and out of the composite. They also demonstrated that it was feasible to take the lead-in and lead-out optical fibers perpendicular to the plane of laminated prepregs. Martin *et al.* [24] used a commercial filament winding machine to carry out the routine embedment of optical fibers and optical fiber sensors. No major problems were encountered with respect to the survival of the optical fibers.

Another point to note is that with some AFRC manufacturing techniques the component has to be trimmed to remove the flash (excess resin around the composite): the option of routing the fibers through the thickness of the laminate may be possible. This may not be a problem for cure-monitoring, but it would certainly pose a serious problem for subsequent health-monitoring using the embedded sensor. It is worth noting that with some sensor designs, a single fiber-end acts as the lead-in and lead-out fiber.

The location of the embedded sensor within the composite can also have an influence on its useful life-span in service. This is also probably more relevant for health-monitoring rather than cure-monitoring. Figure 2.11 shows an ultrasonic C-scan for an impact-damaged cross-ply carbon fiber composite. The sketches illustrate the observed damage modes from right to left in increments of 1 mm. If the optical fiber was located within this damage envelope, the

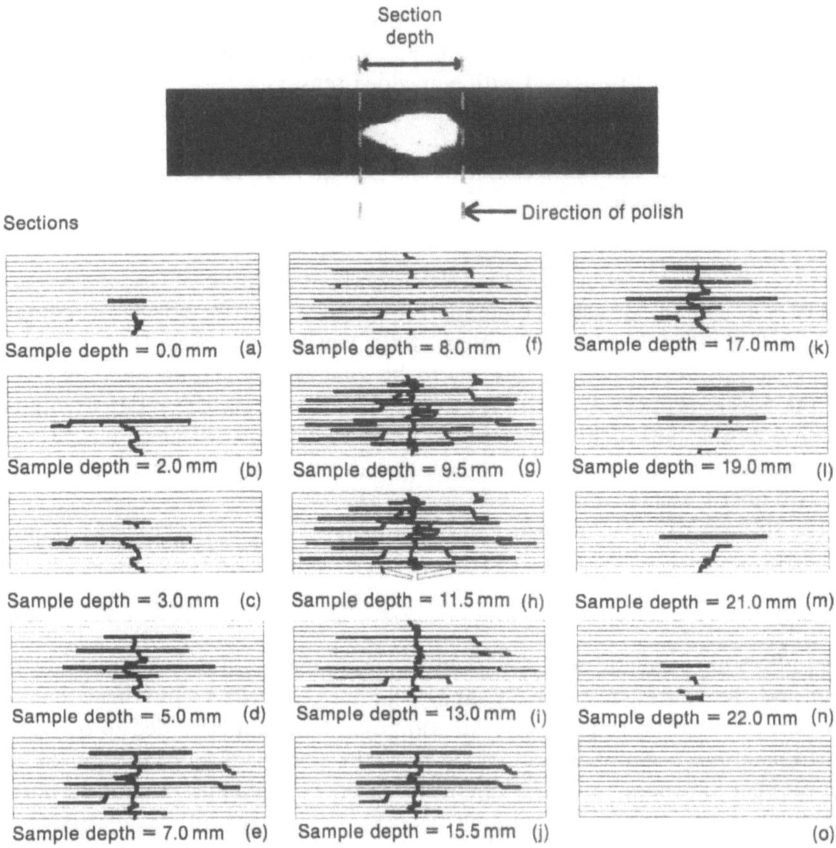


**Fig. 2.11** Details of the observed damage in a cross-ply T33/920 composite system after it was subjected to a 2-Joule impact. The impacted specimen was embedded in a mounting resin and polished to specified distances from one end of the damage, as shown in this C-scan image [6].

probability of survival is questionable, and even if it did survive, the data from the sensor may be unreliable.

On the basis of the evidence available so far, the applied pressure during the processing of common prepregs has not given rise to any concerns other than the induced distortions in the reinforcing fibers. Apart from the robustness of the sensors, the associated equipment for interrogating the sensors must also be compact and rugged for routine use on site.

### 2.3.4 Cost

This tends to be the major issue with most high-volume producers of AFRC components. The cost here is taken to include the cost of the sensor, the associated instrumentation and the expenditure involved in training personnel to operate and maintain equipment. It is not clear at present if certain classes of AFRC which are used for crucial load-bearing applications will have to be re-certified if the structure is instrumented with embedded sensors.

### 2.3.5 Effect of the embedded sensor on the general mechanical properties of the composite

*(a) Tensile properties*

There is a general consensus of opinion to suggest that the effects of a single embedded optical fiber on the quasi-static tensile properties of AFRC are negligible [25, 26]. However, this has to be qualified by stating that the effect on the tensile properties of the AFRC would depend on the relative volume fraction of the optical fiber and its diameter in relation to reinforcing fibers, their orientation relative to each other, and the affinity of the matrix towards the two fiber systems.

The stiffness of the composite can be estimated by using the rule of mixtures (ROM):

$$E_c = E_f V_f + E_m V_m \tag{2.5}$$

where $E_c$ is the axial stiffness of the composite, $E_f$ and $E_m$ are the stiffness of the fiber and the matrix, respectively, and $V_f$ and $V_m$ are the volume fractions of the fiber and matrix, respectively. An additional term can be added to equation 2.5 to take into account the contribution to the composite from the embedded optical fiber in proportion to its volume fraction. This assumes that all the fibers are continuous, with good bonding between the fibers and the matrix, and aligned in the loading direction. The differences in the Poisson's ratio for the different constituents is assumed to be negligible. The probability of conventional optical fibers contributing to a large fiber volume fraction may not be realistic on the basis of costs and the resulting distortion it may cause to the reinforcing fibers.

*(b) Compressive properties*

Large inclusions in the AFRC can have a detrimental effect on the quasi-static compressive properties of the composite as they can act as sites for initiating delaminations. The compressive properties of AFRCs tend to be controlled by the compressive mechanical properties of the fibers, the yield strength of the matrix, the fiber volume fraction and the uniformity of distribution of the fibers, the interfacial bond strength and the degree of misalignment of the reinforcing fibers. The actual test method which is used to determine the compressive properties of the AFRC can have an influence on the results, and therefore it is essential to ensure that identical test methods are used when comparing different sets of data.

Jenson *et al.* [27] reported compressive strength reductions ranging from 1% to 15% and from 34% to 70% for situations where the optical fibers were embedded parallel or perpendicular to the loading direction, respectively. The compressive stiffness reductions ranged from 13% to 20% for optical fibers which were embedded parallel to the loading direction, and from 12% to 20% for optical fibers embedded perpendicular to the loading direction.

Mall *et al.* [28] carried out a detailed study into the effect of (a) the number of optical fibers, (b) the diameter of the optical fibers, (c) the location of the optical fiber within the laminated composite, and (d) the relative orientation of the optical fibers with respect to the reinforcing fibers. They used the IITRI–ASTM D 3410 compression test method with a quasi-isotropic carbon fiber reinforced epoxy composite and reported no degradation of the compressive properties when the optical fibers were positioned parallel to the reinforcing fibers.

The series of micrographs in Figs. 2.12–2.16 [29] illustrate the effect of the embedded optical fibers on the distortions to the laminated composite. Although the presence of a resin-rich region is not apparent when the optical fiber is located in the same direction as the reinforcing fibers, significant ply waviness is seen at the vicinity of the embedded optical fiber. When the optical fiber is embedded in the direction of the reinforcing fibers, the latter are free to move during processing to accommodate the inclusion. In cases where the optical fiber is embedded between a 0° and a 90° prepreg, the latter will have to deform in order to accommodate the inclusion. This will result in the formation of a small resin-rich region between the optical fiber and the transverse ply (Fig. 2.13). Figure 2.14 illustrates the effect of embedding an optical fiber between 0° and 45° prepregs. As before,

**Fig. 2.12** Micrograph of an embedded optical fiber between aligned plies [29].

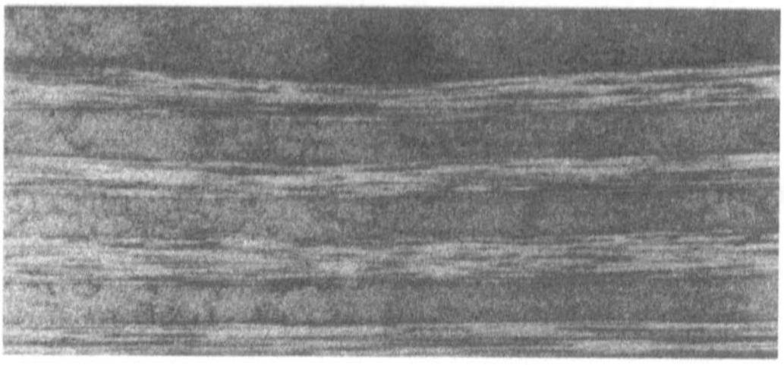

**Fig. 2.13** Micrograph of an embedded optical fiber between one aligned ply and one perpendicular ply [29].

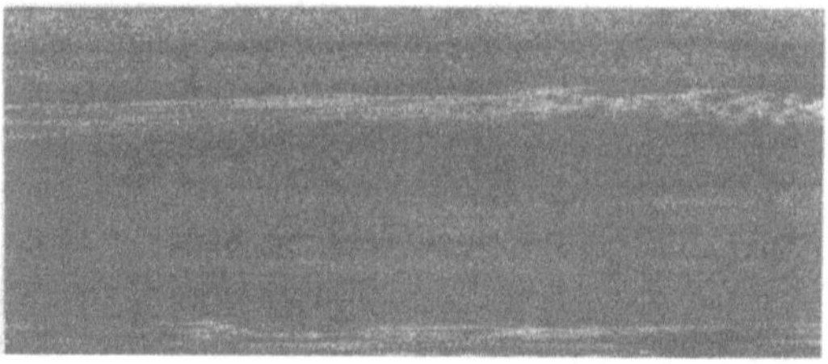

**Fig. 2.14** Micrograph of an embedded optical fiber between one aligned ply and one aligned at 45° [29].

**Fig. 2.15** Micrograph of an embedded optical fiber between two plies aligned at 45° [29].

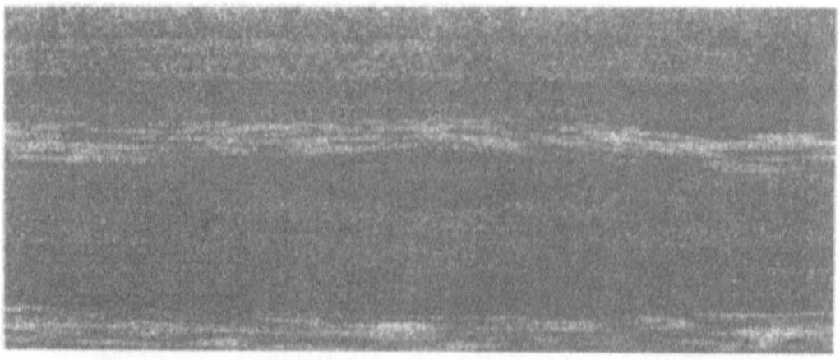

**Fig. 2.16** Micrograph of an embedded optical fiber between a ply oriented perpendicular to the fiber and one aligned at 45° to the other [29].

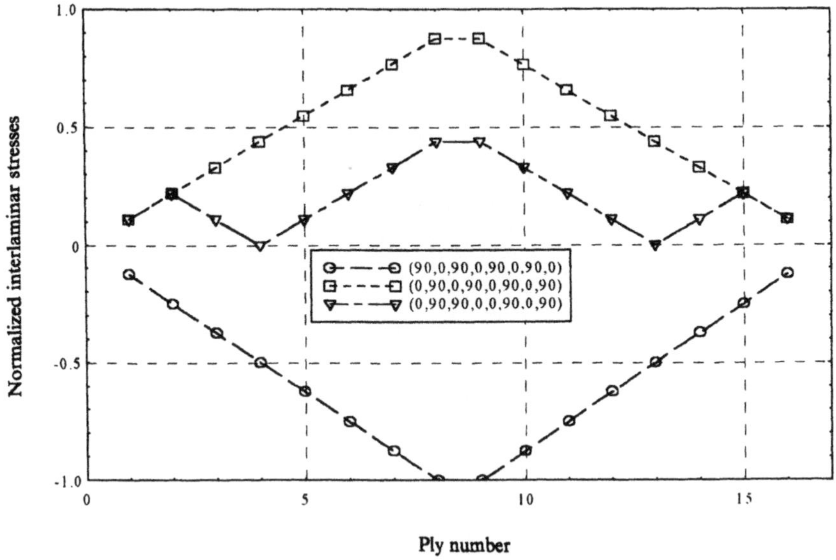


**Fig. 2.17** Effect of the ply stacking sequence on the normalized interlaminar stresses for the cross-ply sequences. The calculations were carried out using the data for a T300/Fibredux 920 resin system [29].

the presence of a small resin-rich region along with the formation of a resin-rich inter-ply layer is observed. Figures 2.15 and 2.16 clearly show the formation of large resin-rich regions when the optical fiber is sandwiched between two plies which are at a different angle to the optical fiber.

The practical point to note is that due to design restrictions or component design, it may not always be possible to embed the optical fibers parallel to the reinforcing fibers. Another issue which has to be considered is the magnitude of the interlaminar stresses between plies for a given stacking sequence. The effect of the ply stacking sequence on the normalized interlaminar stress distribution in the composite is shown in Fig. 2.17. These calculations can be made using a number of commercially available software packages which deal with laminate analysis of fiber-reinforced composites.

Nightingale [30] carried out a study on the location of the embedded optical fiber within a cross-ply composite on its ultimate quasi-static tensile properties. The location of the optical fiber was not found to have any significant influence on the tensile properties. As seen in Fig. 2.18, no major differences were observed in the fracture modes for the plies with and without embedded optical fibers.

Figures 2.18–2.21 represent fracture surfaces where the optical fibers were embedded at specified locations within a cross-ply composite with the following lay-up sequence: (0, $90_2$, $0_2$, 90, 0, 90)s. In Fig. 2.18, the presence of the optical fiber was not seen to have any significant influence on the macroscopic failure

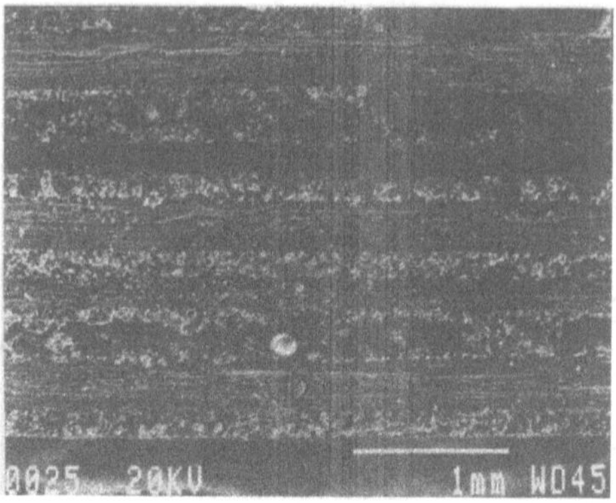


**Fig. 2.18** An embedded optical fiber between two 0° plies (plies 4 and 5).

**Fig. 2.19** Optical fiber embedded between two 0° plies (plies 4 and 5) (this is a different specimen to that shown in Fig. 2.18).

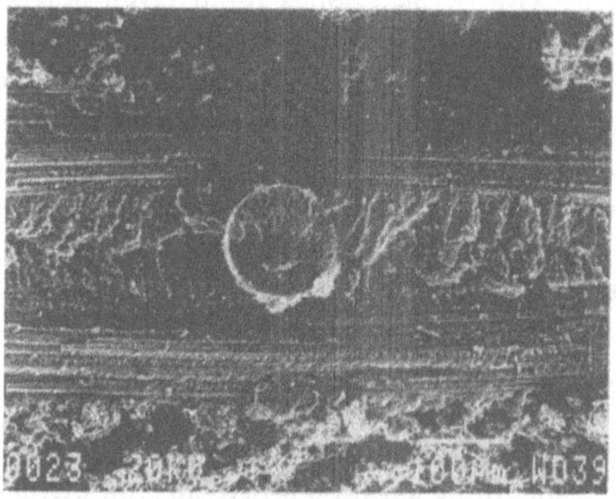

**Fig. 2.20** Fracture surface of a cross-ply composite with the optical fiber between two 90° plies.

**Fig. 2.21** Fracture surface of a cross-ply composite with the optical fiber located between a 0° and a 90° ply. The coating on the optical fiber was retained.

when compared with comparable ply sequences without any embedded optical fibers. However, it is worth noting that post-failure analysis of AFRCs is fraught with difficulties because a significant number of specimens have to be studied in order to make any definite conclusions on failure modes and mechanisms. A close inspection of the surrounding regions near the embedded optical fibers in Figs. 2.19 and 2.20 showed similar fracture modes when compared with comparable plies without any embedded optical fibers. The micrograph shown in Fig. 2.21 represents an optical fiber where the acrylate-based coating was retained. Significant damage is seen in the coating material, but as with previous cases, the macroscopic damage in the vicinity of the embedded sensor was found to be somewhat similar to that in comparable regions without any embedded optical fibers. With respect to the dynamic loading of AFRCs with embedded optical fiber sensors, embedding the optical fibers in areas of high interlaminar stresses may lead to premature debonding and delaminations. This is discussed further in the next section.

### 2.3.6 The long-term durability of the interface between the embedded sensor system and the matrix

A number of excellent reviews have been published on the optimization of the interface of reinforcing fibers [31]. There is general acceptance of the need to surface-treat the optical fibers and sensors prior to embedment. A number of detailed studies have also been carried out on the optimum specifications for the coatings on the optical fibers in order to ensure effective load-transfer from the matrix to the fibers. However, issues relating to the long-term durability of optical fiber sensors in AFRCs have not really been covered in any great detail.

In assessing Fig. 2.22, it is important to appreciate that the nature of fiber damage in a composite will depend on the type of reinforcing fiber. For example, polyaramid fibers can undergo significant surface damage if they are subjected to

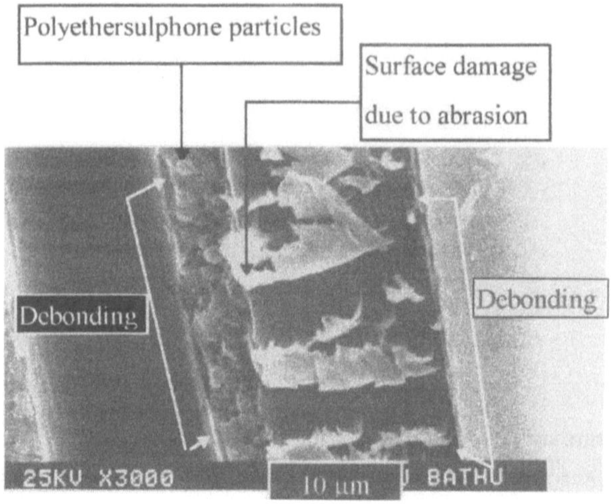


**Fig. 2.22** Micrograph of a fatigue-damaged reinforcing fiber surface of a polyaramid fiber.

prolonged cyclic loading where relatively high-tensile and compressive loads are used. It is envisaged that this would be a cause for concern when the embedded optical fibers cause significant distortion to the reinforcing fibers in terms of fiber waviness or fiber clumping. The periodicity of the fiber surface damage is thought to be due to micro-buckling along the fibers, which in turn results in abrasion damage. This mode of damage to the reinforcing fibers can cause debonding of neighboring fibers, including optical fibers. Another feature shown in Fig. 2.22 is the presence of spherical particles of a thermoplastic material. It is not known at present if there would be preferential precipitation of these particles onto the optical fibers. The particles may interfere with refractive index and evanescent-based cure monitoring.

### 2.3.7 The long-term durability and reliability of the embedded sensor

This is a critical issue, as potential end-users of AFRCs would insist that the lifetime of the sensor be equal to that of the design life of the structure. This may prove to be difficult as AFRCs can sustain significant damage prior to catastrophic failure. For example, the failure sequence for a quasi-isotropic composite when tested in tension is illustrated in Fig. 2.23, where it can be seen that the transverse plies fail first, followed by the 45° and finally the unidirectional plies. This failure sequence is as expected, but what is not known is the effect of this damage on the performance of the embedded sensors, as the propagating crack can have a number of possible interactions. These issues are discussed below.

There is ample evidence in the literature to suggest that measurands such as temperature and quasi-static strain can easily be measured using embedded optical fiber sensors. The complexity of the fracture process in AFRCs was highlighted in Fig. 2.11, and the question of using optical fiber sensors to study

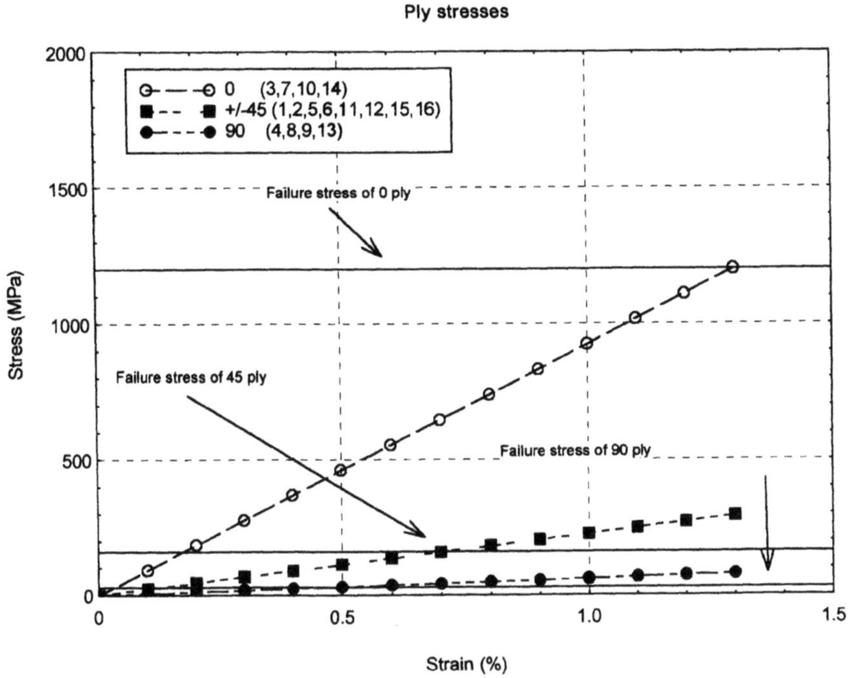


**Fig. 2.23** Calculated failure strains for an incremental applied stress for a quasi-isotropic composite [29].

damage initiation and propagation in these materials has to be considered in detail. In the context of this section, the issue is the reliability of the data from the embedded optical fiber sensor once the composite has sustained a specified level of damage.

Figure 2.24 illustrates a series of potential fracture events which can take place when an unidirectional AFRC (Fig. 2.24(a)) is subjected to tensile loading; in this case, the failure strain of the matrix is assumed to be significantly higher than that of the fibers.

As shown in Fig. 2.24(b), a reinforcing fiber can fracture at its weakest point anywhere along its length. In the vicinity of this fracture, a redistribution of the stress takes place, with additional stresses being imposed on neighboring fibers. Hence there is a higher probability that the fibers in the immediate vicinity of the original fracture would fail. This failure mode is characteristic of most carbon-fiber-reinforced composites. Other failure processes which can take place in the vicinity of the fiber fracture are matrix yielding or interfacial debonding (Fig. 2.24(c)). This mode of failure would depend on the interfacial bond strength. Figure 2.24(d) illustrates the case where the fiber debonding shown in Fig. 2.24(c) has resulted in the development of a longitudinal crack along the fiber and also the formation of a through-thickness crack. The situation is equally

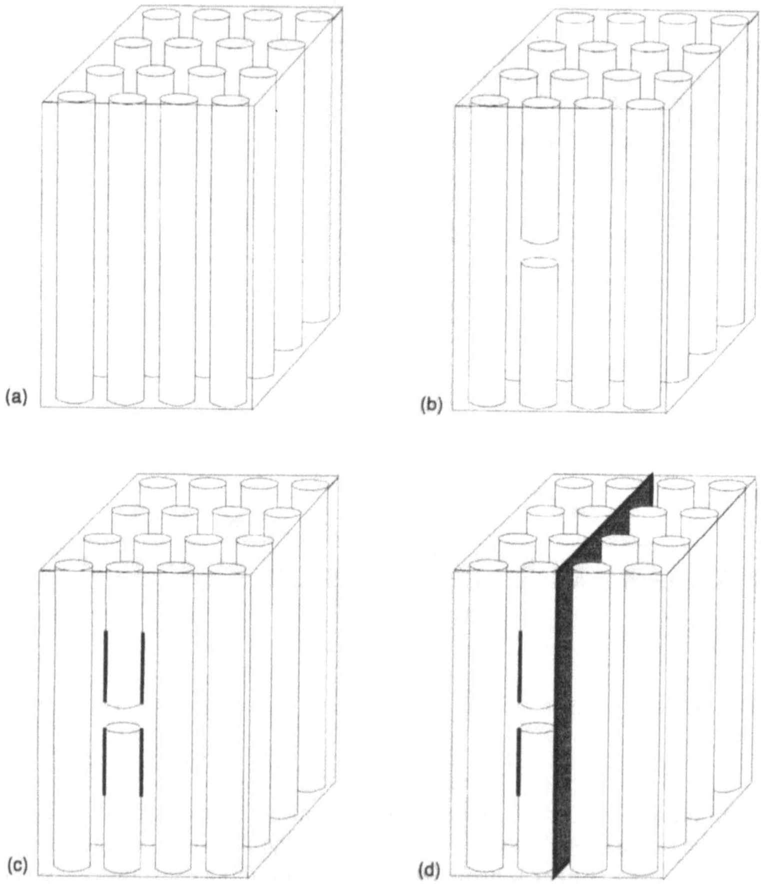


**Fig. 2.24** Schematic illustration of a series of fracture and damage mechanisms which can take place when a unidirectional composite is subjected to a tensile load [32]. See text for details.

complex when the failure strain of the fiber is higher than that of the matrix, as seen in Fig. 2.25.

Figure 2.25 shows that when a load is applied to the composite, a matrix crack forms at a point of weakness (Fig. 2.25(b)) and multiple matrix cracks can form if the applied load is increased, as seen in Fig. 2.25(c). Figure 2.25(d–f) illustrate potential fracture paths for the propagating matrix crack. The stress concentration at the crack-front may be sufficiently high to fracture the fiber (Fig. 2.25(d)) or, if the interfacial bond strength is weak, it can result in debonding along the length of the fiber (Fig. 2.25(e)). The presence of multiple matrix cracks is not shown in these figures for clarity. Figure 2.25(f) illustrates the situation where the propagating matrix crack causes localized fiber–matrix debonding.

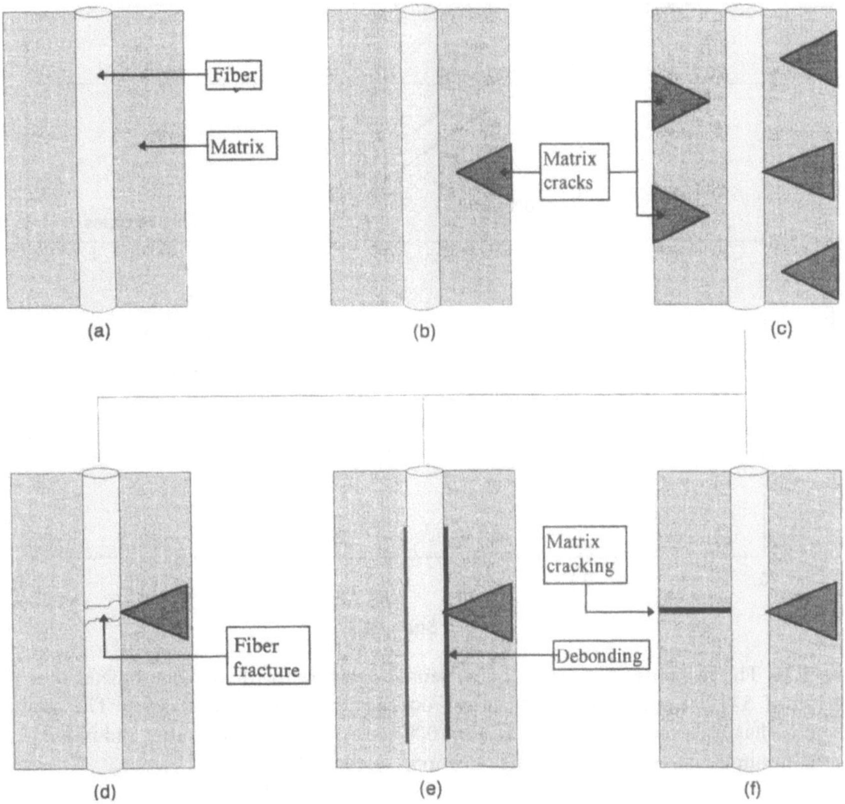


**Fig. 2.25** Illustration of possible fracture modes in an AFRC when the failure strain of the fiber is higher than that of the matrix. See text for details.

The situations with cross-ply and quasi-isotropic AFRCs are even more complex, as the through-thickness stresses can play an important role in determining the fracture mechanisms. In addition to this, cracks propagating in the longitudinal and transverse directions in a cross-ply composite can result in delamination. The scenarios presented above can apply equally to embedded optical fiber sensors. Hence, significant work still needs to be carried out to interpret and correlate the output from embedded sensors and damage mechanisms in AFRCs.

In the final part of this section on the reliability and durability of embedded optical fiber sensors, consideration is given to the effect of dynamic loading on the performance of embedded optical fibers and optical fiber-based sensors. Badcock [29] carried out a study into the feasibility of using embedded, extrinsic Fabry–Perot sensors to monitor the compliance change in a cross-ply carbon-fiber-reinforced epoxy composite. Significant difficulties were encountered in ensuring that the trailing optical fibers were not damaged when in the grips of

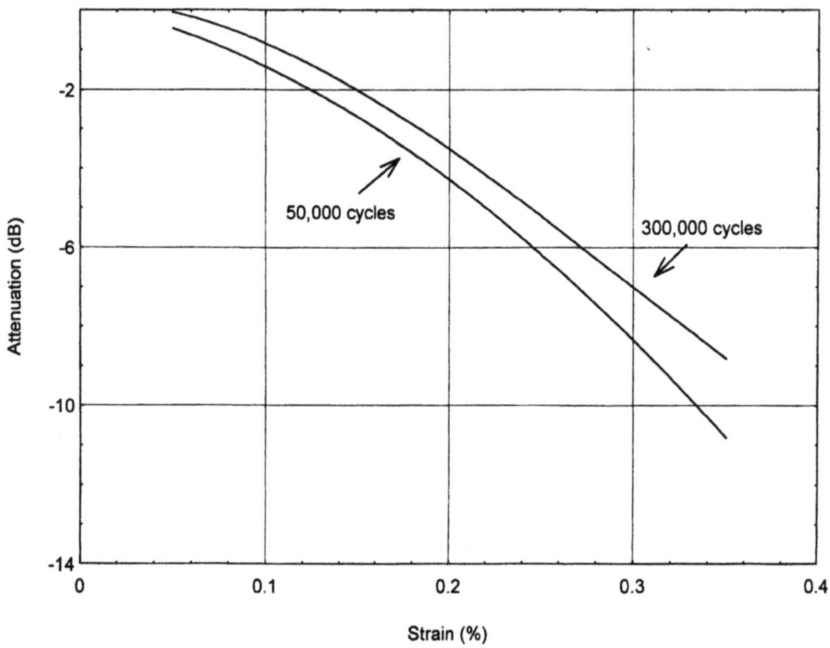


**Fig. 2.26** The response of an optical fiber sensor compared with an extensometer over a period of 570 000 cycles and tested at a frequency of 1.0 Hz and $R = 0.1$. The strain response has been plotted at 50 000 and 300 000 cycles. $R$ is the stress ratio, and is defined as the minimum stress divided by the maximum stress [29].

the fatigue testing machine. With reference to the fatigue data presented in Fig. 2.26, the data suggest a drift from the embedded EFPI sensor as a function of number of cycles. This drift in the signal from the embedded sensor was attributed to a possible failure of the interface between the sensor and the resin matrix, and also to a failure of the EFPI sensor in the vicinity of the fusion splice. A similar observation was reported by Levin and Nilsson [23].

Unlike quasi-static mechanical testing, the interpretation of fatigue data using test specimens with embedded optical fiber sensors is not simple because of difficulties in defining a failure criterion. Firstly, for example, with fatigue-testing machines which do not have amplitude and waveform control, there is no guarantee that the test specimen actually experienced the specified load-range at the frequency. This is a major concern with materials which show a significant change in compliance during the fatigue test. Secondly, fatigue testing of materials which show a rate-dependence (for example, glass fibers) should be carried out at a constant rate of stress application, whereas the majority of the published works in this field tend to use a constant frequency. Thirdly, the dependence on the 'number of cycles to failure' as a failure criterion is not acceptable because of the above-mentioned reasons, and also because it does not impart any information on the mode of damage propagation. For example, a composite may show extensive delamination and surface ply separation prior to failure. Although this may be acceptable in a laboratory-based test, it will not be acceptable if the test specimen is evaluated in-service. Fourthly, the issue of the failure criterion has to be put into context with regards to the actual remit of the embedded optical fiber sensor. For example, the maximum permissible strain for a real structure may only be a small fraction of the ultimate failure strain for the composite. In this case, it is highly unlikely that the damage modes and mechanisms discussed previously would matter as far as the reliability of the embedded optical fiber sensor is concerned. However, if the embedded optical fiber sensor is being used to study damage initiation and propagation in AFRCs, then the embedded sensor must perform up to the failure of the composite. Finally, it is customary to validate the output from the embedded optical fiber sensor against an independent strain/temperature source. With respect to quasi-static strain measurements, surface-mounted strain gauges are frequently used as the reference. With fatigue tests, a surface-mounted extensometer or a fatigue-rated electrical resistance strain gauge is generally used. These techniques are not ideal because if surface delaminations or surface cracking develops during the fatigue test, it would invalidate the data. This is a problem with embedded optical fiber sensors, because there is some evidence to suggest that the introduction of compressive load excursions into the fatigue load can result in premature delaminations [33].

### 2.3.8 Interfacial and environmental interactions

*(a) Cure monitoring*

There is some evidence to suggest that certain compositions of optical fibers can preferentially adsorb or chemisorb components of the resin system onto the fiber surface. For example, Cossins *et al.* [34] found that the spectra recorded using an evanescent sensor design was primarily due to the amine hardener. Further evidence for the physiosorption and condensation of aminosilanes on silica surfaces was presented by Vrancken *et al.* [35]. The possible interactions between the optical fiber sensor and the second phase in the resin system was highlighted previously.

### 2.3.9 Interconnection

This topic has to be the Achilles' heel of embedded optical fiber sensor technology in AFRCs. The subject is of significant practical importance, and deserves far greater attention than it is receiving on the basis of information in the public domain.

It is proposed that four basic techniques could be deployed to power and interrogate the embedded optical fiber sensor: (i) direct coupling; (ii) trailing lead-in and lead-out optical fibres; (iii) embedded connectors; (iv) telemetry.

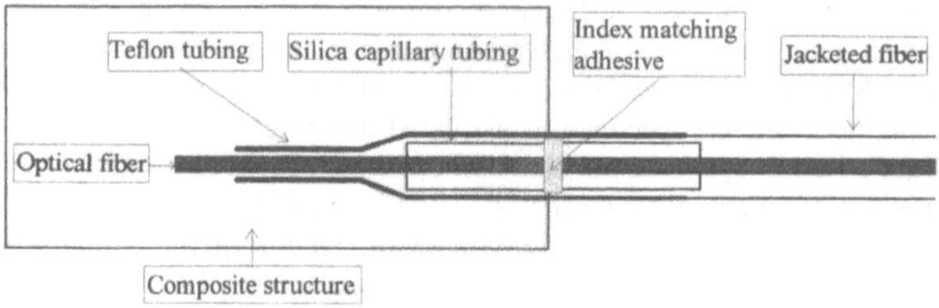


**Fig. 2.27** Illustration of the interconnection method proposed by Talat [36].

(i) Direct coupling. Here the light source and detector are connected directly to the AFRC and become permanent features of the structure. This technique is ideally suited for intensity-based optical fiber sensors.

(ii) Trailing optical fibers. This is probably the most risky technique, as significant protection has to be given to the trailing fibers in order to ensure that they do not fracture at the entry and exit points into and out of the AFRC. Conventional fusion splicing or mechanical connectors can be used to link the sensor to the light source and any associated instrumentation.

(iii) Embedded connectors. Badcock [29] recently reviewed potential interconnection methods for embedded optical fiber sensors and a summary of his findings are presented below.

A schematic illustration of the interconnection design proposed by Talat [36] is presented in Fig. 2.27.

As is shown in Fig. 2.27, the interconnection technique proposed by Talat involves the insertion of a Teflon sleeve over the embedded optical fiber and the location of a silica capillary tube within the Teflon tubing as an alignment-aid. Both the Teflon and the silica capillary are left to protrude from the edge of the composite. Subsequent to curing, a cleaved optical fiber is inserted into the capillary and secured in position using an index-matched adhesive.

The method proposed by Morgan *et al.* [37] (Fig. 2.28) involves a capillary tube being inserted over part of the embedded optical fiber and positioned with a portion of it inside the composite preform. After curing the preform, the protruding optical fiber is cut back to coincide with the end of the capillary and a second capillary tube with a graded index lens (GRIN) is then positioned over the partly embedded capillary tube. The connector relies on a second GRIN lens which is capable of focusing the light from first optical fiber onto a second fiber with a similar connection assembly.

The interconnection technique proposed by Lu and Blaha [38] is shown in Fig. 2.29. Here the fiber is protected from damage by the use of a FC compatible connector assembly near the edge of the composite. A standard ceramic ferule is placed around the optical fiber within a metal insert. This is positioned within the prepreg at the time of laying-up. Once the prepreg is cured, the metal insert is trimmed flush with the edge of the composite. An FC-compatible adapter is

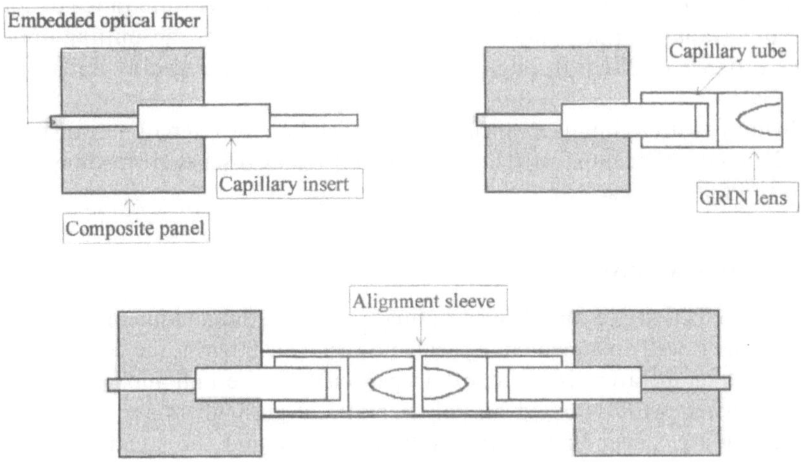


**Fig. 2.28** Illustration of the interconnection design method proposed by Morgan *et al.* [37].

then placed inside the metal assembly. A ceramic split sleeve is then placed over the ceramic ferule, thus facilitating connection to a standard FC connector.

(iv) Telemetry. In the long term, there is a need to develop techniques to power-up and transmit data from the sensors remotely, i.e. without the need for trailing optical fibers. This technology will make it possible to integrate optical fibers into almost any AFRC manufacturing process. A limiting factor

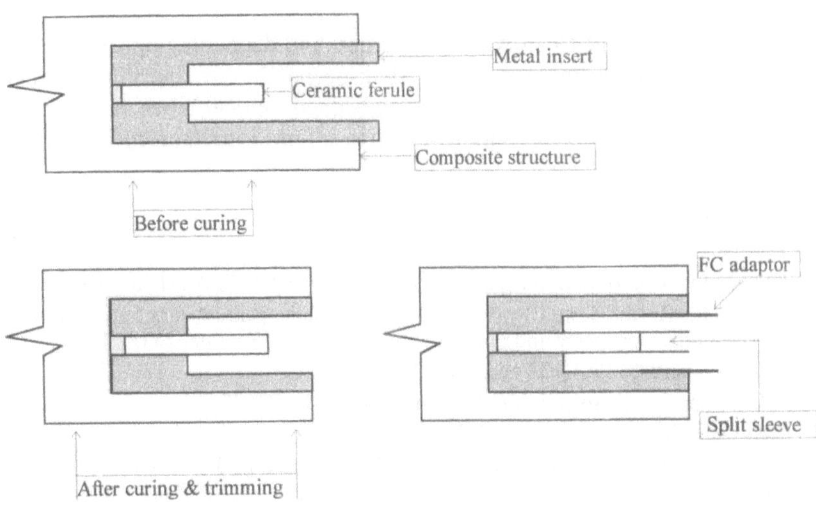


**Fig. 2.29** Schematic illustration of the interconnection system proposed by Lu and Blaha [38].

at present is the relative bulkiness and the operating-temperature range of commercially available transponder systems. The feasibility of transmitting optical data a distance of 20 cm through air was recently demonstrated by Zabaronick *et al.* [39].

Obviously with techniques which rely on embedding the interconnection device into the composite, most of the concerns and issues discussed previously will apply to these devices.

### 2.3.10 Cross-sensitivity

The cross-sensitivity of fiber Bragg gratings towards axial and lateral strain and temperature is well documented. Apart from academic interest, the decoupling of strain from temperature is an important perquisite for certain applications. For example, when AFRCs based on polyaramid and glass fibers are subjected to fatigue loading, depending on the cyclic frequency and the stress amplitude, the surface temperature can rise to in excess of 50°C above ambient.

The ingress of moisture into composites should be considered because with prolonged exposure to moisture the glass transition temperature of the resin system will be lowered, and this can also resulting in swelling. Detailed studies have to be undertaken on the effect of moisture ingress into composites with embedded optical fiber sensors. An area which needs particular attention is the entry and exit ends of the optical fibers in the composite. This is because if these locations are not protected with a water-impermeable coating, moisture may be able to wick along or diffuse through this interface.

### 2.3.11 Repair

In the event of damage to an AFRC structure, and depending on the severity of the damage, repair techniques such as stick-on patches are currently available. However, the repair of AFRCs with embedded optical fiber sensors would pose a very serious problem, and no effective solutions are currently available.

## REFERENCES


1. *Engineering Materials Handbook. Vol. 1. Composites* (1987) ASM International.
2. Klocek, P., Roth, M. and Rock, R. D. (1987) Chalcogenide glass optical fibres and image bundles: property and applications. *Optical Engineering*, **26** (2), 88–95.
3. Martin, A. (1998) Ph.D. Thesis, Brunel University.
4. UK Composites Association & Hunting Engineering (1997).
5. Ciba Geigy Trade Literature (1981) Fibredux 914.
6. Brooks, D. (1997) Department of Materials Engineering, Brunel University, unpublished results.
7. Badcock, R. A. and Fernando, G. F. (1995) An intensity-based optical fibre sensor for fatigue damage detection in advanced fibre-reinforced composites. *Smart Materials and Structures*, **4**, 223–230.

8. Martin, D. A. (1987) Optical fibre coating evaluation for composite embedment applications. *Materials Research Society Symposia Proceedings*, Pittsburgh, PA, **88**, 19–26.
9. DiFrancia, C., Claus, R., Hellgeth, J. W. and Ward, T. C. (1989) Structure/property correlation of several polyimide optical fibre coatings for embedding in an epoxy matrix. *SPIE*, **1170**, 505–512.
10. DiFrancia, C., Ward, T. C. and Claus, R. O. (1996) The single fibre pull-out test. 2. Quantitative evaluation of an uncatalysed TGDDM/DDS epoxy cure study. *Composites Part A*, **27A**, 613–624.
11. Oxford Electronics Ltd. (1997) Trade literature.
12. Sirkis, J. S. and Lu, I. P. (1993) On interphase modelling for optical fibre sensors embedded in unidirectional composite systems. ASME, Aerospace Division, *Proceedings of the 1993 ASME Winter Annual Meeting*, New Orleans, LA, **35**, 419–426.
13. Sirkis, J. S. and Lu, I. P. (1995) On interphase modeling of optical fibre sensors embedded in unidirectional composite systems. *Journal of Intelligent Material Systems and Structures*, **6**, 199–209.
14. Sirkis, J. S. and Grande, R. (1995) Nonlinear analysis of composite strength loss due to embedded ductile metal coated optical fibre sensors. *Proceedings of SPIE*, **2442**, 152–159.
15. Dasgupta, A. and Sirkis, J. S. (1992) Importance of coatings to optical fibre sensors. *AIAA Journal*, **30** (5), 88–99.
16. King, W. W. and Aloisio, C. J., Jr. (1997) Thermomechanical mechanism for delamination of polymer coating from optical fibers. *Journal of Electronic Packaging, Transactions of the ASME*, **119** (2), 133–136.
17. Matthewson, M. J., Kurkian, C. R. and Hamblin, J. R. (1997) Acid stripping of fused silica optical fibres without strength degradation. *Journal of Lightwave Technology*, **15** (3), 490–497.
18. Matejec, V., Hayer, M., Pavlovic, P., Kuncova, M., Kuncova, G. and Guglielmi, M. (1995) Effect of preparation of sol-gel coatings on the strength of optical fibres. *Journal of Sol-Gel Science and Technology*, **5** (3), 193–199.
19. Schultheisz, C. R., Schutte, C. L., McDonough, W. G., Macturk, K. S. and McAuliffe, M. (1996) Effect of temperature and fibre coating on the strength of E-glass fibres and the E-glass/epoxy interface for single-fibre fragmentation samples immersed in water. *ASTM Special Technical Publication*, No. 1290, pp. 103–131, ASTM, Conshohocken, PA.
20. Doremus, R. H. (1995) Diffusion of water in silica glass, *Journal of Materials Research*, **10** (9), 2379–2389.
21. Thomas, J. L. (1995) Interface region in glass fibre-reinforced epoxy resin composites. 2. Water absorption, voids and the interface. *Composites*, **26** (7), 477–485.
22. Evans, T. (1997) High temperature optical fibre sensors. Final Year Project Dissertation, Department of Materials Engineering, Brunel University.
23. Levin, K. and Nilsson, S. (1996) Examination of reliability of fibre optic sensors embedded in carbon/epoxy composites. *Proceedings SPIE – International Society for Optical Engineering*, **2779** pp. 222–229.
24. Martin, A., Fernando, G. F. and Hale, K. (1997) Impact damage detection in filament wound tubes using embedded optical fibre sensors. *Smart Materials and Structures*, **6**, 470–476.
25. Melle, S. M., Liu, K. and Measures, R. M. (1993) Practical fibre-optic grating strain gauge system. *Applied Optics*, **32** (19), 3601–3609.

26. Fernando, G. F., Liu, T., Crosby, P. A., Doyle, C., Martin, A., Brooks, D., Ralph, B. and Badcock, R. A. (1997) A multi-purpose optical fibre sensor design for fibre reinforced composite materials. *Journal of Measurement Science and Technology*, **8**, 1065–1079.

27. Jensen, D. W., August, J. A. and Pascual, J. (1991) Compressive strength and stiffness reductions in graphite/bismaleimide laminates with embedded fibre-optic sensors. *ADPA/AIAA/SPIE Conference on Active Materials and Adaptive Structures*, November, Alexandria, VA. Editor G. J. Knowles, IOP, Bristol, pp. 129–134.

28. Mall, S., Dosedel, S. B. and Holl, M. W. (1996) Performance of graphite–epoxy composites with embedded optical fibres under compression. *Smart Materials and Structures*, **5** (2), 209–215.

29. Badcock, R. A. (1998) Ph.D. Thesis, Brunel University.

30. Nightingale, C. (1996) Final Year Project, Materials Engineering, Brunel University.

31. DiFrancia, C., Ward, T. C. and Claus, R. O. (1996) Single-fibre pull-out test. 1. Review and interpretation. *Composites. Part A. Applied Science and Manufacturing*, **27** (8), 597–612.

32. Badcock, R. A. (1997) DERA Farnborough, SMART Structures Department, Guildford, Private communication.

33. Brooks, D. (1997) Sensors Research Group, Brunel University, unpublished results.

34. Cossins, S., Connell, M., Cross, B., Winter, R. and Kellar, J. (1996) In-situ near-IR cure monitoring of a model epoxy matrix composite. *Applied Spectroscopy*, **50** (7), 900–904.

35. Vrancken, K. C., Van Der Voort, P., Possemieers, K., Grobet, P. and Vansant, E. F. (1994) The physisorption and condensation of aminosilanes on silica gel in *Chemical Modified Surfaces* (eds J. J. Pesek and I. E. Leigh), Royal Society of Chemistry, London, pp. 46–57.

36. Talat, K. (1990) Smart skins and fibre-optic sensors application and issues. *Proceedings SPIE Conference on Fibre Optic Smart Structures and Skins III*, **1370**, 103–114.

37. Morgan, R. E., Ehlers, S. L. and Jones, K. L. (1991) Composite embedded fibre optic data links and related material/connector issues. *Proceedings SPIE Conference on Fibre Optic Smart Structures and Skins IV*, **1588**, 189–197.

38. Lu, Z. J. and Blaha, F. A. (1991) Application issues of fibre optic sensors in aircraft structures. *Proceedings SPIE Conference on Fibre Optic Structures and Skins IV*, **1588**, 276–281.

39. Zabaronick, N., Sherrer, D. W., Claus, R. O., Murphy, K. A., Duncan, P. and Shinpaugh, K. (1996) Remote optical interrogation of embedded optical fibre sensors. *Proceedings SPIE – International Society for Optical Engineering*, **2718**, 234–238.

# 3

# The application of optical fiber sensors in advanced fiber reinforced composites. Part 2: Cure monitoring

P. A. Crosby and G. F. Fernando

## 3.1 INTRODUCTION

Optical-fiber-based cure monitoring in composites or resin systems used in the production of composites can be classified into qualitative or quantitative techniques. In the qualitative approach, correlation is sought between a specified property of the resin as a function of processing time and conditions. For example, the change in the refractive index of the resin system as a function of cure.

The quantitative approach involves obtaining information on the relative chemical concentrations of the active or chemical species which participate in the reaction. This approach usually involves spectroscopy-based optical fiber interrogation techniques.

### 3.1.1 The need for cure monitoring

*(a) On-line process monitoring*

At present there are no reliable tools which can be used to obtain real-time quantitative information on the rate, extent and homogeneity of the cure. The term 'cure' is used to describe a series of complex chemical reactions which take place when the composite is processed, usually by the application of heat. Where advanced fiber-reinforced composites (AFRCs) are used for load-bearing applications, the cure schedule is strictly regulated. The introduction of new resin systems usually requires a lengthy qualification process. The availability of a reliable quantitative cure monitoring technique could result in a faster introduction of new products into the marketplace. In addition to this, on-line process monitoring can be used to carry out multibatch processing of components with different resin systems.

With reference to potential techniques for cure monitoring of AFRCs, a technology for dielectric cure monitoring has been established since the late 1950s,

*Optical Fiber Sensor Technology*, Vol. 3. Edited by K. T. V. Grattan and B. T. Meggitt.
Published in 1998 by Kluwer Academic Publishers, London. ISBN 978-1-4419-4736-9

but some possible disadvantages of this technique are: (i) the operation of the dielectric cure sensor can be affected by electromagnetic magnetic interference; (ii) the dielectric cure sensor can only give qualitative information on the cure kinetics of the resin system; (iii) the sensor system is relatively expensive; (iv) the output from the sensor may be influenced by the concentration of impurities in the resin system, the moisture content and the presence of conducting material in the composite.

On the other hand, optical fiber sensors embedded into the composite preform are immune to electromagnetic interference and can be used to obtain quantitative information on the chemical concentrations of the active functional groups in the resin. Access to real-time data on the chemical nature of the resin system, at specified locations within the composite preform, is a prerequisite in order to develop reliable and quantitative on-line process monitoring and optimization techniques. It is important to appreciate that most aerospace-based thermosetting resins have a finite shelf-life, typically 18 months when stored at −18°C. Obviously the storage conditions, storage time, number of freeze/thaw cycles and cumulative time spent outside the freezer will have an influence on the cure kinetics of the resin system. With the advent of quality assurance procedures, the integrity of the prepreg system is strictly regulated in certain industries. However, there would be significant economic advantages if an *in situ* sensor system could be used to optimize the processing conditions in order to compensate for the variable chemical state of the resin.

### *(b) Real-time temperature measurements*

The cure reaction is itself an exothermic reaction, and hence it is necessary that the heat transfer into and out of the composite preform is controlled accurately. With reference to temperature measurements during processing, the current practice is to use thermocouples at the outer edges of the preform to obtain data on the 'temperature profile'. However, in general, the actual temperature(s) away from the edges is not measured. This is an area of concern for large, complex and thick preforms. A number of optical-fiber-based sensors have been developed in recent years, and a discussion of these techniques is presented in Section 3.3.

### *(c) Residual stresses and dimensional stability*

Residual stresses (fabrication stresses) result in the final structure/component being distorted (warped). The term 'spring-back' is used in the industry to describe this phenomenon. The current practice to combat spring-back is to modify the tooling to compensate for the warping. The optimum tool design is obtained by a costly trial and error process.

### *(d) Effect of the cure schedule on the matrix-dominated properties of AFRCs*

This is an issue of paramount importance with respect to the long-term durability and structural integrity of advanced fiber-reinforced composites. Although the general mechanical properties of AFRCs are dominated by the reinforcing fibers and the nature of the interfacial bond strength, the matrix-dominated properties such as the interlaminar shear strength, resistance to moisture ingress and compressive strength can influence the extent of cure.

## 3.2 THE CURE (CROSS-LINKING) PROCESS

The curing of thermosets involves the transformation of the material from a viscous liquid to a glassy solid via the application of heat. Prepreg-based AFRCs are generally processed in an autoclave [1], but other processing methods such as microwave [2, 3] and electron beam [4] curing are also gaining popularity. Obviously, optical-fiber-based sensor systems are ideal for use in these environments.

The cure process begins with the formation and growth of chains, which then branch and cross-link. As the reaction proceeds, the molecular weight increases rapidly, and eventually several chains become linked together into networks of high molecular weight. Associated with this conversion process are changes in the chemical, physical and mechanical properties of the resin. In order to facilitate the on-line process monitoring and optimization of the cure process, it is necessary to monitor certain system parameters such as the relative concentrations of the active functional groups in the resin system and the temperature. However, other parameters can also be monitored to infer the extent and rate of the reaction, including the heat of reaction, the refractive index, the dielectric constant and the ultrasonic wave propagation velocity through the material. This literature review outlines the current methods that are used to establish the degree of cure of epoxy-resin-based materials, together with optical-fiber-based methods capable of monitoring cure *in situ* during processing.

### 3.2.1 Epoxy resin systems and their applications

Epoxy or epoxide resins can be defined as molecules containing one or more epoxide groups or oxirane rings, the structure of which is shown in Fig. 3.1 [5]. The prefix 'glycidyl' is often used to describe resins with terminal epoxide groups, along with the terms 'ether', 'ester' or 'amine' etc., according to the group attached to the third carbon atom. Epoxy resins can be made by reacting epichlorohydrin and multifunctional acids, amines or alcohols. The epichlorohydrin residue is generally referred to as the glycidyl group. The most common of the commercial epoxy resins is the diglycidyl ether of bisphenol-A (DGEBA) shown in Fig. 3.2. Values of *n* vary from 0 to 12, giving these resins a wide range of physical forms, ranging from low viscosity liquids to tack-free solids.

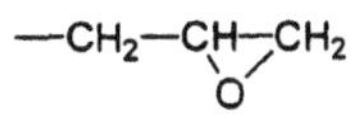


**Fig. 3.1** Structure of oxirane ring.

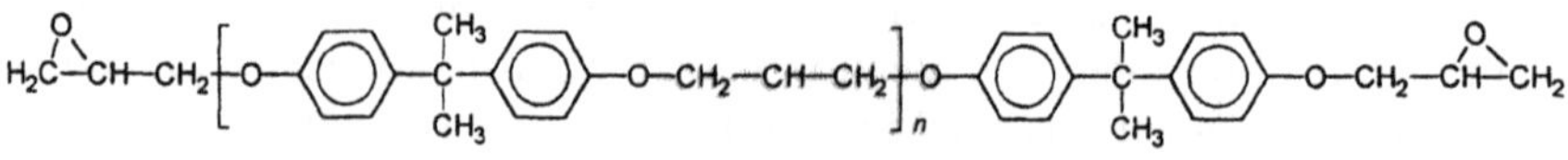

**Fig. 3.2** Structure of diglycidyl ether of bisphenol-A (DGEBA) epoxy resin.

The strained epoxide ring is highly reactive, and this allows the resins to be cross-linked with a variety of hardeners (cross-linking agents). As a result of this versatility, these products have found use in protective coatings, adhesives for most substrates, flooring, tooling compounds for moulds, textiles and fiber-reinforced plastics [6].

For aerospace applications where higher operating temperatures are encountered and high mechanical strengths are necessary, epoxy resins based on tetraglycidyl-4,4′-diaminodiphenyl methane (TGDDM or TGMDA) have found widespread use (Fig. 3.3). These are usually cured with diaminodiphenyl-sulphone (DDS) and/or boron trifluoride monoethylamine ($BF_3$–MEA) for use as a matrix material in AFRCs. Epoxy resins are particularly suitable as matrix resins since they produce easily handled prepregs which require relatively low temperatures for processing [7] (120–180°C), and the extent of shrinkage upon curing is relatively low. Table 3.1 lists common hardeners used as curing agents for epoxy resin systems.

### 3.2.2 Epoxy-amine cure reactions

Multifunctional primary and secondary amines are widely used as curing agents for epoxy resin systems. It is generally accepted that there are two main reactions which can occur during epoxy-amine cure. Primary amines initially present in the reaction mixture react with the epoxide ring to produce secondary amine and hydroxyl groups. Further reaction of the secondary amine groups with the epoxide ring leads to the formation of tertiary amine groups (Fig. 3.4). Additional reactions of epoxy to primary or secondary amine groups are catalyzed by acids such as Lewis acids, phenols and alcohols [8]. Hydroxyl groups formed by these reactions also act as catalysts. Therefore, it can be seen that autocatalysis occurs during the early stages of cure.

Rozenberg [9] states that when the curing agent is an aliphatic amine, it is possible for the tertiary amine formed in the secondary amine-epoxy reaction also to catalyze the epoxy group polymerization. However, this reaction generally

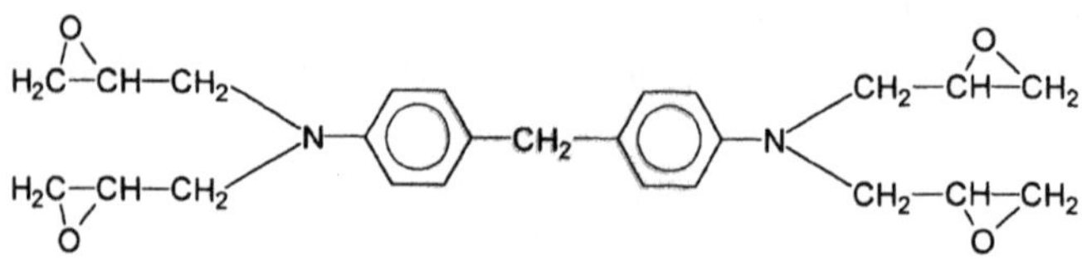

**Fig. 3.3** Structure of tetraglycidyl-4,4′-diaminodiphenyl methane (TGDDM).

**Table 3.1** Common epoxy resin curing agents

| Curing agent | Example structure | Uses |
|---|---|---|
| Aromatic amines and cycloaliphatic amines | 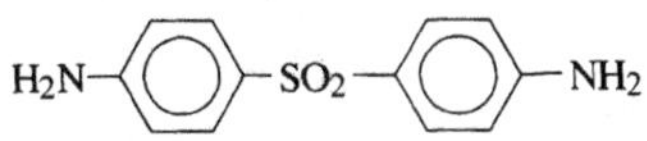 4,4-diaminodiphenylsulphone (DDS) | Composites, coatings, adhesives |
| Dicyandiamide and derivatives | $H_2N-C(=NH)-NH-C\equiv N$ Dicyandiamide | Printed circuit boards, composites, coatings and adhesives |
| Aliphatic amines, polyamide resins and derivatives | $H_2N-(CH_2)_2-NH-(CH_2)_2-NH-(CH_2)_2-NH_2$ Triethylenetetramine | Flooring, paints, adhesives, potting and tooling compounds |
| Acid anhydrides | 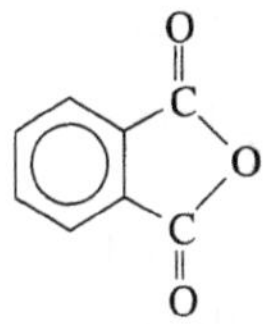 Phthalic anhydride | Electrical castings, coatings and composites |
| Phenol formaldehyde and amino-formaldehyde resins | 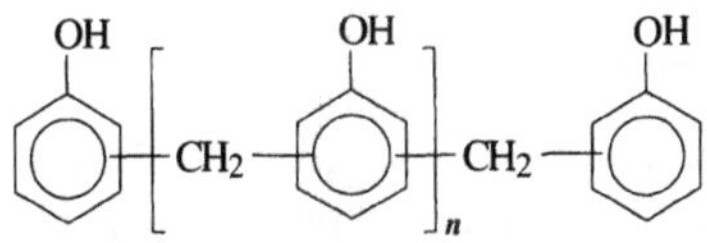 Phenol-novolac resin | Can coatings and drum linings |
| Polymercaptans | $R-[O-(CH_2CH_2CH_2O)_{\overline{n}}CH_2-CH(OH)-CH_2-SH]_3$ $n = 1$ to 2; R = aliphatic hydrocarbon | Fast-curing adhesives |
| Carboxylic acid functional polyesters | 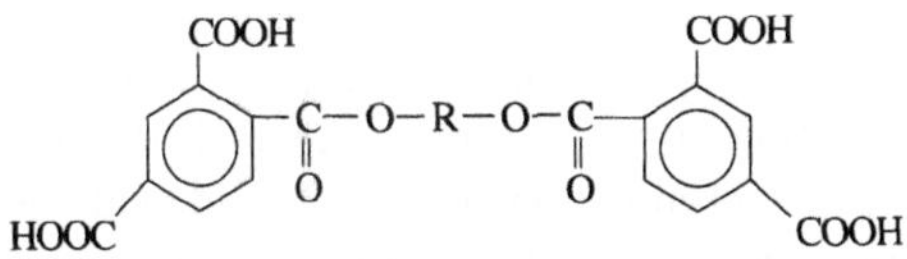 Carboxyl terminated polyester resin where $R$ = polyester chain | Powder coatings |

occurs only at elevated temperatures. Additionally, when aromatic amines are used as curing agents, such reactions do not take place (since the reaction is dependent on the steric accessibility and basicity of the tertiary amine). Generally, for epoxy-amine systems, the tertiary amine-catalysed reaction of epoxy is unfavored compared with amine-addition reactions, and consequently little chain growth occurs via this route unless a good catalyst is used [10]. Shechter and Wynstra [8] and Dusek *et al.* [11] showed that only when epoxide groups were present in excess did the hydroxyl groups formed in the amine-epoxy addition reactions add to the epoxide ring. This reaction is known as etherification (Fig. 3.4).

The term post-curing is used to describe the situation where the composite is subjected to an additional cure schedule but at a higher temperature. The

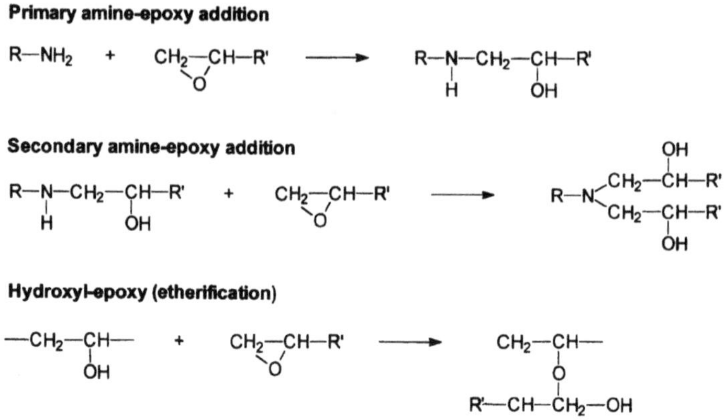


**Fig. 3.4** Generalized reaction scheme of epoxy-amine systems.

functions of the post-cure cycle are to react off any residual active fundamental groups and to increase the glass transition temperature of the resin system.

### 3.2.3 Techniques for cure monitoring

With reference to the generalized reaction schemes shown in Fig. 3.4 for an epoxy/amine resin system, the cross-linking process can be monitored via the techniques described below.

#### *(a) Quantitative analysis*

The reaction between the epoxy and amine functional groups results in the depletion of these two functional groups along with the formation of hydroxyl (OH) groups. Therefore, techniques such as mid- and near-infrared spectroscopy, nuclear magnetic resonance spectroscopy and high-performance liquid chromatography (HPLC) can be used to quantify the relative concentrations of the active functional groups.

#### *(b) Thermal analysis*

Differential scanning calorimetry (DSC) has been used extensively to study the cure kinetics of thermosets. DSC is used to obtain data on the enthalpy of the reaction between the active functional groups in the resin, and it is also used to measure the heat capacity and the glass transition temperature of the cross-linked resin system.

#### *(c) Correlation studies*

The cross-linking reaction scheme shown in Fig. 3.4 also results in changes to the (i) physical, (ii) mechanical, (iii) optical, (iv) chemical, (v) acoustic, (vi) rheological and (vii) electrical properties of the resin system as a function of cure. Hence, a number of corresponding techniques have been demonstrated as being suitable for cure monitoring as a function of the extent of cure: (i) density increase; (ii) increase in the tensile, compressive and flexural properties; (iii) increase in the refractive index; (iv) improved resistance to swelling in solvents; (v) retardation of the ultrasonic wave propagation; (vi) an increase in the viscosity; (vii) an increase in the glass transition temperature; (viii) depletion of active functional groups in the resin system.

We now concentrate on a selected number of cure monitoring techniques which are of relevance to the utilization of optical-fiber sensors. In addition to this, it is necessary to consider in some detail the non-optical-fiber-based techniques which are routinely used to characterize the cure of thermosets. This is necessary because prior to the implementation of any sensor system, the end-users of AFRCs would want to see a direct correlation between the established techniques such as DSC and Fourier transform infrared (FTIR) and the proposed sensor.

### 3.2.4 Near-infrared spectroscopy of epoxy resin systems

The near-infrared spectral region can be defined as the wavelength range from 700 nm to about 2500 nm. Chemists tend to refer to the infrared spectrum in terms of its frequency, but the frequency is not quoted in Hertz but in wavenumbers ($cm^{-1}$): the wavenumber is the number of waves per centimeter and is equal to the reciprocal of the wavelength in centimeters. Absorption bands in the near-IR are due to overtones and combinations of the fundamental mid-IR molecular vibration bands. Figure 3.5 shows the principal vibration modes of a molecule $AX_2$ ($-NH_2$, $-CH_2$, etc.). Overtone bands (harmonics)

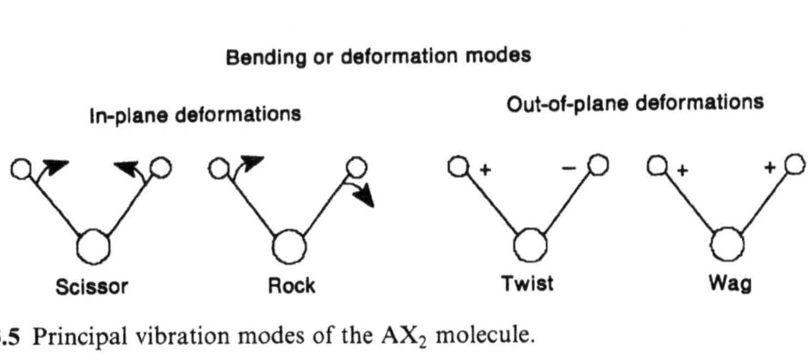


**Fig. 3.5** Principal vibration modes of the $AX_2$ molecule.

appear as approximate integer multiples of fundamental vibrations. For example a strong absorption at $2500\,cm^{-1}$ will give rise to a weaker absorption at $5000\,cm^{-1}$.

Although all fundamental vibration modes can show overtones in the near-IR, the most prominent of those observed are those of O–H, C–H and N–H groups. This is because first overtones of other bands such as C–O occur in the mid-IR, and so only much weaker second or third overtones are seen in the near-IR region. As well as overtones, combinational bands can also occur. These result from simultaneous excitation of two different fundamentals to give beats which are a combination of difference frequencies. For example, absorptions at $x\,cm^{-1}$ and $y\,cm^{-1}$ interact to produce two weaker frequencies at $(x \pm y)\,cm^{-1}$.

These combination bands usually involve O–H, C–H or N–H stretches with one or more bending or rocking modes. Many combinations are possible, which makes band assignments difficult [12]. Several review papers have been published on near-IR spectroscopy and its application to quantitative and qualitative analysis [12–19]. These serve as useful references for the theory, for peak assignments and for data of absorptions of various standard epoxies and amines. Earlier published papers are also useful for peak assignments [20–27].

### 3.2.5 Application of near-infrared spectroscopy to monitor cure of epoxy resin systems

Near-IR spectroscopy has been applied to the quantitative analysis of epoxy resin systems by many authors. For quantitative analysis, the near-IR region has advantages over the mid-IR as many of the bands in the mid-IR region overlap, and this makes quantitative analysis difficult. On the other hand, near-IR spectra are simplified and bands are well resolved, allowing more accurate quantitative measurements. As discussed previously, the functional groups of primary importance are due to vibrations of N–H, O–H and C–H. These bands will be well resolved, and therefore accurate quantitative measurements are possible.

There are three basic types of epoxy resin which have generally been used in the literature to investigate the techniques of near-IR spectroscopy cure monitoring: (i) 1,2-epoxy-3-phenoxypropane, a simple mono-functional epoxy more generally known as phenyl glycidyl ether (PGE), which is used for model reaction studies [28–31]; (ii) diglycidyl ether of bisphenol-A (DGEBA) [26, 32–37], a commercially available bi-functional resin; (iii) tetraglycidyl 4,4′-diaminodiphenylmethane (TGDDM) [10, 38–41], a commercially available tetra-functional resin. However, the near-IR spectra of these three types of resin differ very little, with major absorption peaks being found in similar locations. Table 3.2 shows the position of near-IR absorption bands for DGEBA resins.

For epoxy resin systems using diamine hardeners, band positions are dependent on whether the amine is aromatic or aliphatic. Much of the literature on near-IR analysis of epoxy resin systems concerns aromatic amine curing agents [10, 24, 25, 28, 32, 33, 35, 38, 40, 41]. Publications concerning aliphatic amines

**Table 3.2** Near-infrared absorption bands of diglycidyl ether of bisphenol-A (DGEBA)

| Group | Assignment | Wavelength | Comments |
|---|---|---|---|
| Epoxy | 2nd overtone of the C–H stretching vibration | 1.159 μm (8627 $cm^{-1}$) | Weak band [26, 42] |
| | 1st overtone of C–H stretch in epoxy group | 1.650 μm (6060 $cm^{-1}$) | Can be overlapped by adjacent bands [26, 42] |
| | Combination of the C–H stretching fundamental at 3050 $cm^{-1}$ with the C–$H_2$ deformation band at 1460 $cm^{-1}$ [20, 26, 33, 42] (Chike *et al.* [34] assigns this band as a combination of the C–O fundamental stretch (~900 $cm^{-1}$) and the fundamental C–H stretch (~3050 $cm^{-1}$)) | 2.206 μm (4532 $cm^{-1}$) | Most suitable for quantitative analysis [26, 32–34, 42, 43] |
| Hydroxyl | Combination of O–H stretching and deformation vibrations [33, 42] | 2.090 μm (4785 $cm^{-1}$), 2.050 μm (4878 $cm^{-1}$) | |
| | 1st overtone of the O–H stretching fundamental | 1.432 μm (6983 $cm^{-1}$) | Hydrogen-bonded peak [26, 32, 33, 42] (hydrogen bonding causes a shift to longer wavelengths) [42] |
| C–H (aromatic) | 1st overtone of C–H stretching vibration | 1.668 μm (5995 $cm^{-1}$) | Internal calibration peak [26, 33, 42] |
| | Combination band of the aromatic conjugated C=C stretch (1625 $cm^{-1}$) with the aromatic C–H fundamental stretch (3050 $cm^{-1}$) [33, 34] | 2137 nm (4680 $cm^{-1}$), 2164 nm (4622 $cm^{-1}$) | Used as an internal standard by Min *et al.* [32] |
| C–H (aliphatic) | –CH overtones of –$CH_2$, –$CH_3$ [27, 33] | 1.700 μm (5882 $cm^{-1}$) | The series of bands at 1770 nm, 1734 nm, 1670 nm and 1690 nm can all be attributed to aliphatic C–H groups [27] |
| | 2nd harmonic of a C–H stretch of the methyl group [26] | 1.188 μm (5232 $cm^{-1}$) | |
| | –$CH_3$ combination band [20, 34] | 2.3 μm (4350 $cm^{-1}$) | |
| | –$CH_2$ combination band of the methylene group [16, 34] | 2.4 μm (4170 $cm^{-1}$) | |
| | Terminal methylene band due to unsaturation [20] | 2.1 μm (4760 $cm^{-1}$) | |

**Table 3.3** Near-infrared band assignments for aliphatic and aromatic amines

| Group | Assignment | Wavelength (nm) | Comments |
|---|---|---|---|
| Aromatic primary amine | Combination of bending and stretching vibrations | 1974 nm [38–41], 1970–1995 nm [32], 1989 nm [28] | DDS hardener |
| | Overtone of symmetric and asymmetric stretches | 1496 and 1528 nm [38–41], 1509 nm (max. of two overlapping peaks) [28], 1494–1520 nm [32] | DDS hardener |
| | | 2205 nm [32] | Unknown band attributed to primary amine [32] in DDS hardener |
| Aromatic secondary amine | Overtone of N–H stretch | 1496 nm [38–41], 1494–1520 nm [32], 1502 nm [29] | DDS hardener |
| Aliphatic primary amine | N–H combination | 2025 nm [23] | Decylamine |
| | 1st overtone of symmetric N–H stretch | 1526 nm [23] | Decylamine |
| Aliphatic secondary amine | 1st overtone of N–H symmetric stretch | 1543 nm [23] | Di-*n*-hexylamine |

may also be found in the literature [23, 26, 34]. Table 3.3 lists the peaks present in the near-IR region for aromatic and aliphatic amines and assigns their origin.

There are generally two near-IR spectroscopy methods which have been used in the literature to follow the cure of epoxy resin systems: (i) *in situ* methods [26, 28–31, 34, 38–41] whereby the cure reaction occurs in the spectrometer and the spectra are collected in real time; (ii) *ex situ* methods [32, 33, 35, 44] whereby the resin is cured externally from the spectrometer, with samples being removed periodically to obtain spectra.

## 3.3 *IN SITU* CURE MONITORING METHODS

An early paper in this area of research was presented by Dannenberg [26]. In his work, near-IR spectroscopy was used to follow the cure of a diglycidyl ether of bisphenol-A resin with ethylenediamine at room temperature over a period of 12 h. The epoxy peak height at 2205 nm was used to determine the epoxy value after ratioing with the nonchanging aromatic C–H absorbance band at 1668 nm. The data presented for the epoxy value was seen to drop quite steeply from the first to the second data point. After this initial decrease, the shape of the epoxy value against time curve showed the 's-shape' typical of these systems. Other peaks analysed by the authors included the NH bands at 2025 and 1540 nm and the OH peak at 1428 nm.

Chike *et al.* [34] used a near-IR spectroscopy method to investigate the reaction between diglycidyl ether of bisphenol-A and diethylamine at room temperature. Only the epoxy band at 2210 nm was analysed using a corrected peak height method, and the results were compared with those obtained for Raman studies over the range 600–1350 $cm^{-1}$. An excellent correlation was seen for quantitative data obtained by the two methods.

Fu and Schlup [28] describe an *in situ* near-IR spectroscopy method to study the reaction between phenyl glycidyl ether and different aromatic amines. Samples were cured in a heated transmission cell with data acquisition throughout the reaction. Although a number of peaks were identified as being suitable for quantitative analysis, only the epoxy peak at 2206 nm was monitored. The intensity of the epoxy peak was ratioed with the aromatic C–H band intensity and subsequently converted into an epoxy conversion rate format. The results for one isothermal temperature (90° C) were then fitted to the rate model proposed by Ryan and Dutta [45]. In further papers by the same research group, Xu *et al.* [46] describe the application of near-IR spectroscopy to the study of the reactions of phenyl glycidyl ether (PGE) and methylaniline [29], and the cure of PGE with aniline and *meta*-phenylenediamine (m-PDA) [46].

Mijovic and co-workers [30, 31] used near-IR spectroscopy to investigate the cure of a model and a multifunctional epoxy-amine system. In their first paper [30], the cure of a stoichiometric mixture of PGE with aniline was investigated. Results for the extent of cure were obtained from the epoxy absorption band at 2208 nm and were compared with data from HPLC and mid-IR spectroscopy. Good correlation was observed between HPLC and near-IR data. However, a discrepancy was observed between mid- and near-IR results which was attributed to the mid-IR epoxy absorption band at 915 $cm^{-1}$ being affected by hydrogen bonding and an overlapping C–O–C absorption at 920 $cm^{-1}$. In a further publication [31], the authors compared results obtained from near-IR spectroscopy with those from dielectric spectroscopy for the cure of PGE with aniline and N-methylaniline as well as a DGEBA resin with 4,4′-methylenedianiline (MDA). Results obtained for dielectric spectroscopy, near-IR spectroscopy and HPLC showed good agreement for the PGE + aniline and PGE + N-methylaniline systems. Good agreement was also seen between results from dielectric spectroscopy and near-IR for the DGEBA/MDA system. Since both HPLC and near-IR spectroscopy techniques produce information based upon the chemical state of the material, it would be expected that the reaction profiles should compare closely. Dielectric spectroscopy measurements, however, are an indirect method of monitoring cure. Since the results from dielectric spectroscopy were shown to be similar to results from HPLC and near-IR spectroscopy, the authors indicated that the dielectric response was driven by the same underlying mechanism. Other authors who have compared the results for Fourier transform (FT)–near-IR spectroscopy with other techniques include Lachenal *et al.* [47] for a DGEBA resin cured with poly(oxypropylene)triamine.

The cure of a DGEBA epoxy resin with dicyandiamide and diuron was investigated using near-IR spectroscopy by Poisson *et al.* [48]. Although this system was heterogeneous, the authors found no significant problems in obtaining spectra due to light-scattering effects. Results obtained during cure were seen to compare favorably with data from size-exclusion chromatography and chemical titration experiments.

Near-IR spectroscopy of TGDDM + 4,4-diaminodiphenylsulphone (DDS) reactions have been reported in several papers by George, St John and co-workers [38–41]. The reaction kinetics of a stoichiometric system at 160°C were discussed by St John and George [39], who used near-IR spectroscopy results as evidence for the formation of ether-links from the reaction of hydroxyl groups with epoxy. Additionally, the authors also reported that the presence of a chlorohydrin impurity in the commercial TGDDM epoxy resin caused a cyclization reaction onto the aromatic ring.

In a further paper, DeBakker *et al.* [40] discuss the use of FT–Raman spectroscopy to follow the cure of TGDDM + DDS. Raman spectra were obtained by curing samples for different lengths of time in a heating block at 160°C remote from the spectrometer. After various time-intervals, samples were removed from the heating block and scanned by Raman spectroscopy. The results were compared with *in situ* near-IR spectroscopy data and a good correlation was seen for the limited data set.

A novel application of near-IR spectroscopy cure monitoring involving simultaneous differential scanning calorimetry (DSC) analysis is reported by DeBakker *et al.* [41] for a TGDDM + DDS system. The technique involved collimating and focusing an external tungsten lamp into a DSC microscopy cell. A sample of resin was placed in a glass pan on the DSC sensor, and radiation from the lamp source was set up to pass through the resin and the glass pan. After the radiation had passed through the sample, a mirror directed the light towards the sample compartment of the spectrometer. Two isothermal temperatures were used for cure monitoring: 150 and 170°C. By obtaining instantaneous reaction rates for primary amine, tertiary amine and ether groups from near-IR spectroscopy together with DSC heat flow data, reaction enthalpies were calculated for each of the reactions taking place during cure.

Recently, Xu and Schlup [49] described a method where near-IR spectra of epoxy resins were collected using a ZnSe attenuated total internal reflection (ATR) crystal ($n = 2.4$). The spectra presented appeared to have low absorbance values, but the absorbance could be improved by choosing a crystal with a refractive index closer to that of the resin system to be analyzed. This ATR technique was demonstrated as being applicable to optical fibers.

### 3.3.1 Refractive index cure monitoring

During the cure of an epoxy resin there is a decrease in volume as the reaction progresses. Associated with this change in density of the sample is a change in the refractive index, which can serve as a measure of the extent of cure. Dannenberg [50] published an early paper on the use of refractometry for cure monitoring. In his work, an Abbe refractometer was used to follow the cure of an Epon 828 (a diglycidyl ether of bisphenol-A resin) with the following amine hardeners: (i) ethylenediamine; (ii) hexamethylenediamine; (iii) *m*-phenylenediamine; (iv) piperidine; (v) triethylamine.

Vaz and Montgomery [51] used a Jelley refractometer to study the curing reactions of a DGEBA resin and an aliphatic triglyceride (EPON 812) cured with Devcon 5-min Part B hardener. Different stoichiometric combinations of the two epoxy resins and hardener were made up, and the refractive index at 589 nm was measured throughout cure at 25° C. A model for the change in refractive index during cure was developed by the authors. For the reacting epoxy system, it was assumed that at any time, $t$, during cure, the mixture would consist of resin, hardener and cured epoxy, which have time-dependent refractive indices $n_r$, $n_h$ and $n_p$ and time-dependent volumes $V_r(t)$, $V_h(t)$ and $V_p(t)$, respectively. When the resin and hardener react, the volume of each of the starting components decreases with time, whilst the volume of the resulting polymer increases. It was proposed that the reactant volume would decay exponentially in time with its own characteristic time constant $\tau_i$ resulting in the equation

$$V_i(t) = V_i(0)e^{-t/\tau_i}. \tag{3.1}$$

An expression for the refractive index of the system at any time during cure was then defined as

$$n_t = n_p - \frac{V_r(0)}{V}(n_p - n_r)e^{-t/\tau_r} - \frac{V_h(0)}{V}(n_p - n_h)e^{-t/\tau_h}, \tag{3.2}$$

where $V$ is the total volume of the system, and is simplified for this case to be independent of time. Since the polymer refractive index would also change during cure, a further expression was defined assuming that the polymer index starts at a value $n_{pi}$ and approaches its final value $n_{pf}$ exponentially in time, with a time-constant $\tau_p$:

$$n_p(t) = n_{pf} - (n_{pf} - n_{pi})e^{-t/\tau_p}. \tag{3.3}$$

A Simplex algorithm was used to obtain a least-squares fit to the experimental data, and it was found that the model reproduced the shape of the experimental curves quite well. However, the model appears somewhat oversimplified in assuming that the epoxy reaction is a one-step process where no change in total volume occurs during cure.

Fridman and Shchurov [52] describe a method whereby the refractive index of the epoxy can be calculated solely from its chemical structure. The method is based on the Lorentz–Lorenz equation, which uses an additive principle in determining the molecular refraction and the eigenvolume of repeating units of polymer, assuming that the molecular packing coefficient is constant. This leads to the expression

$$\frac{n^2 - 1}{n^2 + 2} = K_{av}\frac{R}{N_A \sum_i \Delta V_i}, \tag{3.4}$$

where $N_A$ is the Avogadro number. $R$ and $\sum_i \Delta V_i$ are a Lorentz–Lorenz type of molecular refraction and the eigenvolume of the repeating unit of the polymer, which are equal to the sums of the refractions and volume increments, respectively, for atoms and atomic groups forming part of the repeating unit of the polymer. $K_{av}$ is the mean value of the molecular packing coefficient for a given type of polymer, whereupon

$$K = \frac{1}{M} dN_A \sum_i \Delta V_i \tag{3.5}$$

where $d$ is the polymer density and $M$ is the molecular weight of the repeating unit.

By this method the refractive index of epoxy resins cured with phthalic anhydride was calculated. Results for the calculations were compared with experimental refractive index values obtained using a microscopic apparent-depth method [53]. A high degree of accuracy was obtained between the calculated and experimental values.

An alternative method of calculating the refractive index for linear co-polymers was also described by the authors. In this method, the molecular weight, $M$, the molecular refraction and the eigenvolume of the repeating unit of the epoxy polymer network is represented by additive expressions in the form of molecular average quantities:

$$\begin{aligned} \bar{M} &= f_e M_e + f_a M_a \\ \bar{R} &= f_e R_e + f_a R_a \sum_i \Delta V_i + f_e \sum_i \Delta V_{i_e} + f_a \sum_i \Delta V_{i_a}, \end{aligned} \tag{3.6}$$

where $M_e$ and $M_a$, $R_e$ and $R_a$, and $\sum_i \Delta V_{i_e}$ and $\sum_i \Delta V_{i_a}$ are the molecular weights, molecular refractions and eigenvolumes of epoxy and anhydride fragments, respectively, and $f_e$ and $f_a$ are their molar concentrations. The refractive index can then be calculated from the Lorentz–Lorenz expression, where the quantities $M$ and $R$ have been replaced by $\bar{M}$ and $\bar{R}$:

$$\frac{n^2 - 1}{n^2 + 2} = d\frac{\bar{R}}{\bar{M}}. \tag{3.7}$$

Values for group contributions to molar refraction and molar volumes can be found in reference tables [54].

The wavelength-dependence of the refractive index of a cured epoxy resin system was studied by Cooper [55]. In this work, the author found that the relationship between the refractive index and wavelength was nonlinear over the range 500–1600 nm. Other useful references for monitoring the refractive index during cure can be found in articles by Galiatsatos *et al.* [56] and in the *Encyclopedia of Polymer Science and Engineering* [57].

Other important methods for monitoring the refractive index during cure are based on optical-fiber sensors. These will be discussed later in a separate section.

### 3.3.2 Optical-fiber sensor cure monitoring methods

The use of optical-fiber sensors for cure-monitoring of epoxy resin systems has gained considerable interest in recent years. This is in part due to the increased interest in on-line and *in situ* techniques which can determine the cure state and mechanical properties of the resin during cure as an aid to processing. Additionally, their small dimensions and compatibility with the resin matrix makes them ideal candidates for embedment within the resin material. Techniques as diverse as near-IR [38], Raman [58] and fluorescence [59] spectroscopy, refractive index [60] and ultrasound [61] have all been applied as monitoring methods. General reviews have been published concerning optical-fiber chemical sensors by Grey and Partin [62] and Roberts and Davidson [63].

#### *(a) Near-infrared spectroscopy cure sensors*

The use of near-IR spectroscopy as a method of monitoring the cure of epoxy resin systems has already been discussed. Optical fibers are of particular interest for near-IR spectroscopy cure monitoring since standard telecommunications grade optical fiber transmits near-IR wavelength light, allowing data to be collected at a distance from the spectrometer. Sensor configurations have been described using transmission spectrosopy [38, 64–66], reflection [67] and evanescent spectroscopy [68–70] techniques.

A seminal work in cure-monitoring via remote near-IR transmission spectroscopy was published by George *et al.* [38]. The authors constructed an optical-fiber sensor from two plastic-clad silica (PCS) fibers inserted into a glass capillary with an internal diameter slightly greater than the external diameter of the fiber. A sample of epoxy (TGDDM resin + DDS hardener) was placed inside the capillary, and the fibers were arranged with a gap of approximately 1 mm between their end-faces. Light from a Fourier transform near-IR spectrometer was coupled into one of the fibers and returned from the second fiber to the spectrometer detector. The spectrum of TGDDM/DDS obtained through the optical fiber was compared with that obtained by conventional near-IR spectroscopy [50]. The main difference between the two spectra was the presence of a large band at 1408 nm due to the silicone resin cladding of the fiber.

Recently, Mijovic *et al.* [64] used a similar type of optical fiber near-IR transmission sensor to monitor the cure of several epoxy resin systems. The systems investigated were PGE/aniline, PGE/N-methylaniline, and DGEBA, DGEBF and triglycidyl ether of para-aminophenol (TGEPA) resin with DDM hardener. The sensor was prepared from 600/630 μm low-OH silica optical fiber aligned in a glass capillary. The size of the sensor suggests that this arrangement is not particularly suitable for composite cure monitoring. The authors also showed that the optical fiber near-IR transmission sensor can work with bismaleimide resin systems [71].

Calvert *et al.* [67] described a method whereby near-IR spectra are collected during cure of a TGDDM/DDS system using only one optical fiber. This technique involved embedding an optical fiber approximately 2 mm from the surface of the epoxy specimen and launching light from a remote quartz–halogen source directly at the specimen so that it is collected by the optical fiber and led to a detector. A diagram of the sensor arrangement is shown in Fig. 3.6. These authors also demonstrated the ability of the sensor to detect the ingress of water into the specimen from spectra recorded after immersion of the sample in water for a period of 53 h.

Evanescent spectroscopy using an optical fiber with its cladding stripped away to produce the sensing element has been described by several authors [68–70]. Cossins *et al.* [68, 72] described an evanescent sensor manufactured from standard fused silica core optical fiber whose glass cladding was stripped away over a distance of 1 cm. Using this sensor, the cure of a specialty low-refractive-index epoxy resin was monitored over the range 12 000–4000 $cm^{-1}$. Tapanes *et al.* [73] described an evanescent sensor also using standard silica optical fiber to monitor

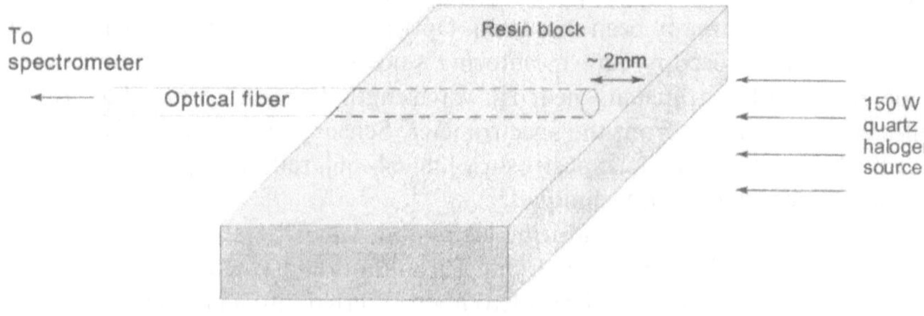


**Fig. 3.6** Diagram of single-ended near-infrared cure sensor used by Calvert *et al.* [67].

**Table 3.4** Characteristics of infrared transmitting fibers [76]

| Property | Chalcogenide | Fluoride | Sapphire | AgBr/Cl |
|---|---|---|---|---|
| Wavelength range | 3300–1000 $cm^{-1}$ | 20 000–2325 $cm^{-1}$ | 50 000–2500 $cm^{-1}$ | 3000–667 $cm^{-1}$ |
| Attenuation | 0.5 dB/m at 6 μm | 0.02 dB/m at 2.6 μm | 20 dB/m at 3 μm | 0.7 dB/m at 10.6 μm |

the cure of a commercial epoxy-amine resin system. This was achieved by etching the cladding to a depth of <1 μm over a distance of 3 cm. A series of papers by Crosby, Powell and co-workers [69, 70] described the use of a high-refractive-index glass-core optical fiber ($n = 1.65$) with a portion of its cladding stripped away to monitor the cure of a DGEBA/hexanediamine system via evanescent wave spectroscopy.

*(b) Mid-infrared spectroscopy cure sensors*

Optical fiber cure sensors based on mid-IR spectroscopy cannot use standard silica optical fibers since these can generally only guide light at wavelengths up to 2500 nm (4000 $cm^{-1}$). Therefore specialty fibers have been developed based on materials such as chalcogenide (arsenic germanium selenide) glass [74], fluoride glass [74], sapphire glass [75, 76] or silver halide glass [77]. The optical characteristics of these fibers are given in Table 3.4.

An early paper by Compton *et al.* [78] described the use of chalcogenide fibers for monitoring the IR spectrum from 4500 to 500 $cm^{-1}$ during the cure of a polyimide composite using an evanescent wave spectroscopy sensor with a 3–6 cm stripped region embedded in the sample. A diagram of the sensor arrangement used for evanescent spectroscopy is shown in Fig. 3.7, together with a schematic showing the penetration of the evanescent wave from the fiber core into a medium of lower refractive index surrounding the core.

Druy and Elandjian [74] developed a chalcogenide evanescent wave optical-fiber sensor with a stripped region to follow the cure of an epoxy-resin-based prepreg material. The absorption band of the –NH bending mode at 3367 $cm^{-1}$ was followed during the reaction. Other authors who have used chalcogenide optical fibers to follow epoxy resin cure include Ge *et al.* [79], who used an evanescent wave technique with a 15-cm-long unclad sensor region to monitor the cure of an epoxy-based prepreg.

Bunimovich *et al.* [77] described the use of silver halide mid-IR transmitting fibers to monitor cure from 2400 to 2100 $cm^{-1}$ using an evanescent wave technique. These authors found that a stripped cladding region of between 10 and 20 cm was necessary for adequate absorption.

Authors who have used sapphire optical fibers to monitor cure include Druy *et al.* [75], who used a sapphire optical fiber operating in a transmission mode to monitor the cure of an epoxy-based composite system (Fiberite 934), and Driver *et al.* [80], who used an evanescent wave technique with an 80 mm stripped length sensor.

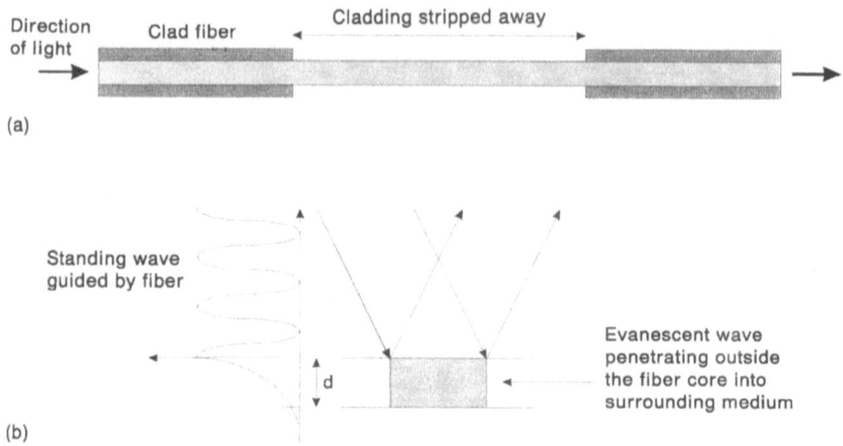


**Fig. 3.7** (a) Typical sensor arrangement used for evanescent wave spectroscopy [74]; (b) schematic diagram showing the penetration of the evanescent wave from the fiber core into a medium of lower refractive index surrounding the core [77].

Mijovic and Andjelic [81] discussed a novel technique of collecting transmission mid-IR spectra during epoxy-amine cure using a pair of hollow gold-coated nickel waveguides tipped with sodium chloride windows. Results obtained using this technique showed spectra comparable with standard FTIR spectroscopy, but the size of the nickel tubes (inside diameter 1.56 mm) meant that this method is not practical for embedment within composite materials.

Other authors who have used mid-IR transmission sensors include Fildes *et al.* [66] and Milkovich *et al.* [82], who investigated the cure of T300/934 prepreg using fluoride glass fiber together with neural network processing.

### *(c) Raman spectroscopy cure sensors*

Lyon *et al.* [58] used optical-fiber Raman spectroscopy to monitor the cure of PGE resin with a cycloaliphatic primary amine, *n*-amino-ethylpiperazine. The cure sensor consisted of two 250 μm silica fibers bonded side-by-side in an SMA connector placed in the resin sample. Light from an 820-nm laser diode was launched into one fiber end, and the scattered light was collected by the second optical fiber. Light emerging from the second optical fiber was passed to a spectrometer. Data obtained from the Raman sensor during cure were compared with transmission near-IR spectroscopy results with good correlation. Other authors who have used similar types of dual-fiber Raman probes include Stellman *et al.* [83], who described the use of a two-fiber Raman probe to monitor the microwave cure of an epoxy system, and Hong *et al.* [84], who described a two-fiber Raman probe to follow the reaction of a DGEBA resin with diethylamine.

Recently, Aust *et al.* [85, 86] described an optical-fiber Raman probe technique where a single fiber is used for both delivery of the excitation wavelength and collection of the Raman scattered light. This system was used to monitor the cure of an epoxy-amine system and an epoxy-based composite.

Several authors have conducted comparative studies on different types of optical-fiber Raman sensor designs. These serve as useful references for qualitative and quantitative Raman analysis [87–90].

### *(d) Fluorescence spectroscopy cure sensors*

For standard fluorescence spectroscopy epoxy resin cure monitoring, two methods are generally used together with optical fibers for *in situ* monitoring: (i) addition of an extrinsic fluorophore, which has a similar reactivity to the diamine curing agent; (ii) intrinsic reactive fluorophore monitoring.

Paik and Sung [59] used intrinsic fluorescence to investigate the cure of both a neat DGEBA/DDS formulation and a commercial graphite-fiber-reinforced composite (Hexcel T-190-12-F263) using an optical fluorescence sensor. The sensor was prepared from a 240-μm silica fiber, with a freshly cleaved distal end embedded in the sample. Excitation spectra obtained during cure showed the excitation peak, which was initially at 361 nm, shifting to longer wavelengths. The shift in the excitation spectra was found to be linear with DDS concentration, allowing the fall in concentration of DDS to be monitored. Results were compared with DSC and transmission mid-FTIR spectroscopy with excellent correlation.

Extrinsic fluorescence spectroscopy via Fresnel reflection was described by Dang and Sung [91] using several types of fluorophore to monitor the cure of a DGEBA/DDS resin system. The authors used a single-ended sensor prepared from standard silica fiber which was embedded in the curing resin. This acted as both the excitation source (carrying the excitation light of 456 nm to the sample) and for collection of the fluorescence intensity (emission peak at around 540 nm).

Extrinsic fluorescence spectroscopy via evanescent wave interaction of the light from the optical fiber with the epoxy resin has been investigated by several authors [92–94]. Woerdeman and co-workers [92, 95] describe an evanescent spectroscopy fluorescence method to monitor the cure of a commercial DGEBA epoxy resin with Jeffamine D400 (Texaco) curing agent. The fluorophore 4-(N,N-dimethyl amino)-4′-nitrostilbene (DMANS) was added to the epoxy-hardener mixture and using a high-refractive-index (1.62) optical fiber, the shift in the fluorescence peak was monitored over the range 495–700 nm following excitation at 488 nm. A similar single-probe evanescent sensor was described by Fuchs and Sung [93] for monitoring the cure of a DGEBA/DDS resin system.

### *(e) UV-vis spectroscopy sensors*

Rice [96, 97] developed an optical-fiber cure sensor based on changes in the transmission spectra of a commercial epoxy composite in the range 450–800 nm. The sensor was prepared by aligning two silica optical fibers in a machined channel to form a transmission cell. The gap between the two optical fibers allowed interaction of the light with the resin. The path-length between the two optical fibers was altered between 1 and 10 mm. Light from a broadband xenon lamp was introduced into one of the optical fibers, following which it was transmitted through the resin and collected by the second fiber and introduced to a monochromator and detector. Results from the sensor showed that the intensity fell as a function of the cure process. No sharp absorption bands were found over the range 450–950 nm, but the authors were able to monitor the cure reaction from a shift in the absorption edge to higher wavelengths as a function of temperature.

### *(f) Refractive-index-based optical fiber cure monitoring sensors*

The method of refractive index monitoring during epoxy cure using optical fibers has been investigated extensively by Afromowitz and co-workers [60, 98–100]. Two methods were proposed to determine the state of cure during the reaction. The first method involved constructing an 'epoxy optical fiber' from a pre-cured sample of the epoxy resin to be monitored [98]. The epoxy optical fiber was then butt-connected to conventional optical fibers to allow light to be launched into the epoxy fiber and returned to a detector. The optical fiber sensor set-up was then embedded in the sample of resin to be monitored. As the resin cured, the intensity of light reaching the detector was seen to fall as the refractive index of the resin approached that of the epoxy optical fiber. A similar type of sensor arrangement was used by Zimmermann *et al.* [101] for a DGEBA resin cured with an unknown hardener, by Zimmermann *et al.* [102] for a neat resin system and an epoxy composite system, by Lou *et al.* [103] to monitor the cure of a commercial carbon-fiber–epoxy-resin prepreg, and by May *et al.* [104] for a multifunctional sensor which was used for both cure monitoring and strain monitoring. The sensor used by May *et al.* [104] consisted of an adaptation of an extrinsic Fabry–Perot interferometer sensor whereby a single-mode fiber was used as the input/output fiber and an unclad high-refractive-index fiber was used as the reflector. When the sensor arrangement was embedded in curing epoxy, a decrease in the light intensity from the sensor was observed as the curing epoxy resin approached the refractive index of the reflection fiber.

In their second method for monitoring the refractive index during epoxy cure, Lam and Afromowitz [60, 99] calculated the refractive index during cure using a multimode optical fiber immersed in the curing resin with an 850 nm LED source. The refractive index was calculated from the reflected light returning from the fiber–resin interface using Fresnel's equation for light incident normally to a dielectric interface:

$$R = \frac{(n_{12} - 1)^2}{(n_{12} + 1)^2} \tag{3.8}$$

where $R$ is the reflection coefficient and $n_{12} = n_1/n_2$. The system that was used for cure-monitoring included Epon 828 (a DGEBA-type resin) and *m*-phenylenediamine cured between 60 and 150°C. From the results, it was found that within the range of cure temperatures studied, the refractive index decreased linearly with increasing temperature, and also that the refractive index increased linearly with conversion to the gel-point. Thereafter a deviation from linearity occurred.

Ganesh and co-workers [105, 106] investigated the cure of epoxy resins using an optical fiber Fresnel reflectometer and artificial neural networks. Neural networks were used in this study in order to deconvolute degree-of-cure and temperature effects from the reflected signal, as well as normalizing the different signal levels obtained from different sensors.

Fawcett *et al.* [107] discussed the design and performance of an optical-fiber refractive index sensor based on a side-polished single-mode fiber with a dielectric waveguide overlay. When different refractive index materials were placed over the structure, it was found that a polarization-dependent resonance peak was observed whose position was found to vary in wavelength as a result of the material's refractive index. A similar method of cure-monitoring was also demonstrated by Michie *et al.* [108] and by Levesque *et al.* [109] to determine the refractive index using an ATR crystal waveguide.

Fildes *et al.* [66] and Milkovich *et al.* [82] monitored the cure of graphite-fiber-reinforced epoxy prepregs using a commercially available optical fiber refractive index sensor from Photonetics/MetriCor [110]. This sensor was based on a Fabry–Perot interferometer, and consisted of a cavity with a mirrored surface at the far end with a slot in the cavity to allow the resin to enter. When the cavity was illuminated with an LED source via an optical fiber, changes in the reflected signal occurred as a function of the refractive index of the material.

Several authors have described optical-fiber refractive index sensors based on a fiber with its cladding removed over a small region [111, 112]. A series of papers by Crosby, Powell and co-workers [69, 70, 112, 113] detailed experiments that were carried out to track the cure of a DGEBA resin with hexanediamine hardener using a high-refractive-index optical fiber ($n = 1.65$) with a 20-mm portion of its cladding removed. Other authors who have used similar optical fiber refractive index sensors to characterize epoxy cure include Klosterman and Saliba [111], who described a method of using a refractive-index-based optical-fiber sensor to detect the presence of voids during processing, and Fei *et al.* [114], who described a method based on an optical fiber with its cladding removed (although very few experimental details are given).

### *(g) Additional optical fiber cure sensors*

May and Claus [115] reported a novel method of undertaking point rheology measurements during cure using an extrinsic Fabry–Perot interferometer strain sensor with a magnetorestrictive actuator coating to apply stress to the matrix. A diagram showing this arrangement is shown in Fig. 3.8. Other authors who have monitored strain changes during cure using optical fibers include Dunphy

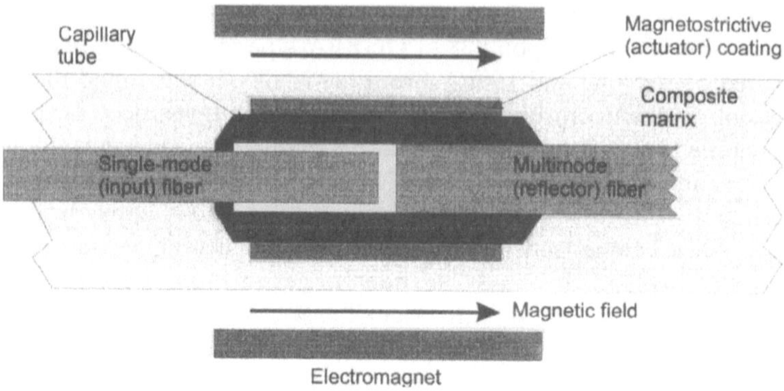


**Fig. 3.8** Schematic diagram of the sensor arrangement used for resin viscoelasticity-based cure monitoring [115].

*et al.* [116], who used Bragg grating sensors to predict the onset of vitrification, completion of vitrification and the build-up of compressive strain in the sample on cooling.

Several authors have used optical-fiber sensors together with ultrasound to follow epoxy resin cure [61, 117]. Ohn and co-workers [61, 118, 119] described a technique where ultrasound pulses were generated in epoxy resin materials using a Q-switched laser or a piezoelectric transducer. The ultrasonic pulses served to interrogate the mechanical properties of the resin during cure and were detected using an optical-fiber Michelson interferometer. Dorighi *et al.* [117] used an intrinsic fiber-optic Fabry–Perot sensor to monitor an ultrasonic pulse that was introduced into an epoxy resin during cure. From their results of the ultrasonic attenuation and velocity, the authors were able to infer the viscosity of the system.

Dielectric methods of monitoring the progress of epoxy cure have become increasingly popular in recent years. Kranbuehl *et al.* [120–122] used a commercial dielectric sensor to monitor the cure of an epoxy-amine resin system and found that ionic and dipolar mobilities were related to the viscosity, the degree of cure and the build-up to the glass transition temperature. Choi and Lee [123] carried out dielectric measurements on an epoxy-resin-based composite material by placing the specimen between two electrodes. An alternating current was applied to the specimen and the dissipation factor was used as an index of cure. Other authors who have used dielectric sensor techniques to monitor epoxy cure include Stephan *et al.* [124], Steele *et al.* [125], Rice and Lee [126], Kim and Lee [127, 128], Maistros and Partridge [129] and Mathieu *et al.* [130].

An alternative method of monitoring the dielectric properties of the resin during cure was proposed by Lee *et al.* [131]. In their method, the authors determined the direct current resistance of the composite material during cure by measuring the voltage across a known reference resistor placed in series with the material. This technique provided similar information to standard dielectric measurements recorded simultaneously.

Biermann *et al.* [132] described a method whereby a pulse–echo ultrasonic technique was used to measure the time for an ultrasonic pulse to pass through a graphite epoxy laminate during cure. The propagation characteristics of the waves were related to the material properties during cure, which allowed key cure events (such as the end of cure) to be determined. Other authors who have used ultrasonic techniques to monitor cure include Saliba *et al.* [133].

A novel method of monitoring the thickness and compaction during composite cure using a Hall-effect sensor is described by Saliba *et al.* [133]. The sensor consisted of a magnet and a Hall transducer aligned on the heated press used to cure the composite. Since the Hall voltage was proportional to the magnetic field strength, any movement of the Hall transducer from the magnet was related to the change in thickness of the composite. The use of this type of sensor has also recently been described by Rice and Lee [126].

Other types of sensors that have been demonstrated for epoxy cure monitoring include eddy-current proximity sensors [134], force-sensing resistor sensors for pressure monitoring [134] and heat flux sensors [135].

## REFERENCES


1. P. R. Ciriscioli and G. S. Springer, *Smart autoclave cure of composites*. Technimic Publishing, Pennsylvania, ISBN: 87762-802-5 (1990).
2. G. S. Springer, Microwave cure of polymeric matrix composites. *Proc. ACS Division of Polymeric Materials Science and Engineering*, **66**, 420–421 (1992).
3. C. J. de Baker, P. M. Fredericks and G. A. George, Fibre optic FT–Raman spectroscopy for the thermal and microwave cure of advanced composite resins. Preliminary studies. *14th Int. Conf. Raman Spectroscopy*, **524**, 702–703 (1994).
4. A. S. Castro, R. Y. Kim and B. P. Rice, Electron beam cure of composites for aerospace structures. **42**(1), 487–497 (1997).
5. Y. Tanaka, Synthesis and characterization of epoxides. *Epoxy Resins, Chemistry and Technology*, Editor C. A. May, 2nd edn. Marcel Dekker, New York, pp. 9–284 (1988).
6. C. A. May, Introduction to epoxy resins. *Epoxy Resins, Chemistry and Technology*, Editor C. A. May, 2nd edn. Marcel Dekker, New York, pp. 1–8 (1988).
7. D. C. L. Greenfield, The cure characteristics and physical properties of glycidyl ether epoxy resins. PhD. Thesis, Brunel University (1986).
8. L. Shechter and J. Wynstra, Glycidyl ether reactions with amines. *Industrial and Engineering Chemistry*, **48**, 94–97 (1956).
9. B. A. Rozenberg, Kinetics, thermodynamics and mechanism of reactions of epoxy oligomers with amines. *Advances in Polymer Science*, **75**, 113–165 (1985).
10. N. A. St John and G. A. George, Diglycidyl amine-epoxy resin networks: kinetics and mechanisms of cure. *Progress in Polymer Science*, **19**, 755–795 (1994).
11. K. Dusek, M. Bleha and S. Lunak, Curing of epoxide resins: model reactions of curing with amines. *Journal of Polymer Science: Polymer Chemistry Edition*, **15**, 2393–2400 (1977).
12. L. G. Weyer, Near-infrared spectroscopy of organic substances. *Applied Spectroscopy Reviews*, **21**(1 & 2), 1–43 (1985).

13. E. Stark, K. Luchter and M. Margoshes, Near-infrared analysis (NIRA): a technology for quantitative and qualitative analysis. *Applied Spectroscopy Reviews*, **22**(4), 335–399 (1986).

14. K. B. Whetsel, Near-infrared spectrophotometry. *Applied Spectroscopy Reviews*, **2**(1), 1–67 (1968).

15. R. F. Goddu, Near-infrared spectrophotometry. *Advances in Analytical Chemistry and Instrumentation*, **1**, 347–425 (1960).

16. C. E. Miller, Near-infrared spectroscopy of synthetic polymers. *Applied Spectroscopy*, **26**(4), 277–339 (1991).

17. H. E. Howell and J. R. Davis, Qualitative identification of polymeric materials using near-infrared spectroscopy. *Advances in Chemistry Series*, **263**, 263–285 (1993).

18. G. Lachenal, Dispersive and Fourier transform near-infrared spectroscopy of polymeric materials. *Vibrational Spectroscopy*, **9**, 93–100 (1995).

19. B. Chabert, G. Lachenal and C. Vinh Tung, Epoxy resins and epoxy blends studied by near infrared spectroscopy. *Macromolecular Symposia*, **94**, 145–158 (1995).

20. R. F. Goddu and D. A. Delker, Determination of terminal epoxides by near-infrared spectrophotometry. *Analytical Chemistry*, **30**(12), 2013–2016 (1958).

21. O. H. Wheeler, Near infrared spectra – a neglected field of spectral study. *Journal of Chemical Education*, **37**(5), 234–236 (1960).

22. R. F. Goddu and D. A. Delker, Spectra-structure correlations for the near-infrared region. *Analytical Chemistry*, **32**(1), 140–141 (1960).

23. F. H. Lohman and W. E. Norteman Jr., Determination of primary and secondary aliphatic amines by near-infrared spectrophotometry. *Analytical Chemistry*, **35**(6), 707–711 (1963).

24. K. Whetsel, W. E. Roberson and M. W. Krell, Near-infrared analysis of mixtures of primary and secondary aromatic amines. *Analytical Chemistry*, **30**(10), 1594–1597 (1958).

25. K. B. Whetsel, W. E. Roberson and M. W. Krell, Near-infrared spectra of primary aromatic amines. *Analytical Chemistry*, **30**(10), 1598–1604 (1958).

26. H. Dannenberg, Determination of functional groups in epoxy resins by near-infrared spectroscopy. *SPE Transactions*, January, 78–88 (1963).

27. C. Tosi and A. Pinto, Near-infrared spectroscopy of hydrocarbon functional groups. *Spectrochimica Acta*, **28A**, 585–597 (1972).

28. J. H. Fu and J. R. Schlup, Mid- and near-infrared spectroscopic investigations of reactions between phenyl glycidyl ether (PGE) and aromatic amines. *Journal of Applied Polymer Science*, **49**, 219–227 (1993).

29. L. Xu, J. H. Fu and J. R. Schlup, In-situ near-infrared spectroscopic investigation of epoxy resin–aromatic amine cure mechanisms. *Journal of the American Chemical Society*, **116**, 2821–2826 (1994).

30. J. Mijovic and S. Andjelic, A study of reaction kinetics by near-infrared spectroscopy. 1. Comprehensive analysis of a model epoxy/amine system. *Macromolecules*, **28**, 2787–2796 (1995).

31. J. Mijovic, S. Andjelic and C. F. Winnie Yee, A study of reaction kinetics by near-infrared spectroscopy. 2. Comparison with dielectric spectroscopy of model and multifunctional epoxy/amine systems. *Macromolecules*, **28**, 2797–2806 (1995).

32. B. G. Min, Z. H. Stachurski, J. H. Hodgkin and G. R. Heath, Quantitative analysis of the cure reaction of DGEBA/DDS epoxy-resins without and with thermoplastic polysulfone modifier using near-infrared spectroscopy. *Polymer*, **34**(17), 3620–3627 (1993).

33. V. Strehmel and T. Scherzer, Structural investigation of epoxy amine networks by mid- and near-infrared spectroscopy. *European Polymer Journal*, **30**(3), 361–368 (1994).

34. K. E. Chike, M. L. Myrick, R. E. Lyon and S. M. Angel, Raman and near-infrared studies of an epoxy resin. *Applied Spectroscopy*, **47**(10), 1631–1635 (1993).

35. D. W. Schiering, J. E. Katon, L. T. Drzal and V. B. Gupta, An infrared spectroscopic investigation of the curing of the EPON 828/meta-phenylenediamine system. *Journal of Applied Polymer Science*, **34**, 2367–2375 (1987).

36. K. A. Kozielski, G. A. George, N. A. St John and N. C. Billingham, Kinetic studies by FT–NIR of the curing reactions of two glycidyl ether epoxy resins mixed with stoichiometric quantities of 4,4′-DDS. *High Performance Polymers*, **6**, 263–286 (1994).

37. N. Poisson, G. Lachenal and H. Sautereau, Near- and mid-infrared spectroscopy studies of an epoxy reactive system. *Vibrational Spectroscopy*, **12**, 237–247 (1996).

38. G. A. George, P. Cole-Clarke, N. St John and G. Friend, Real-time monitoring of the cure reaction of a TGDDM/DDS epoxy resin using fibre optic FTIR. *Journal of Applied Polymer Science*, **42**, 643–657 (1991).

39. N. A. St John and G. A. George, Cure kinetics and mechanisms of a tetraglycidyl-4,4′-diaminodiphenylmethane/diamindiophenylsulphone epoxy resin using near i.r. spectroscopy. *Polymer*, **33**(13), 2679–2687 (1992).

40. C. J. DeBakker, G. A. George, N. A. St John and P. M. Fredericks, The kinetics of the cure of an advanced epoxy resin by Fourier transform Raman and near-IR spectroscopy. *Spectrochimica Acta*, **49A**(5), 739–752 (1993).

41. C. J. DeBakker, N. A. St John and G. A. George, Simultaneous differential scanning calorimetry and near-infrared analysis of the curing of tetraglycidyldiaminodiphenylmethane with diaminodiphenylsulphone. *Polymer*, **34**(4), 716–725 (1993).

42. T. Scherzer, Characterization of diol-modified epoxy resins by near- and mid-infrared spectroscopy. *Journal of Applied Polymer Science*, **51**, 491–502 (1994).

43. M. C. Paputa Peck, R. O. Carter III and S. B. A. Qaderi, Near-infrared measurements of terminal epoxides in polymer resin systems. Part 1. Analytical considerations. *Journal of Applied Polymer Science*, **33**, 77–86 (1987).

44. R. J. Varley, G. R. Heath, D. G. Hawthorne and J. H. Hodgkin, Toughening of a trifunctional epoxy system. 1. Near infrared spectroscopy study of homopolymer cure. *Polymer*, **36**(7), 1347–1355 (1995).

45. M. E. Ryan and A. Dutta, *Polymer*, **20**, 203–206 (1979).

46. L. Xu, J. H. Fu and J. R. Schlup, In-situ near-infrared spectroscopic investigation of the kinetics and mechanisms of reactions between phenyl glycidyl ether (PGE) and multifunctional aromatic amines. *Industrial and Engineering Chemistry Research*, **35**, 963–972 (1996).

47. G. Lachenal, A. Pierre and N. Poisson, FT–NIR spectroscopy: trends and application to the kinetic study of epoxy/triamine system (comparison with DSC and SEC results). *Micron*, **27**(5), 329–334 (1996).

48. N. Poisson, G. Lachenal and H. Sautereau, Near- and mid-infrared spectroscopy studies of an epoxy reactive system. *Vibrational Spectroscopy*, **12**, 237–247 (1996).

49. L. Xu and J. R. Schlup, Application of near-infrared attenuated total reflectance spectroscopy for monitoring epoxy resin/amine cure reactions. *Applied Spectroscopy*, **50**(1), 109–114 (1996).

50. H. Dannenberg, Refractive index method for determining cure rates of epoxy resins. *SPE Journal*, October, 875–880 (1959).

51. N. A. Vaz and G. P. Montgomery Jr., Refractive indices of polymer-dispersed liquid film materials: Epoxy-based systems. *Journal of Applied Physics*, **62**(8), 3161–3172 (1987).

52. Y. B. Fridman and A. F. Shchurov, On relations between the refractive indices of epoxy polymer networks and their chemical structure. *Polymer Science USSR*, **25**, 1702–1707 (1983).

53. ASTM Standards on Plastics, *Standard Test Methods for Index of Refraction of Transparent Organic Plastics*. D542-50. Philadelphia (1960).

54. D. W. Van Krevelen and P. J. Hoftyzer, *Properties of Polymers – Their Estimation and Correlation with Chemical Structure*, 2nd edn. Elsevier Scientific, Amsterdam (1976).

55. P. R. Cooper, Refractive-index measurements of paraffin, a silicone elastomer, and an epoxy resin over the 500–1500-nm spectral range. *Applied Optics*, **21**(19), 3413–3415 (1982).

56. V. Galiatsatos, R. O. Neaffer, S. Sen and B. J. Sherman, Refractive index, stress-optical coefficient and optical configuration parameter of polymers. *Physical Properties of Polymers*, Editor James E. Mark. American Institute of Physics Press, New York (1996).

57. R. E. Pepper and R. J. Samuels, Refractometry. *Encyclopedia of Polymer Science and Engineering*, **14**, 261–298 (1992).

58. R. E. Lyon, K. E. Chike and S. M. Angel, In situ cure monitoring of epoxy resins using fiber-optic Raman spectroscopy. *Journal of Applied Polymer Science*, **53**, 1805–1812 (1994).

59. H.-J. Paik and N.-H. Sung, Fiberoptic intrinsic fluorescence for in-situ cure monitoring of amine-cured epoxy and composites. *Polymer Engineering and Science*, **34**(12), 1025–1032 (1994).

60. K. Y. Lam and M. A. Afromowitz, Fiber-optic epoxy composite cure sensor. I. Dependence of refractive index of an autocatalytic reaction epoxy system at 850 nm on temperature and extent of cure. *Applied Optics*, **34**(25), 5635–5638 (1995).

61. M. M. Ohn, A. Davis, K. Liu and R. M. Measures, Embedded fiber optic detection of ultrasound and its application to cure monitoring. *SPIE Fiber Optic Smart Structures and Skins V*, **1798**, 134–143 (1992).

62. A. E. Grey and J. K. Partin, Fiber optic chemical sensors – an overview. *Leak Detection for Underground Storage Tanks*. ASTM STP 1161, Editors: P. B. Durgin and T. M. Young. American Society for Testing and Materials, Philadelphia, pp. 105–114 (1993).

63. S. S. J. Roberts and R. Davidson, Cure and fabrication monitoring of composite materials with fibre-optic sensors. *Composites Science and Technology*, **49**, 265–276 (1993).

64. J. Mijovic, S. Andjelic and J. M. Kenny, In situ real-time monitoring of epoxy-amine kinetics by remote near infrared spectroscopy. *Polymers for Advanced Technologies*, **7**, 1–16 (1996).

65. P. A. Crosby, G. R. Powell, T. Liu, X. Wu and G. F. Fernando, In-situ cure monitoring of an epoxy/amine resin system using an optical-fiber transmission sensor. *Proceedings of the Society of Photo-optical Instrumentation Engineers (SPIE)*, **2895**, 109–115 (1996).

66. J. M. Fildes, S. M. Milkovich, R. Altkorn, R. Haidle and J. Neatrour, In situ infrared spectroscopy and neural network analysis for composite cure monitoring. *25th International SAMPE Technical Conference*, October (1993).

67. P. Calvert, G. George and L. Rintoul, Monitoring of cure and water uptake in a freeformed epoxy resin by an embedded optical fiber. *Chemistry in Materials*, **8**, 1298–1301 (1996).

68. S. Cossins, M. Connell, B. Cross, R. Winter and J. Kellar, In-situ near-IR cure monitoring of a model epoxy matrix composite. *Applied Spectroscopy*, **50**(7), 900–905 (1996).

69. G. R. Powell, P. A. Crosby, G. F. Fernando, C. M. France, R. C. Spooncer and D. N. Waters, Optical-fiber evanescent-wave cure monitoring of epoxy-resins. *Proceedings of the Society of Photo-optical Instrumentation Engineers (SPIE)*, **2718**, 80–92 (1996).

70. P. A. Crosby, G. R. Powell, G. F. Fernando, C. M. France, R. C. Spooncer and D. N. Waters, In-situ monitoring of epoxy-resins using optical-fiber sensors. *Smart Materials & Structures*, **5**(4), 415–428 (1996).

71. J. Mijovic and S. Andjelic, Study of the mechanism and rate of bismaleimide cure by remote in-situ real time fiber optic near-infrared spectroscopy. *Macromolecules*, **29**, 239–246 (1996).

72. S. L. Cossins, M. E. Connell, W. M. Cross, R. M. Winter and J. J. Kellar, Evanescent wave spectroscopy for in-situ cure monitoring. *Proceedings of the Society of Photo-optical Instrumentation Engineers (SPIE)*, **2836**, 147–156 (1996).

73. E. E. Tapanes, J. R. Goode, P. L. Rossiter and A. J. Hill, Evanescent wave spectroscopy as an interface probe. *Materials Science Forum*, **189–190**, 205–210 (1995).

74. M. A. Druy and L. Elandjian, Composite cure monitoring with infrared transmitting optical fibers. *Proceedings of the Society of Photo-optical Instrumentation Engineers (SPIE)*, **986**, 130–134 (1988).

75. M. A. Druy, P. J. Glatkowski and W. A. Stevenson, Embedded optical fiber sensors for monitoring cure cycles of composites. *ADPA/AIAA/ASME/SPIE Conference on Materials and Adaptive Structures*, **42**, 805–808 (1992).

76. M. A. Druy, P. J. Glatkowski and W. A. Stevenson, Applications of remote fiber optic spectroscopy using IR fibers and Fourier transform infrared (FTIR) spectrometers. *Proceedings of the Society of Photo-optical Instrumentation Engineers (SPIE)*, **1591**, 218–224 (1991).

77. D. Bunimovich, R. Kellner, R. Krska, A. Mesica, I. Paiss, U. Schiesl, M. Tacke, K. Taga and A. Katzir, A system for monitoring and control of processes based on IR fibres and tunable diode lasers. *Journal of Molecular Structure*, **292**, 125–132 (1993).

78. D. A. C. Compton, S. L. Hill, N. A. Wright, M. A. Druy, J. Piche, W. A. Stevenson and D. W. Vidrine, In-situ FT–IR analysis of a composite curing reaction using a mid-infrared transmitting optical fiber. *Applied Spectroscopy*, **42**(6), 972–979 (1988).

79. Z. Ge, C. W. Brown and M. Brown, Monitoring of lamination processes in an autoclave with fiber-optic infrared spectroscopy. *Journal of Applied Polymer Science*, **56**, 667–675 (1995).

80. R. D. Driver, J. N. Downing and G. M. Leskowitz, Evanescent-wave spectroscopy down infrared transmitting optical fibers. *Proceedings of the Society of Photo-optical Instrumentation Engineers (SPIE)*, **1591**, 168–179 (1991).

81. J. Mijovic and S. Andjelic, Monitoring of reactive processing by remote mid-infrared spectroscopy. *Polymer*, **37**(8), 1295–1303 (1996).

82. S. M. Milkovich, R. I. Altkorn, R. H. Haidle, M. J. Neatrour and J. M. Fildes, In situ sensors for intelligent process control for fabrication of polymer-matrix composite materials. *Proceedings of the Society of Photo-optical Instrumentation Engineers (SPIE)*, **2191**, 349–360 (1994).

83. C. M. Stellman, J. F. Aust and M. L. Myrick, In situ spectroscopic study of microwave polymerization. *Applied Spectroscopy*, **49**(3), 392–394 (1995).

84. K. C. Hong, T. M. Vess, R. E. Lyon, K. E. Chike, J. F. Aust and M. L. Myrick, Remote cure monitoring of polymeric resins by laser Raman spectroscopy. *38th International SAMPE Symposium*, pp. 427–435 (1993).

85. J. F. Aust, K. S. Booksh, C. M. Stellman, R. S. Parnas and M. L. Myrick, Precise determination of percent cure of epoxide polymers and composites via fiber-optic Raman spectroscopy and multivariate analysis. *Applied Spectroscopy*, **51**(2), 247–252 (1997).

86. J. F. Aust, K. S. Booksh and M. L. Myrick, Novel in situ probe for monitoring polymer curing. *Applied Spectroscopy*, **50**(3), 382–387 (1996).

87. T. F. Cooney, H. T. Skinner and S. M. Angel, Comparative study of some fiber-optic remote Raman probe designs. Part 1. Model for liquids and transparent solids. *Applied Spectroscopy*, **50**(7), 836–848 (1996).

88. T. F. Cooney, H. T. Skinner and S. M. Angel, Comparative study of some fiber-optic remote Raman probe designs. Part 2. Tests of single-fiber, lensed, and flat- and bevel-tip multi-fiber probes. *Applied Spectroscopy*, **50**(7), 849–860 (1996).

89. J. F. Maguire and P. L. Talley, Remote Raman spectroscopy as a sensor technology in composite materials processing. *Journal of Advanced Materials*, **January**, 27–40 (1995).

90. I. R. Lewis and P. R. Griffiths, Raman spectrometry with fiber-optic sampling. *Applied Spectroscopy*, **50**(10), 12A–30A (1996).

91. W. Dang and N.-H. Sung, In-situ cure monitoring of diamine cured epoxy by fiber-optic fluorimetry using extrinsic reactive fluorophore. *Polymer Engineering and Science*, **34**(9), 707–715 (1994).

92. D. L. Woerdeman, K. M. Flynn, J. P. Dunkers and R. S. Parnas, The use of evanescent wave fluorescence spectroscopy for control of the liquid molding process. *Journal of reinforced plastics and composites*, **15**, 922–943 (1996).

93. A. Fuchs and N.-H. Sung, Composite interphase study by evanescent fiberoptic fluorescence. *Antec 95*, 2437–2441 (1995).

94. A. Fuchs and N. Sung, Epoxy/fiber interphase probing via evanescent fluorimetry using sapphire optical fiber, *Polymeric Materials Science and Engineering*, **71**, 439–440 (1994).

95. D. L. Woerdeman and R. S. Parnas, Cure monitoring in RTM using fluorescence. *Plastics Engineering*, **October**, 25–27 (1995).

96. B. P. Rice, Monitoring of composite resin cure with a tool-mounted UV–vis–NIR fiber optic sensor. *38th International SAMPE Symposium*, **May** (1993).

97. B. P. Rice, Composite cure monitoring with a tool-mounted UV–vis–NIR fiber optic sensor. *39th International SAMPE Symposium*, **April**, 893–904 (1994).

98. M. A. Afromowitz, Fiber optic polymer cure sensor. *Journal of Lightwave Technology*, **6**(10), 1591–1594 (1988).

99. M. A. Afromowitz and K. Y. Lam, The optical properties of curing epoxies and applications to the fiber optic cure sensor. *Sensors and Actuators*, **A21–A23**, 1107–1110 (1990).

100. K. Y. Lam and M. A. Afromowitz, Fiber-optic epoxy composite cure sensor. 2. Performance characteristics. *Applied Optics*, **34**(25), 5639–5644 (1995).

101. B. Zimmermann, C. DiFrancia, K. Murphy, A. Vengsarkar and R. Claus, Optical fiber methods for autoclave and epoxy cure monitoring. *Review of Progress in Quantitative Nondestructive Evaluation*, **9**, 2047–2053 (1990).

102. B. Zimmermann, M. de Vries and R. Claus, Resin sensors for composite cure monitoring. *ADPA/AIAA/ASME/SPIE Conference on Active Materials and Structures*, **18**, 313–316 (1992).

103. K. A. Lou, B. Zimmermann and G. Yaniv, Combined sensor system for process and in-service health monitoring of composite structures. *Proceedings of the Society of Photo-optical Instrumentation Engineers (SPIE)*, **2191**, 32–45 (1994).

104. R. G. May, J. M. Sanderson and R. O. Claus, Combined fiber optic strain sensor and composite cure monitor for smart structure applications. *Proceedings of the Society of Photo-optical Instrumentation Engineers (SPIE)*, **2191**, 46–57 (1994).

105. C. Ganesh, J. P. H. Steele, H. Zhang, D. Mishra and J. Jones, Predicting degree-of-cure of epoxy resins with fiber optic sensors and artificial neural networks. *39th International SAMPE Symposium*, **April**, 883–892 (1994).

106. Y. M. Liu, C. Ganesh, J. P. H. Steele and J. E. Jones, Fiber optic sensor development for real-time in-situ epoxy cure monitoring. *Journal of Composite Materials*, **31**(1), 87–102 (1997).

107. G. Fawcett, W. Johnstone and W. L. K. Yim, Design and experimental optimization of an evanescent-field fibre-optic refractometer. *Conference on Lasers and Electro-optics (Amsterdam, 1994)*, pp. 372–373 (1994).

108. W. C. Michie, G. Thursby, W. Johnstone and B. Culshaw, Optical techniques for determination of the state of cure of epoxy resin based systems. *Proceedings of the Society of Photo-optical Instrumentation Engineers (SPIE)*, **1798**, 11–18 (1992).

109. L. Levesque, B. E. Paton and S. H. Payne, Precise thickness and refractive index determination of polyimide films using attenuated total reflection. *Applied Optics*, **33**(34), 8036–8040 (1994).

110. L. C. Gunderson, Fiber optic sensor applications using Fabry–Perot interferometry. *Proceedings of the Society of Photo-optical Instrumentation Engineers (SPIE)*, **1267**, 194–204 (1990).

111. D. A. Klosterman and T. E. Saliba, Development of an in situ fiber optic sensor for on-line void detection and control. *38th International SAMPE Symposium*, **May**, 1322–1332 (1993).

112. G. R. Powell, P. A. Crosby, G. F. Fernando, C. M. France, R. C. Spooncer and D. N. Waters, In-situ cure monitoring of advanced fibre reinforced composites. *Proceedings of the Society of Photo-optical Instrumentation Engineers (SPIE)*, **2444**, 386–396 (1995).

113. P. A. Crosby, G. R. Powell, G. F. Fernando, D. N. Waters, C. M. France and R. C. Spooncer, A comparative study of optical fibre cure monitoring methods. *Proceedings of the Society of Photo-optical Instrumentation Engineers (SPIE)*, **3042**, 141–153 (1997).

114. L. Fei, M. Aidong and L. Jianyi, Optical fiber sensing method for the in-situ cure monitoring of composite materials. *Proceedings of the Society of Photo-optical Instrumentation Engineers (SPIE)*, **2839**, 413–419 (1996).

115. R. G. May and R. O. Claus, In-situ fiber optic sensor for composite cure monitoring through characterization of resin viscoelasticity. *Proceedings of the Society of Photo-optical Instrumentation Engineers (SPIE)*, **2948**, 24–34 (1996).

116. J. R. Dunphy, G. Meltz, F. P. Lamm and W. W. Morey, Multi-function, distributed optical fiber sensor for composite cure and response monitoring. *Proceedings of the Society of Photo-optical Instrumentation Engineers (SPIE)*, **1370**, 116–118 (1990).

117. J. Dorighi, S. Krishnaswamy and J. D. Achenbach, A fiber optic ultrasound sensor for monitoring the cure of epoxy. *Review of Progress in Quantitative Nondestructive Evaluation*, **15**, 657–664 (1996).

118. A. Davis, M. M. Ohn, K. Liu and R. M. Measures, Composite cure monitoring with embedded optical fiber sensors. *Proceedings of the Society of Photo-optical Instrumentation Engineers (SPIE)*, **1489**, 33–43 (1991).

119. A. Davis, M. M. Ohn, K. Liu and R. M. Measures, A study of an opto-ultrasonic technique for cure monitoring. *Proceedings of the Society of Photo-optical Instrumentation Engineers (SPIE)*, **1588**, 264–274 (1991).

120. D. Kranbuehl, D. Hood, Y. Wang, G. Boiteux, F. Stephan, C. Mathieu, G. Seytre, A. Loos and D. McRae, In situ monitoring of polymer processing properties. *Polymers for Advanced Technologies*, **8**, 93–99 (1997).

121. D. Krambuehl, D. Hood, A. Kriss and R. Barksdale, In-situ FDEMS sensing and modeling of the epoxy infiltration, viscosity and degree of cure during resin transfer molding of a textile preform. *Polymeric Materials Science and Engineering*, **74**, 28–29 (1996).

122. D. E. Kranbuehl, P. Kingsley and S. Hart, In situ sensor monitoring and intelligent control of the resin transfer molding process. *Polymer Composites*, **15**(4), 299–305 (1994).

123. J. H. Choi and D. G. Lee, Expert cure system for the carbon fiber epoxy composite materials. *Journal of Composite Materials*, **29**(9), 1181–1200 (1995).

124. F. Stephan, A. Fit and X. Duteurtre, In-process control of epoxy composite by microdielectrometric analysis. *Polymer Engineering and Science*, **37**(2), 436–449 (1997).

125. J. P. H. Steele, C. Ganesh, K. Liu, H. Zhang, D. Mishra and J. Jones, Predicting degree of cure of epoxy resins using dielectric sensor data and artificial neural networks. *38th International SAMPE Symposium*, **May**, 1333–1345 (1993).

126. B. P. Rice and C. W. Lee, Novel, low-cost sensors for intelligent process control. *41st International SAMPE Symposium*, **March**, 1518–1529 (1996).

127. J. S. Kim and D. G. Lee, Measurement of the degree of cure of carbon fiber epoxy composite materials. *Journal of Composite Materials*, **30**(13), 1436–1457 (1996).

128. J.-S. Kim and D. G. Lee, On-line cure monitoring and viscosity measurement of carbon fiber epoxy composite materials. *Journal of Materials Processing Technology*, **37**, 405–416 (1993).

129. G. M. Maistros and I. K. Partridge, Dielectric monitoring of cure in a commercial carbon-fibre composite. *Composites Science and Technology*, **53**, 355–359 (1995).

130. C. Mathieu, G. Boiteux, G. Seytre, R. Villain and P. Dublineau, Microdielectric analysis of the polymerization of an epoxy-amine system. *Journal of Non-Crystalline Solids*, **172–174**, 1012–1016 (1994).

131. C. W. Lee, B. Rice and M. Buczek, Direct current resistance based resin state sensors. *41st International SAMPE Symposium*, **March**, 1511–1517 (1996).

132. P. J. Biermann, J. H. Cranmer, C. A. Lebowitz and L. M. Brown, End-of-cure sensing using ultrasonics for autoclave fabrication of composites. *Proceedings of the Society of Photo-optical Instrumentation Engineers (SPIE)*, **2948**, 72–83 (1996).

133. T. E. Saliba, D. Hofmann and P. Smolinski, Development of an in situ Hall-effect sensor for on-line monitoring of thickness and compaction during composite curing. *Composites Science and Technology*, **51**, 1–9 (1994).

134. Y. Bar-Cohen, A. Chatterjee and M. West, Sensors for cure monitoring of composite materials. *Review of Progress in Quantitative Nondestructive Evaluation*, **12**, 1039–1046 (1993).

135. M. J. Perry, L. J. Lee and C. W. Lee, On-line cure monitoring of epoxy/graphite composites using a scaling analysis and a dual heat flux sensor. *Journal of Composite Materials*, **26**(2), 274–292 (1992).

4

# The application of optical fiber sensors in advanced fiber reinforced composites. Part 3: Strain, temperature and health monitoring

T. Liu and G. F. Fernando

## 4.1 INTRODUCTION

The virtues of advanced fiber-reinforced composites (AFRCs) were highlighted in the previous chapters. A major concern with the use of AFRCs is the detection of damage within these materials. AFRCs can sustain significant internal damage before the nature of the damage is visible on the surface. In general, this is because it is the surface which is opposite to the impact surface, i.e. the tensile face which sustains the bulk of the damage. With reference to aerospace structures, impact damage can be induced by bird-strikes, damage from debris and accidental impacts from tools being dropped on the structure. In the case of fatigue damage, the design criteria which are used for these materials ensure that the operating loads are well below those required to initiate damage. However, an issue to note is that aerospace structures may be under a state of dynamic stress when the impact event takes place. There is sufficient evidence now to suggest that the damage mechanics when a composite is impacted under load are different, and the degree of damage is more significant, when compared with conventional impact testing [1]. It is also worth bearing in mind that expensive sensor systems may not be a prudent investment if the sensor is damaged when it receives the first impact.

From a materials point of view, the availability of reliable and robust sensor systems which can impart real-time information on the structural integrity of AFRCs will facilitate: (i) the rapid evaluation and introduction of new materials into the market place; (ii) the *in situ* evaluation of damage mechanisms within the composite; (iii) the inspection of AFRC structures only when required rather than adhering to an arbitrarily defined maintenance schedule.

*Optical Fiber Sensor Technology*, Vol. 3. Edited by K. T. V. Grattan and B. T. Meggitt.
Published in 1998 by Kluwer Academic Publishers, London. ISBN 978-1-4419-4736-9

## 4.2 TECHNIQUES FOR DAMAGE DETECTION IN AFRCS

The nature and the extent of damage within AFRCs can be assessed using destructive [2] and nondestructive [3] techniques.

There are four main destructive testing techniques. (1) Testing the AFRC to failure using appropriate test methods [4], for example, tensile, compression, flexural, interlaminar shear, shear and fracture toughness. (2) Sectioning, where the damaged or partially damaged AFRC is mounted in a thermosetting resin and then sectioned and polished. Due care and attention has to be paid with this technique to ensure that: (a) the sample is representative of the macroscopic damage mode; (b) the act of sectioning and polishing does not induce any additional damage; (c) the polishing is carried out in a step-wise and sequential fashion towards the primary damage zone. (3) A de-plying technique, which is used to study the nature of fiber fractures in the AFRC and involves pyrolyzing the resin in a furnace. Most epoxy/amine-based thermosets tend to pyrolyze above 300–400°C. This technique is ideally suited to glass-fiber-reinforced composites. Carbon-fiber-based composites are usually de-plied using hot concentrated sulphuric acid, as pyrolysis is not generally recommended because the fiber tends to oxidize. (4) Fractography, which is the post-failure analysis of fracture surfaces using optical and scanning electron microscopy. This technique can impart useful information on actual modes of damage initiation and propagation in AFRCs [5, 6].

Typical examples of the more popular nondestructive techniques which are used to assess the structural integrity of AFRCs include: (1) ultrasonic; (2) X-ray radiography; (3) thermography; (4) acoustic emission; (5) edge-replication; and (6) optical-based nondestructive techniques including embedded optical fiber sensors.

1. An ultrasonic C-scan is capable of detecting defects such as porosity, density variations and delamination in the plane normal to the wave. This technique is used extensively in the composites industry as a quality control tool to assess the porosity in cured laminates, and to quantify the extent of delamination in AFRCs which have been subjected to impact and fatigue loading. A limitation of this technique is the need for a coupling medium between the transducer and the AFRC. However, the use of air-coupled ultrasonic C-scanning systems has been reported.
2. X-ray radiography is used primarily to detect matrix cracks and delaminations. This technique requires the use of low-energy sources (10–20 kV) in conjunction with a dye-penetrant. The dye is required because even with the use of low-energy X-rays, both the matrix and the fibers (glass and carbon) are transparent. An aqueous solution of zinc iodide in conjunction with a surfactant is commonly used. The dye is introduced into the test specimen whilst the sample is held under a relatively small tensile load (to open out the cracks). Prolonged exposure of the AFRC to the dye-penetrant can degrade the matrix-dominated properties.

3. Thermography is the mapping of isothermal contour lines from a surface, and can be achieved with the use of a video-thermographic camera [7] or simply surface-mounted thermocouples [8]. Both hysteretic heating and friction-based heating can give rise to a steady increase in the surface temperature of AFRCs during dynamic loading. In general, a rapid rise in the temperature is observed just prior to catastrophic failure [8].
4. Acoustic emission has been used extensively to characterize the onset and nature of damage development in AFRCs [9]. Conventional acoustic emission monitoring requires the use of a couplant between the transducer and the surface of the specimen. The use of embedded optical-fiber-based acoustic emission sensors in AFRCs has also been reported [10–12].
5. Edge-replication is a simple technique, and strictly it cannot be classified as a nondestructive technique, but it is useful as an aid to assessing the extent of transverse ply cracking and edge-delaminations. Edge replication involves the use of a strip of an acetate tape which is partially secured onto the edge of the AFRC. Acetone is then introduced between the acetate and the edge of the specimen. Gentle pressure is applied to the partially plasticized acetate tape in order to transfer the edge profile of the specimen to the acetate. Despite the simplicity of this technique, accurate information can be obtained on the state of damage of the specimen.
6. Optical-based nondestructive techniques which have gained popularity in recent years include electronic speckle pattern interferometry [13], TV holography [14], holographic interferometry, shearing interferometry [15] and speckle photography [16]. Included in this category is optical-fiber-based nondestructive testing of AFRCs.

The remainder of this chapter is dedicated to a review of optical-fiber-based sensor systems for health monitoring in AFRCs.

## 4.3 POTENTIAL STRATEGIES FOR HEALTH MONITORING IN AFRCs USING OPTICAL-FIBER SENSORS

### 4.3.1 Complexity of the problem

The details of potential damage modes in AFRCs were discussed in Chapters 2 and 3. The actual nature of the induced damage will be a function of: (i) the fiber type and associated properties; (ii) the composition of the matrix and its failure strain and fracture toughness; (iii) the stacking sequence and the dimensions of the test specimen; (iv) the relative magnitudes of the coefficient of thermal expansion for the matrix and the fibers; (v) the magnitude of the fabrication stresses; (vi) the test temperature; (vii) other factors such as the void content, the degree of fiber waviness and the fiber volume fraction. The actual test method (tension, compression, flexural etc.) in conjunction with the duration and magnitude of the applied stress or strain will dictate the nature of the damage mechanisms for quasi-static experiments. The rate of stress application is an important factor for specimens which exhibit viscoelastic behavior. With reference to impact testing, the clamping method and the level of free deformation available to the test specimen will have a significant influence on the nature and extent of damage. Therefore, it is important to appreciate the complexity of the damage initiation and propagation processes in AFRCs prior to comparing data from different sources.

### 4.3.2 Potential strategies

(i) Crack detectors. This technique relies on the fracture of surface-mounted or embedded optical fibers to infer the presence of damage in the AFRC [17]. The feasibility of etching the optical fibers to control their failure strain has also been demonstrated. On the basis of the complexity of the damage processes in AFRCs, the interpretation of the data from crack-detector-based sensors has to be approached with a degree of caution.
(ii) Self-sensing composites. Here a proportion of the reinforcing fibers are converted to act as light guides. This technique offers the potential for the whole area of the AFRC to be monitored [18].
(iii) Strain-based damage assessment. The majority of the optical-fiber-based techniques which are used for damage detection in AFRCs rely on strain sensors. The strain measured by the optical fiber sensor is then correlated to the damage state in the composite. Typical examples of optical-fiber-based strain sensors include intrinsic and extrinsic Fabry–Perot sensors, fiber Bragg gratings and intensity-based sensors. With regard to applying these sensors to on-site applications on engineering structures, a fundamental issue has to be addressed, namely, what do the strain data actually mean? The measured strain in an AFRC structure may be affected by one or more of the following factors: (a) temperature, (b) stress relaxation, (c) creep, (d) moisture content, (e) rate of mechanical loading, (f) extent of reinforcing fiber fractures, (g) extent of debonding and matrix cracking, (h) delaminations, (i) relative orientation and alignment of the optical-fiber sensor in relation to the reinforcing fibers, (j) Poisson's contraction, (k) the integrity of the optical fiber and the interface between the matrix and the sensor system.

At present, optical fiber strain sensors reported in the literature have mostly been used in specimens tested in the laboratories or subjected to controlled and simple loading conditions. Thus most of the factors which may be encountered in a real engineering environment, as listed above, have not been considered in any detail.

The strain measured by the sensor will also depend on its relative location with respect to the location of the damage in the AFRC. This problem can be overcome to some extent by the use of multiple sensors. It is conceivable that an expert system can be developed to distinguish and categorize the consequences of the above-mentioned events.
(iv) Stiffness-based damage detection. It is well established that the extent of global damage in an AFRC can be related to its stiffness [2]. The data required

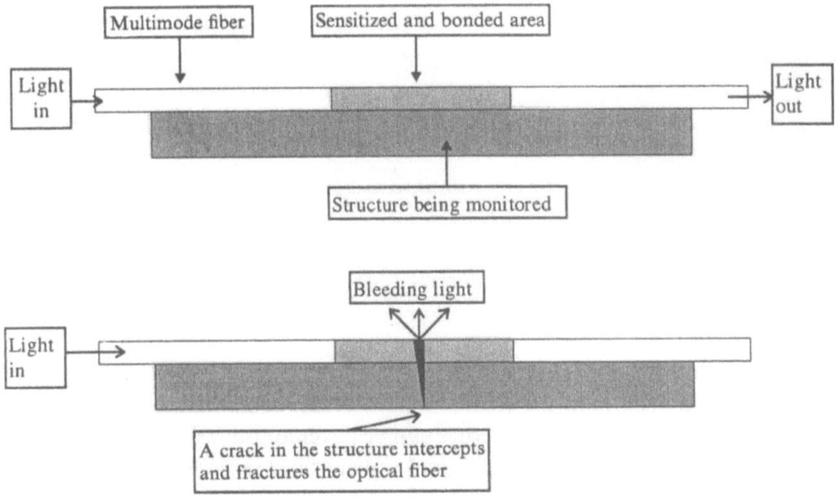


**Fig. 4.1** Schematic illustration of a crack-detector system proposed by Hale [17].

to calculate the stiffness are the strain and the stress. For laboratory-based tests, the load can be measured independently using a load cell. However, this may not be easy to achieve on site with a real structure. The limitations mentioned in the above section with respect to embedded sensors also apply here for strain monitoring.

### 4.3.3 Crack detectors

A schematic illustration of the crack detector concept is shown in Fig. 4.1. This technique requires perfect bonding between the substrate and the sensor. It is also necessary for the failure strain of the sensitized area of the fiber to be similar to that of the substrate. The feasibility of using custom-made 30 μm optical fibers as embedded crack-detectors in filament wound tubes was recently demonstrated by Martin *et al.* [19].

### 4.3.4 Reinforcing fiber light guides (self-sensing composites)

These self-sensing composites are glass reinforcing composites where some of the reinforcing fibers are converted to light guides and used as sensors [18]. The surface treatment on the reinforcing fibers involves the initial removal of the protective coating on the reinforcing fibers, followed by a cladding process. These authors demonstrated the feasibility of using epoxy-, silicone-, fluoropolymer- and sol-gel-based claddings. The concept of the reinforcing fiber light-guide (RFLG) is ideal for impact damage detection and has a number of advantages:

1. there is minimal induced distortion of the reinforcing fibers when the RFLGs are embedded in the AFRC;

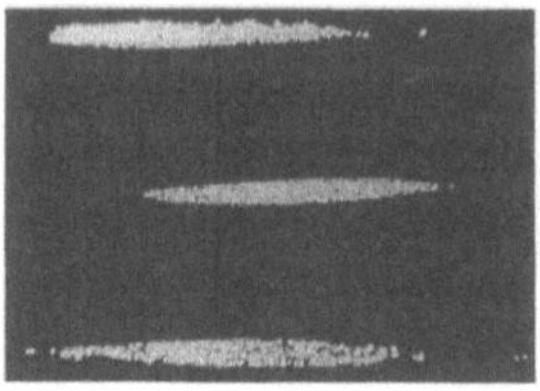

(a)

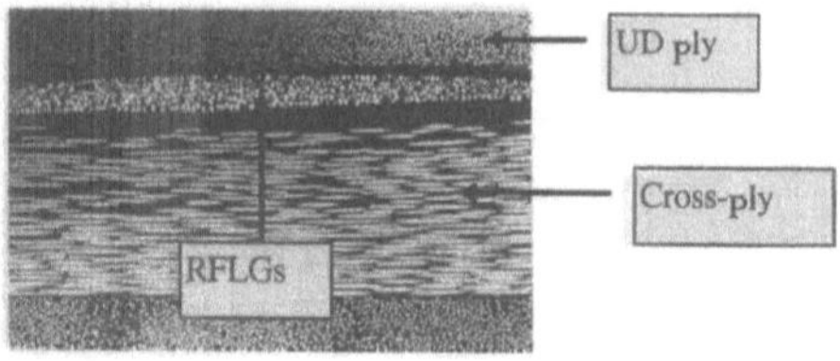


(b)

**Fig. 4.2** Micrographs of embedded reinforcing fiber light guides in a 16-layer cross-ply composite.

2. RFLGs eliminate the presence of the resin-rich regions;
3. the RFLGs contribute to the reinforcement of the composite structure;
4. large areas of the composite structure can be covered using the RFLGs, which is a major advantage over conventional optical-fiber sensors;
5. the sol-gel-based cladding technique offers the potential for custom-designing the coating to make it specific for a given parameter of interest, for example, thermochromic, photochromic and cladding materials which can undergo color change as a function of pH, salt ingress, moisture ingress etc.

The illumination of Fig. 4.2(a) was achieved by transmitting light through RFLG fiber bundles. Figure 4.2(b) illustrates the presence of the embedded RFLGs using reflectance microscopy. The apparent resin-rich region in Fig. 4.2(b) is misleading, because this region also has the embedded RFLGs but they are out of focus.

It is important to appreciate that RFLGs are not conventional optical fibers, and as a consequence, the composition and optical quality of these reinforcing fibers will influence their particular transmission characteristics. Typical compositions for three types of reinforcing fibers are presented in Table 4.1.

The compositions of quartz, E- and S-glass are presented in Table 4.1. E-glass has a calcium aluminoborosilicate composition and is used extensively as a general purpose reinforcing fiber. S-glass, on the other hand, contains magnesium aluminosilicate and is used in applications where high tensile strength is required.

**Table 4.1** Typical compositional ranges (in weight %) for glass fibers used in composites

| | Glass type | | |
|---|---|---|---|
| | E-glass | S-glass | Quartz |
| Silica | 52–56 | 63–65 | >99.95 |
| Alumina | 12–16 | 25 | – |
| Calcium oxide | 16–20 | – | – |
| Magnesium oxide | 0–5 | 10 | – |
| Boron oxide | 5–10 | 0 | – |
| Miscellaneous | <2 | <1 | – |

**Table 4.2** Specific properties of selected fiber types. Optical refers to typical commercially available 50/125 multimode optical fibers

| Property | Fiber type | | | |
|---|---|---|---|---|
| | Quartzel | E-glass | S-glass | Optical |
| Diameter (microns) | 9 | 12 | 9 | 125 |
| Coefficient of thermal expansion ($\times 10^{-6}$) | 0.54 | 5 | 4 | 0.54 |
| Refractive index (589 nm: 23°C) | 1.4585 | 1.547 | 1.524 | 1.46 |
| Young's modulus ($GN\,m^{-2}$) | 73 | 76 | 86 | 73.1 |
| Tensile strength ($GN\,m^{-2}$) | 3.6 | 3.2 | 4.4 | 3.6 |
| Maximum service temperature (°C) | 1050 | 650 | 700 | 600 |

Table 4.2 gives the specific properties of different glass fibers. E-glass and Quartzel™ fibers are seen to have comparable mechanical properties. The quartz fibers have a low dielectric constant, offering high heat resistance and a high resistance to thermal shock. As a consequence of this, they are used extensively in the aircraft industry for radomes, electromagnetic windows and thermal insulation applications.

A wide range of 'cladding' materials can be used to manufacture RFLGs, but these must be compatible with the matrix system in the AFRC. A summary of selected properties for cladding materials is presented in Table 4.3.

The RFLGs can be located on the outer skins of the prepreg during laminating and then processed in the conventional manner. They can be interrogated by illuminating them with an appropriate light source, and bleeding light (for surface-mounted RFLGs) will indicate the presence of impact damage. This will then indicate that a closer and more detailed inspection should be carried out in the vicinity of the bleeding light. RFLGs can also be embedded, and in this instance, the attenuation or the lack of transmitted light can be correlated to a specified state of damage within the AFRC. The small diameter of the reinforcing glass fiber also results in a high sensitivity to impact damage. A 2-J impact event has been successfully detected with RFLGs embedded in the

**Table 4.3** Selected properties for potential cladding materials

| Property | Resin code | | | |
|---|---|---|---|---|
| | Epoxide[a] | Sylgard.182[b] | Xylan[c] | Sol-gel[d] |
| Polymer type | Epoxy | Silicone | Fluoro | Silica |
| Refractive index (589 nm: 23°C) | 1.45 | 1.41 | 1.35 | 1.42 |
| Coefficient of thermal expansion ($\times 10^{-6}$) | 58 | 96 | – | – |
| Young's modulus ($GN\,m^{-2}$) | 2 | 0.006 | 0.006 | – |
| Cure time and temperature | 150°C, 1 h | 150°C, 15 min | Solvent cast | 23°C, 2 days |
| Maximum service temperature (°C) | 120–180 | 200 | 330 | 1200 |

[a] Aldrich Chemicals.
[b] Dow Corning (UK).
[c] Whitford Plastics Ltd. (UK).
[d] Sol-gel-based coating.

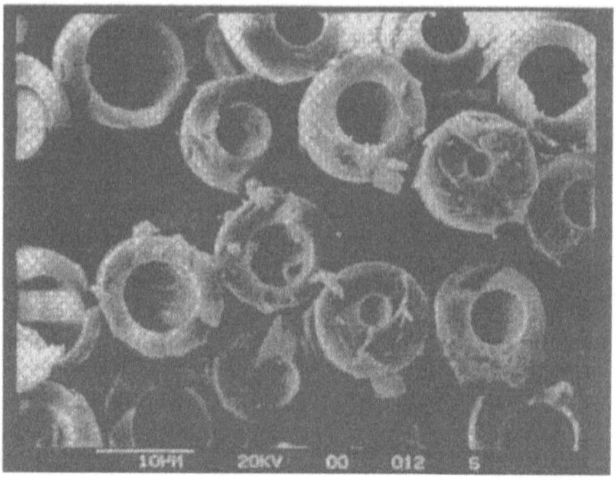

**Fig. 4.3** Micrograph of hollow glass fibers as supplied by Owens Corning (UK).

middle of 16-ply carbon fiber composites, which represented a significant improvement over the 6-J limit typically obtained by the fiber-fracture-based sensors using standard 125 μm optical communication fibers.

An extension of the self-sensing concept is the use of hollow reinforcing glass fibers [21]. A micrograph of hollow reinforcing glass fibers is shown in Fig. 4.3. There is no technical reason why these fibers cannot be converted to self-sensing fibers, as discussed above. In addition to this, the core capillary can be filled with chemicals such as thermochromic dyes to measure internal temperature, or even fluorescent dyes to detect large-area impact damage of surface layers.

The concept can obviously be extended to realistic self-repair composites [22] where the hardener and the resin are contained in individual fiber bundles which are co-mingled. The use of these fiber light guides for photo-curing in order to allow self-repair merits further investigation. Practical issues relating to infiltration of the capillary, and introducing fluids into the hollow core fibers with the facility for launching and detecting the transmitted light will have to be addressed.

### 4.3.5 Strain and stiffness-based damage detection in AFRCs

Monitoring the stiffness of a composite as a function of static or dynamic loading is an effective means of monitoring the progressive nature of damage in cross-ply and quasi-isotropic AFRCs [2]. Monitoring the stiffness tends to be more difficult with unidirectional (UD) carbon-fiber composites because the ultimate failure is sudden once a critical level of damage has been sustained. As mentioned previously, computation of the stiffness requires the applied stress in the vicinity of the sensor to be known. Since this is not always possible, optical-fiber-based strain monitoring techniques generally assume a perfect load-transfer from the matrix to the sensing element.

The operation of optical-fiber strain sensors is based on the modulation of one or more of the properties of the light wave which propagates in the optical fibers.

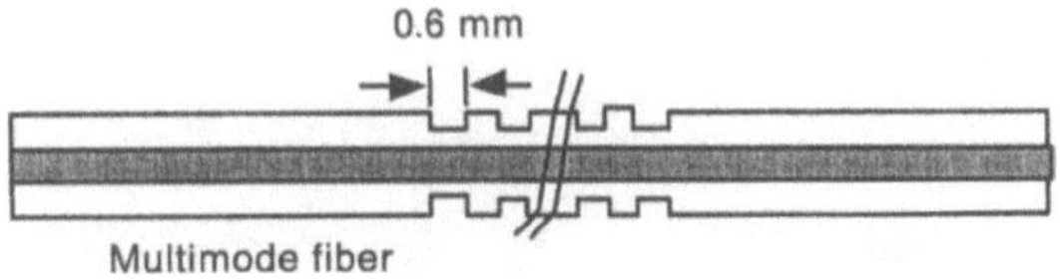


**Fig. 4.4** Schematic illustration of an etched intensity-type strain sensor [23].

A variety of optical parameters have been used for strain measurements, including optical intensity, polarization, optical path lengths (elongation and refractive index), wavelength, time of flight and optical scattering effects. The optical fibers used in the construction of sensors include standard telecommunication single mode or multimode fibers, polarization-maintaining fibers and other specialty fibers.

*(a) Intensity-based strain sensors*

In situations where simplicity, robustness and low cost are primary concerns, intensity-based strain sensors are most attractive. Vaziri and Chen [23] demonstrated that etched multimode fibers can be used as strain gauges. A sensor, as illustrated in Fig. 4.4, was made by etching the cladding of 50/125 μm multimode fibers using hydrofluoric acid. The thickness of the cladding removed was approximately 30 μm, and the period of the etched pattern was approximately 1.20 mm. The relationship between the attenuation of the sensor and the strain was linear, and the maximum strain was 4 m$\varepsilon$. The sensor had a fatigue life greater than 13 000 cycles.

Badcock and Fernando [24] described an intensity-based strain sensor based on a multimode fiber extrinsic Fabry–Perot cavity. The sensor consisted of two multimode fibers separated with an air-gap and housed in a 15-mm-long, 128/300 μm glass capillary. The fibers were fusion-spliced onto the capillary tube. The transmission loss was approximately 1 dB/m$\varepsilon$. The sensors were embedded in carbon-fiber-reinforced composites for static and dynamic strain measurements.

Martin *et al.* [25] described another intensity-based multimode fiber strain sensor. The sensor was made of a profile with a reduced cross section which resembled a symmetrical taper. The fiber diameter was drawn-down using an electric arc produced from an optical-fiber fusion splicer. A micrograph of the sensor is shown in Fig. 4.5. When the tapered fiber is under strain, its numerical aperture varies and hence so does the attenuation of light. The strain sensitivity is a function of the angle of the taper. A linear relationship between strain and attenuation was observed. The sensors were embedded into glass-fiber-reinforced composites and exhibited a typical failure strain of 2.3%.

Intensity-based optical-fiber sensors are characterized by simplicity, low cost and fast response time, but they generally suffer from interference of fiber bending loss and light source intensity fluctuation. Many techniques have been used in attempts to obtain effective referencing. In spatially referenced systems two

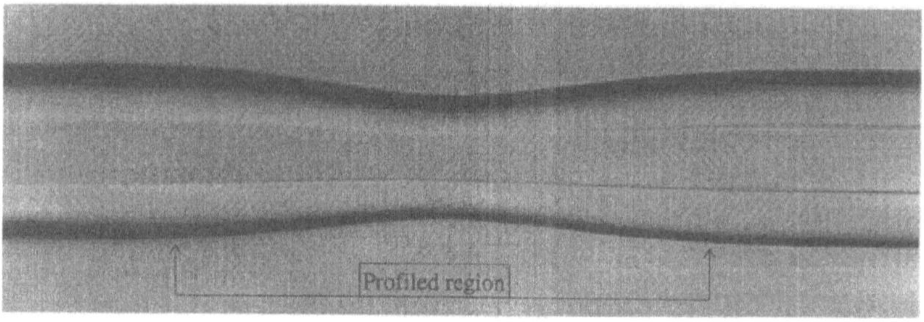


**Fig. 4.5** Micrograph of a typical profile sensor fabricated on a 50/125 μm multimode fiber.

optical paths are provided, one for the reference signal and one for the sensing signal, differing only in that the reference path does not contain the sensor itself [26]. This technique requires that the sensing and transmission parts of the system be duplicated exactly, one for sensing and the other to provide a reliable optical power reference [27]. Temporal referencing involves forcing a single short pulse to make circuits of an optical loop, with part of its power being removed and measured at each cycle, perhaps with the aid of a partially reflecting mirror [28]. If the sensing and reference signals comprise two separate wavelength bands, the system is said to be wavelength-referenced. The two wavelength bands could be provided by two sources [29] or a single broadband source with wavelength demultiplexing just before the detectors [30]. An elegant method of wavelength referencing was presented by Senior *et al.* [31], who used two spectral slices from a single light-emitting diode (LED), which had the effect of eliminating source thermal effects. Wang *et al.* [32] described a self-referencing technique for the multimode extrinsic Fabry–Perot-type intensity-based sensors by coating the tip of the lead-in/lead-out fiber with an interference edge filter. The filter reflects a portion of the spectrum above its edge wavelength and transmits the portion of the spectrum which is below the edge wavelength. The sensor signal at shorter wavelengths is modulated by the air gap. The reference signal at longer wavelengths is independent of the air gap. Hence the ratio between the reference and the sensor signal is in direct proportion to the attenuation modulated by the cavity length. The sensor was demonstrated to be feasible for strain, temperature and pressure sensing.

### *(b) Polarization-based strain sensors*

The polarimeter-type strain sensor is a compromise between the highly sensitive interferometry-based and the low-sensitivity intensity-based sensors. The principle of the polarimeters is based on the fact that the propagation constants of the two orthogonal eigenmodes of light propagating in an optical fiber change differently with strain. Therefore strain can be measured from the difference between the optical path lengths of these two orthogonal eigenmodes within one fiber. In order to minimize the environmental effect on the measurements, high birefringence (hi-bi)

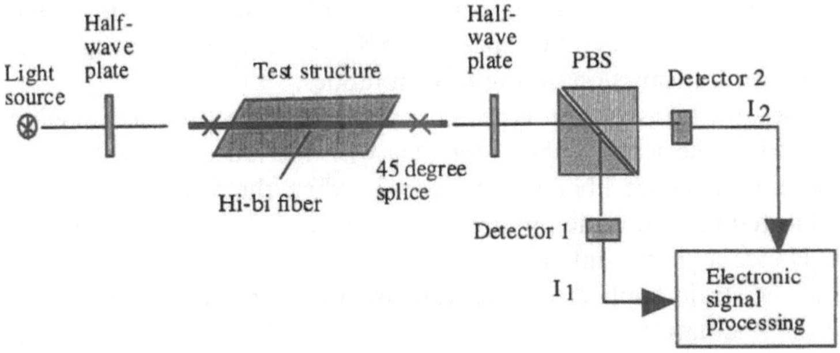


**Fig. 4.6** Schematic illustration of a polarimetric sensing system using high-birefringent fibers [33].

fibers are used as both lead-in/lead-out fibers and the sensor components. A typical transmission-type polarimeter strain-sensing system is illustrated in Fig. 4.6.

In Fig. 4.6, linearly polarized light is coupled into a high birefringence lead-in fiber through a half-wave plate, which is used to adjust the polarization states so that only one eigenmode is guided in the fiber. The lead-in fiber is spliced onto a section of sensing hi-bi fiber with their eigenaxes oriented at 45°. At the other end of the sensing fiber, a lead-out fiber is also spliced onto hi-bi fiber with their eigenaxes oriented at 45°. The two orthogonal components are separated via the polarization beam splitter (PBS) and detected with photodetectors. The half-wave plate in front of the PBS is used to align the eigenmodes of the fiber at 45° to the PBS. An alternative to the combination of half-wave plates and the polarization beam splitter is to use the polarizer and a twist on the hi-bi fiber [34]. Light reaching the two detectors is the result of interference between these two orthogonal modes. Their phases can be related to strain.

The sensing system can also be configured in the reflection mode and all-fiber mode [33]. Meltz and Dunphy [35] have demonstrated the use of hi-bi fiber embedded in unidirectional graphite epoxy composite for detecting sub-surface strain. Waite *et al.* [36] described the use of such a polarimeter for strain measurements in woven glass epoxy composites.

### *(c) Modal interferometric strain sensors*

Like polarimeter-type strain sensors, modal interferometric strain sensors also retain the advantages of both the interferometer- and the intensity-type sensors. They are based on the interference between two spatial modes propagating within one optical fiber. The first modal interferometric sensor was reported by Layton and Bucaro [37], and had a 28-m-long multimode fiber for the detection of sound. However, elliptical-core two-mode fibers have been the most well-developed modal interferometric sensors. A typical sensor system consists of a lead-in/lead-out fiber which supports the lowest order transverse mode ($LP_{01}$). A section of elliptical-core fiber that permits two transverse modes ($LP_{01}$ and $LP_{11}$) to propagate is fusion-spliced to a single-mode fiber [38]. The splice is made with the cores slightly offset so that both spatial modes are excited. The signal can be detected in both transmission and reflection modes.

In the case of a transmission-type sensor, a second section of a single-mode fiber is spliced to the sensing fiber. Whereas with the reflection-type sensor, the end of the elliptical-core fiber was mirrored, when the sensing section of the fiber is subjected to strain, the phases of the two spatial modes vary by different amounts. Hence a spatial intensity modulation is generated, which is a cosine function of the strain applied to the elliptical-core fiber. Hi-bi fibers are used to eliminate the fluctuation of polarization states in the lead-in/lead-out fibers.

The main advantage of a modal interferometer is the simplicity of signal recovery. Although modal interferometers are less sensitive than the single-mode interferometers (potentially about 20 dB less sensitive [39]), they are more sensitive than most of the other sensor types, such as intensity and speckle interferometric types [40]. Their main disadvantage is the complexity of splicing the polarization-maintaining fibers.

### *(d) Interferometry-based strain sensors*

Fiber optic interferometric-type strain sensors offer the highest possible sensitivity for strain measurements [41]. An optical fiber interferometer involves the superposition of electromagnetic fields of two or more beams or modes. They can be configured by using two single-mode fibers (Michelson and Mach–Zehnder type), one single-mode fiber (Fabry–Perot type), one multimode fiber (Fabry–Perot type) or one elliptical-core highly birefringent fiber (two-mode sensor). Phase change in the interferometer can be modulated via the physical elongation, the change in the refractive index of the core, or birefringence of the optical fiber.

Michelson [42–44] and Mach–Zehnder [45] types are two well-known dual-fiber interferometers, and they have been demonstrated to be capable of determining strain- and damage-induced acoustic emission signals in composites. However, these dual-fiber sensors suffer from the following disadvantages:

1. they may introduce more perturbation into the host than single-fiber sensors;
2. stress-induced birefringence in the sensing and referencing fibers affects the interference signal and results in an interference fading which is difficult to correct;
3. it is difficult to maintain the reference fiber in the strain-free condition without introducing significant deterioration in the host;
4. they are sensitive to integrated strain rather than local strain.

Optical-fiber Fabry–Perot-type sensors are among the most popular single-ended strain sensors and can overcome these problems. In particular, they are immune to environmental effects on the lead-in/lead-out fiber. According to the configuration of the Fabry–Perot (FP) cavity, they can be classified as (i) intrinsic FP, (ii) in-line fiber etalon (ILFE) type, or (iii) extrinsic FP interferometric type (EFPI) strain sensors.

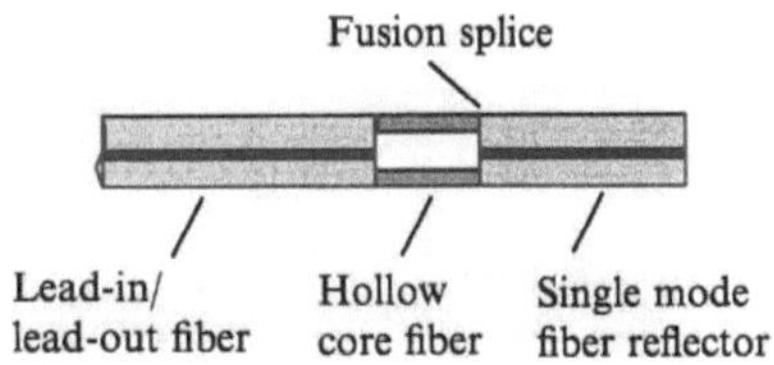


**Fig. 4.7** Schematic illustration of an in-line fiber Fabry–Perot strain sensor.

(i) Intrinsic fiber FP-type strain sensor. An intrinsic FP strain sensor consists of a section of single-mode fiber with both ends coated with semi-reflective film and fusion-spliced onto a lead-in/lead-out fiber.

The reflective film can be made of a number of materials and using different techniques. Lee *et al.* [46] fabricated high-reflectivity (86%) reflectors by depositing seven layers of quarter-wave $TiO_2/SiO_2$ stack on the cleaved fiber ends using a d.c. planar magnetron sputtering system. A finesse of 21 was obtained from the FP interferometers made of such reflector mirrors. Thin aluminum films have been evaporated onto the core of the fiber in a diameter of 15 μm [47]. Silver plating and gold coating by electron beam deposition [48] have also been used as fiber reflectors. Ultra-high finesse of 500 was achieved with multilayer dielectric mirrors attached to single-mode fiber [49]. Apart from axial strain measurements, intrinsic FP sensors have also been used for sensing temperature [50], ultrasound detection [51] and pressure in a combustion engine [52].

The disadvantages of intrinsic single-mode fiber FP sensors are:

1. cross-sensitivity between strain and temperature;
2. cross-sensitivity between axial and transverse strain;
3. the gauge length of the sensor is the same as the cavity length;
4. the power coupling efficiency to single-mode fibers is low;
5. intrinsic FP sensors are generally weak at the joint between the metallic coating and the glass.

(ii) In-line fiber etalon FP strain sensor. One of the techniques developed to improve the mechanical strength of the FP-type sensors is the introduction of an in-line fiber etalon (ILFE) type sensor [53]. As shown in Fig. 4.7, the ILFE consists of a section of hollow-core fiber (80/125 μm) spliced between the ends of two single-mode fibers (9/125 μm). The cavity consists of an air gap in the hollow fiber. The cavity length of the ILFE, which is the length of the hollow-core fiber, is in the range 10–150 μm. A failure strain of 1.96% was achieved with this type of sensor. The ILFE strain sensor maintains the advantage of the small size which IFFP sensors offer, but features a higher mechanical strength and low cross-sensitivity to transverse strain.

(iii) Extrinsic fiber Fabry–Perot (EFPI) strain sensors. These represent another important type of strain sensor. A schematic of the EFPI sensor is shown in Fig. 4.8. A lead-in/lead-out fiber and a reflector fiber are housed in a hollow-core fiber (128/130 μm). The fibers are either fusion-spliced onto the hollow fiber or fixed

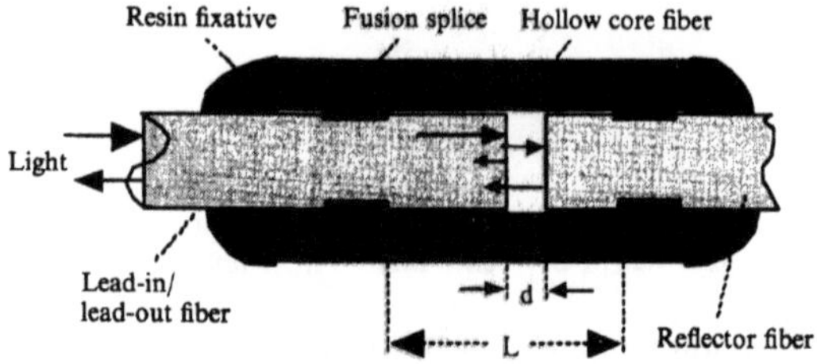


**Fig. 4.8** Schematic illustration of an EFPI sensor design.

with epoxy adhesive or high-temperature adhesive. Both single-mode [54] and multimode [55, 56] fibers have been used as the lead-in/lead-out fiber. The reflector can be either cleaved multimode fiber, single-mode fiber or polished metal wire [57]. Interference occurs between the reflected light at the end of the lead-in/lead-out fiber and the reflector.

Low-finesse EFPI sensors are made of cleaved optical fibers. High-finesse EFPI sensors are made of fibers with metallic coating on their cleaved ends [58]. High finesse offers high strain sensitivity but small operation range. The single-mode EFPI sensors offer high strain sensitivity but suffer from a low power-coupling efficiency. In contrast, multimode EFPI sensors have lower sensitivity but offer high power-coupling efficiency, and hence the cost of the light source is lower. Apart from an air cavity, EFPI sensors have also been constructed with silicon wafers [59] for pressure and temperature measurements. Thin polymer films have also been used to fabricate EFPI sensors for ultrasound detection [60].

The main advantages of EFPI sensors include: (i) the sensor is sensitive to axial strain but not transverse strain; (ii) the temperature sensitivity is negligible because the optical fibers and the quartz hollow-core fiber have similar thermal expansion coefficients but in opposite directions, and the refractive index of air is much more stable than that of silica glass; (iii) the mechanical strength of EFPI sensors is higher than that of IFFP sensors. Up to 1.4% failure strain was obtained with fusion-spliced EFPI sensors. The main disadvantage of EFPI sensors is their relatively large diameter and the perturbation introduced as a consequence.

EFPI strain sensors have been used in a number of applications, including measurement of fabrication residual strain in carbon-fiber-reinforced composites [61], and static and dynamic fatigue testing [62, 63].

### *(e) Fiber Bragg-grating-based strain and temperature sensors*

The fiber Bragg grating (FBG) is an intrinsic device 'photo-imprinted' into the fiber which exhibits a wavelength-encoded response to strain and temperature. The first FBG was developed by Hill *et al.* [64]. The mechanism of an FBG can be summarized as follows. Conventional telecommunications-grade germanium-doped silica-glass optical fiber has been shown to exhibit a significant photosensitivity response to UV radiation in the region of 248 nm, which is the

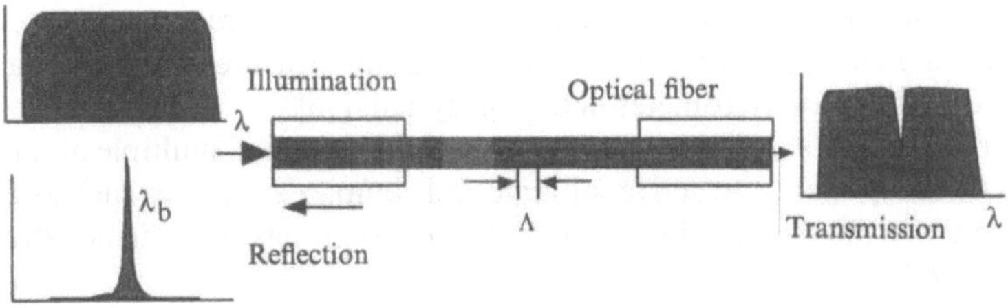


**Fig. 4.9** Schematic illustration of the characteristics of an FBG sensor.

absorption band in the glass associated with Ge/$SiO_2$ bonding [65]. Absorption of UV light in this region modifies the absorption characteristics of the glass, and consequently shifts the refractive index of the glass at wavelengths removed from the absorption region through the Kramers–Kronig relationship. The UV wavelength can readily be obtained from a number of sources: KrF excimer lasers, frequency-doubled $Ar^+$ lasers, dye lasers and quadrupled Nd:YAG lasers. Changes of index are in the order of $10^{-3}$ for normal fiber, and up to $10^{-2}$ can be achieved with hydrogen-loaded fibers [66]. Low-reflectivity FBGs were fabricated using a single 20-ns pulse with a power density of 1 J $cm^{-2}$ from KrF excimer laser radiation [67].

The germanosilicate glass fiber also has a weak absorption band at around 334 nm. This wavelength has been successfully used for fabricating weak long-period gratings on germanosilicate fibers ($\Delta n \sim 10^{-5}$) [68]. Strong Bragg gratings in hydrogen-loaded fibers ($\Delta n \sim 10^{-3}$) were also made using near-UV radiation at this wavelength [69]. The main advantage of using the near-UV wavelength is that Bragg gratings can be written on the fiber without stripping off the polymer jacket [70]. Two main methods have been used in the fabrication of FBGs. These are holographic side-exposure [71] and phase-mask techniques [72].

The transmission and reflection characteristics of an FBG are illustrated in Fig. 4.9. When a wide-band spectrum is injected in an FBG, a narrow band of the spectrum is reflected back. The peak reflection wavelength ($\lambda_b$) is dependent on the refractive index ($n$) of the core, and the spatial period of $\Lambda$ and can be described by

$$\lambda_b = 2n\Lambda. \tag{4.1}$$

The spatial period of an FBG can be modulated via thermal expansion and strain on the fiber. The refractive index of the core responds to temperature and strain via thermo-optic and photo-elastic effects. Hence the FBG reflection wavelength ($\lambda_b$) is a function of both strain and temperature. For fused silica fiber, the wavelength shift due to a 1°C change of temperature is approximately equivalent to 10 $\mu\varepsilon$ of tensile strain.

FBG sensors have been used for strain and temperature measurements [73] in a wide range of applications including fiber-reinforced composites [74] and concrete structures [75]. The strain response of an 850 nm FBG is typically about 0.65 nm $m\varepsilon^{-1}$. The failure strain of an FBG is up to 2%. The accuracy of the sensor depends on the detection system. A static strain accuracy of up to 1 $\mu\varepsilon$ and a higher accuracy for dynamic strain measurements can easily be achieved with an interferometer detection system [76].

Apart from the small size, FBG sensors offer superior multiplexibility. This makes them especially attractive for large and complex structures such as bridges, dams and marine-based vehicles. A 60-FBG system has been demonstrated by Davis *et al.* [75].

Long-period gratings (LPGs) were developed as band-rejection filters [77], but they have found increasing application in strain, temperature, refractive index and other chemical sensing applications. Similar to FBG sensors, LPGs are intrinsic and wavelength-encoded devices. They are also fabricated by illuminating interference patterns of UV radiation onto single-mode fibers. The period of an LPG is much longer than that of an FBG and is normally in the order of a few hundred μm. The periodic modulation of refractive index in the fiber results in power coupling from the fundamental guided mode to the discrete cladding modes. These modal couplings result in a series of distinct resonant bands in the transmission spectrum of an LPG. The absorption wavelengths of an LPG can be expressed by

$$\lambda_L = (n_c - n_{cL}^m)\Lambda, \tag{4.2}$$

where $n_c$ is the refractive index of the core for the fundamental guided modes and $n_{cL}^m$ is the refractive index of the $m$th cladding mode. The absorption peak wavelength $\lambda_L$ is primarily a function of strain, temperature and the ambient refractive index surrounding the cladding, which affects the value of $n_{cL}^m$. LPGs have been used for strain, temperature and refractive index measurements. The typical strain response of a 1257 nm resonance band is 1.52 nm m$\varepsilon^{-1}$, which is about twice the sensitivity of an FBG sensor at the same wavelength [78]. The temperature response at 1300 nm is 0.154 nm per °C, which is about ten times greater than for FBGs at the same wavelength.

The main advantage of LPG sensors is their high strain sensitivity. By carefully choosing the resonant wavelengths, FBGs can exhibit minimal cross-talk between strain and temperature. LPGs also offer the potential of multifunction sensing, e.g. strain, temperature and chemical sensing. The main drawback of LPGs is their susceptibility to bending.

### *(f) Strain measurements based on the transit time of light*

Johnson and Ulrich [79] demonstrated that strain measurement can be achieved by measuring the absolute length of a long multimode fiber using the transit time of a circulating light pulse in the fiber. For a length of 10–1000 m, the absolute length can be measured with an accuracy of ±0.2 mm. The performance can be improved by using a gradient-index (GRIN) multimode fiber.

The accuracy of transit time can be improved by measuring the phase shift of a continuous intensity-modulated optical wave. This technique was successfully demonstrated by Kashyap and Reeve [80] using a 277.7 MHz sinusoidally modulated GaAs laser source. The strain sensitivity was enhanced by using partial reflection splices between segmented fibers and looping the segments along the measurement region so that light traveled many times (up to 35) backward and forward [81]. The looped sensor system was surface-mounted onto a 1-m-long aluminum cantilever beam for static and dynamic strain measurements.

*(g) Strain- and stress-sensing based on Raman spectroscopy*

The Raman effect is a scattering process. Raman spectra are characteristic of variations in the molecular polarizability and dipole moments. If radiation of frequency $\nu_0$ illuminates a transparent specimen, it is found that weak scattering occurs at the same frequency as well as very weak scattering at frequencies $(\nu_0 + \nu')$ and $(\nu_0 - \nu')$. The radiation at $(\nu_0 - \nu')$ is called Stokes radiation, and that at $(\nu_0 + \nu')$ is called anti-Stokes radiation. Stokes scattering is much stronger than anti-Stokes scattering, and hence most Raman spectroscopy work uses Stokes scattering. An argon ion laser is the most commonly used light source. The main application of Raman spectroscopy is in the identification of chemical species. However, it has been used successfully for nondestructive stress/strain measurements [82]. Most of the Raman backbone vibrational modes of crystalline fibers shift to lower values in tension and to higher values in compression [83]. This is because the bond extension or contraction changes the bond stiffness and hence the atomic vibrational frequency.

*(h) Long-gauge-length strain sensors*

Long-gauge-length strain sensors are used for measuring the integrated strain along a length of structure. Two beam interferometers, such as the Michelson [84] or Mach–Zehnder type, measure the integrated strain along one arm of the interferometer. However, they suffer from polarization-induced fading, the inconvenience of requiring a reference arm and difficulties associated with the location of the sensing region. For low sensitivity and a large operation range, a polarimeter operating in the reflection mode is an ideal choice. Long-gauge-length intrinsic FP sensors have been developed for strain measurement with very high strain resolution [48]. These sensors have a 1-m-long cavity length made of high-quality cleaved single-mode fiber ends (cleaved angle of $<0.25$–$0.5°$). The fiber-end faces are gold-coated to obtain high reflectivity, and the finesse of the FP sensor is about 55. Strain measurement is obtained with a resolution in the order of n$\varepsilon$. For long-gauge-length intrinsic FP sensors, thermal effects are a major problem. Special fibers made of glasses which have negative thermo-optic coefficients, e.g. ZBLA and ZBLAN, were investigated with the overall aim of producing a thermally insensitive system. Another type of localized long-gauge-length strain sensor was made of a pair of low-reflectivity FGB reflectors [85]. A 1.6-m intrinsic FP was made using this method. The length of the FBGs was very short (0.5 mm) so that the reflection profile was not affected by the strain gradient. The reflectivities of the FBGs were about 0.25% and the FWHM was about 1 nm. A measurement error of 4 $\mu\varepsilon$ was obtained. The FBG reflectors offer much higher strength than the fusion splices used in making the intrinsic FP sensors, and the need to prepare a high-quality cleave is also eliminated.

*(i) Measurement of multiple-strain components*

The sensors described above are assumed to be subject only to axial strain. This is adequate when the structure is subject to simple loading conditions and only one strain component is required. For a general three-dimensional structure under complex loading, there may be six independent components of strain at a given location. Therefore it is necessary to develop sensors capable of multiple-strain-state measurements.

Gauges are often used at locations subject to stress in more than one direction. If the test point is on a free surface, the condition is called biaxial. The outer surface of a cylindrical pressure vessel is a good example. Biaxial strain is a combination of tensile and Poisson's ratio effects.

$$\varepsilon_x = (\delta_x - \nu\delta_y)/E, \tag{4.3}$$

$$\varepsilon_y = (\delta_y - \nu\delta_x)/E. \tag{4.4}$$

It can be seen from equations 4.3 and 4.4 that the structural transversal and axial strains are always linked together. To measure two strain components, a combination of two independent sensors is required. Sirkis and Lo [86] reported the use of combined Michelson and polarimetric interferometric sensors and passive signal processing methods. Bhatia *et al.* [54] reported a two-strain components sensor based on two quadrature phase-shifted EFPI sensors which were integrated in one sensor head. Fiber rosette sensors consisting of three fibers have been demonstrated for three-axis strain measurements. The basic configuration is shown in Fig. 4.10. Individual sensors in the rosette can be the Michelson interferometric type [42], the intrinsic FP type [47], the EFPI type [87] or FBGs.

The three-fiber rosette configurations suffer from large perturbation to the composite due to the number of fibers involved. To overcome this problem, a dual-wavelength, high-birefringence fiber Bragg grating scheme has been reported for three-axis strain and temperature measurement [88, 89].

Since all the intrinsic fiber sensors exhibit sensitivity to temperature, to obtain three-axis strain measurement in practical applications, at least four sensors will be required. However, the errors incurred by individual sensors will be

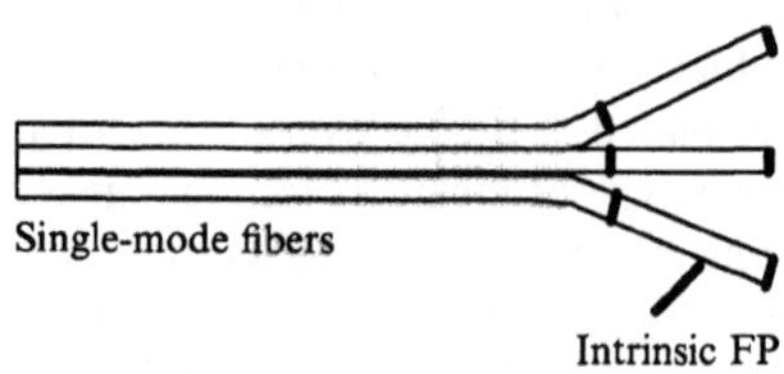


**Fig. 4.10** Intrinsic fiber Fabry–Perot rosette.

accumulated in the process of solving the simultaneous equations, and therefore this still remains a technical challenge.

*(j) Distributed-strain sensors*

Distributed-strain sensors have been developed for monitoring the distribution of strain along a fiber which will be embedded in a large structure such as an aircraft, a dam or a bridge. Distributed-strain sensors have all the advantages of point sensors and provide the spatial distribution of strain. The important criteria for a distributed sensing system are (i) to locate the sensing region, and (ii) to convert strain into a measurable optical parameter.

Optical time-domain reflectometry (OTDR) is one of the main techniques for spatial location. It has been used routinely in the optical telecommunications industry to characterize the loss, joint and fracture of fibers. A narrow pulse of laser light is launched in the fiber, and at a distance $z$ away from the launching point the back-scattered light arrives at the detector with a time-delay of $t$, which can be described by $t = 2z/V_g$, where $V_g$ is the group velocity. The power of back-scattered light from position $z$ normalized to the power launched in the fiber can be expressed in the equation by Healey [90]:

$$P(z) \propto V_g \alpha_s(z) S(z) \exp\left(-\int_0^z [\alpha_f(x) + \alpha_b(x)]\, dx\right) \tag{4.5}$$

where $\alpha_s$ is the scattering component of the fiber attenuation coefficient, $S$ is a capture ratio of the fiber representing the fraction of reverse scattered light coupled into the fiber and the total scattered power at point $z$, and $\alpha_f$ and $\alpha_b$ are forward and backward total scattering coefficients, respectively.

For strain measurements, special fibers which exhibit high microbending loss are commercially available [91]. Polarization modulation is another mechanism for sensing strain. Rogers [92] developed polarization optical time-domain reflectometry for the measurement of polarization states along the fiber, and this can be used for distributed strain sensing. The main drawback is the dependence of polarization on strain and EMF.

Horiguchi *et al.* [93] observed that the frequency shift of Brillouin scattering in optical fibers is a function of strain and temperature. They developed a technique based on Brillouin optical time-domain analysis (BOTDA) to measure both the Brillouin spectra and their location along the fiber [94]. The technique was based on the Brillouin interaction between pulse-pump and continuous-probe light waves counterpropagating in optical fibers. The BOTDA technique was used for the first time to measure distributed strain along a submarine optical fiber cable due to laying and recovering. The resolution for strain measurement was better than 200 $\mu\varepsilon$, and the spatial resolution was 100 m over a 3.7-km length of fiber cable [95]. Bao *et al.* [96] reported experimental results of distributed temperature and strain sensing over a length of 51 km of fiber based on a similar principle. Parker *et al.* [97] studied the dependence of spontaneous Brillouin back-scattering power and frequency shift on the strain and temperature. They demonstrated the feasibility of simultaneously resolving distributed strain and temperature based on the power and frequency of spontaneous Brillouin back scattering. Strain and temperature resolutions of 4°C and 100 $\mu\varepsilon$, respectively, were obtained with a spatial resolution of 40 m over 1200 m.

### *(k) Multiplexed strain sensors*

Optical-fiber-based strain sensors can be multiplexed along a single fiber (in series) or using multiple fibers (in parallel) or a combination of both. Multiplexing the fibers in series offers the advantage of economy, because both the optical fiber and the power are used with high efficiency. However, a disadvantage of this approach is that a fault sensor or a fiber fracture point in the network can partially or completely interrupt the whole sensing system. This disadvantage can be overcome by multiplexing the fibers in parallel, but this can be a more expensive option due to the added costs of the fibers, detectors and light sources. A comprehensive review of multiplexing techniques for optical fiber sensors (OFS) was given by Kersey [98].

With reference to interferometric-type strain sensors, coherence multiplexing has been a popular choice. The technique is based on the use of a low coherence light source. Sensors having optical path lengths longer than the coherence length of the source are interrogated by scanning a reference interferometer [99]. Multiplexed fiber-optic Michelson-type strain sensors have been used for civil structure health monitoring [84].

In the case of FBG sensor systems, wavelength division multiplexing using a tunable FP filter has been a popular approach. A tunable FP with a free spectral range of 45 nm can be used to multiplex 12 sensors in series spaced at >3 nm [100]. The feasibility of obtaining surface-displacement maps was demonstrated by multiplexing 16 FBG strain sensors attached to a cantilever plate [101]. In another application, wavelength division multiplexing was used to attach 12 FBG sensors to the hull of a glass-fiber-reinforced composite boat. Transient loads were detected in service using these sensors [102]. Rao *et al.* [103] demonstrated that the number of sensors multiplexed in one system could be increased by combining a wavelength-division technique with time-division or spatial-division techniques.

### *(l) Apparent thermal strain due to expansion and thermo-optic effects*

Intrinsic OFSs generally have a smaller size than extrinsic-type sensors, and hence they are more suitable for embedding in composites. However, since the fiber optic waveguide characteristics exhibit sensitivity to environmental temperature, they are also more sensitive to temperature than the extrinsic type, especially for the widely used interferometric-type FP and FBG sensors. A thorough theoretical treatment of the phase–strain–temperature relationship for interferometric-type sensors has been given by Sirkis [104, 105]. Another excellent review of this subject was published by Measures [39]. The following section outlines the sources contributing to thermally induced apparent strain in an intrinsic FP strain sensor.

The apparent thermal strain, defined as the strain per °C of temperature change in an intrinsic FP sensor, is similar to that in bulk glass, and can be described by [48]:

$$\varepsilon_T = (n-1)(1+\nu)\alpha + \frac{n^3\alpha E}{4}(q_{11}+q_{12}) + \frac{\mathrm{d}n}{\mathrm{d}T} \tag{4.6}$$

where $n$ is the refractive index, $\nu$ is Poisson's coefficient, $\alpha$ is the linear thermal expansion coefficient, $E$ is Young's modulus, and $q_{11}$ and $q_{12}$ are the stress-optic coefficients. The parameters of Ge-doped silica fiber core are very similar to those of bulk fused silica. Typical parameters are $n = 1.47$, $\nu = 0.17$, $\alpha = 0.55 \times 10^{-6}\,\mathrm{C}^{-1}$, $E = 7.4 \times 10^4\,\mathrm{MPa}$, $q_{11} = 0.42 \times 10^{-12}\,\mathrm{Pa}^{-1}$, $q_{12} = 2.7 \times 10^{-12}\,\mathrm{Pa}^{-1}$, $\mathrm{d}n/\mathrm{d}T = 11 \times 10^{-6}\,\mathrm{C}^{-1}$. The thermal apparent strain in silica fiber is 11.4 $\mu\varepsilon$ per °C, where the thermal expansion, stress and thermo-optic components are 0.29, 0.094 and 11.0 $\mu\varepsilon$ per °C, respectively. Therefore, the thermo-optic component is the dominant factor.

It has to be borne in mind that when an intrinsic FP sensor is embedded or surface-mounted in a structure, the thermal expansion coefficient will be determined by the host or substrate material. For example, typical linear thermal expansion coefficients of aluminum and carbon steel are $24 \times 10^{-6}$ and $12 \times 10^{-6}$, respectively [106], which results in a thermal expansion component of a surface-mounted intrinsic FP sensor of 15.6 and 7.8 $\mu\varepsilon$ per °C, respectively. Therefore, for high-sensitivity strain measurements, thermal apparent strain is an important consideration.

## 4.4 VIBRATION MONITORING

Vibration monitoring can be used to detect impact events, the resonance characteristics of the structure [107] and reconstruction of surface deformation. Modal interference in multimode fiber has been used for vibration monitoring [108]. When coherent light is launched into a multimode fiber, modal interference will result in an output speckle pattern. When the multimode fiber is subject to vibration, the total power from the fiber is relatively stable, but the speckle patterns will vary. A photodetector array can be used to detect these changes. Statistical analysis can be used to correlate the speckle patterns with characteristics of the vibration applied to the fiber.

Two-mode elliptical-core fiber sensors provide another effective means for vibration monitoring. The gauge lengths range from a few centimeters to tens of meters. These sensors do not suffer from lead-in/lead-out fiber intensity perturbations [109]. Murphy *et al.* [110] described a Michelson-type interferometer which was surface-mounted on metal bars for the characterization of acoustic effects. The sensor consists of two very short fiber arms to eliminate thermal drift. One of the fiber ends is coated with a metallic reflective coating and acts as the reference arm, and the other is used to illuminate and receive light scattered from the surface of the specimens. The system was found to be stable for a few hours. The separation between the fiber tip and the surface of the specimen was limited to less than 100 mm. The direction of the fiber was also very critical, and had to be maintained to less than 0.5°.

## 4.5 IMPACT DAMAGE DETECTION

As discussed previously, qualitative measurement of impact damage in AFRCs can be achieved by light bleeding from the embedded fibers. In this section, attention is paid to impact-location and the assessment of the magnitude of the impact damage. The feasibility of using bleeding light for the detection of low-energy impact damage in the leading-edge of an aircraft was demonstrated by Measures [111]. Doyle *et al.* [112] used this technique to study low-velocity impact damage in filament wound tubes. Detailed analysis on the nature and extent of low-velocity impact damage in AFRCs may be found in the work of Martin [113] and Sirkis and co-workers [114, 115].

Spillman and Fuhr [116] demonstrated a technique for impact detection and location by detecting the impact-generated acoustic emission signal using speckle interferometric fiber optic vibration sensors embedded in a fiber glass panel. Relative arrival time was used to locate the impact. Light intensity was used to correlate the impact magnitude. Based on the modal interference principle, Fuhr [117] demonstrated a single-fiber system for vibration monitoring, impact location and communications. Fan *et al.* [118] reported the use of laser-excited fluorescence in rare-earth-doped fibers to determine the location of impact damage. The position of a fracture in an erbium-doped optical fiber was determined by measuring the backward fluorescence power. Spatial accuracy of 3 mm within a sensing length of 43 cm was achieved. White light polarimetry has been used for the detection of impact events [119, 120]. Greene *et al.* [121] demonstrated real-time impact detection and location in a ten-ply graphite-fiber-reinforced composite panel with dimensions 121 cm × 79 cm × 0.325 cm using four embedded EFPI sensors. The EFPI sensors detect the impact-generated acoustic signal and the arrival time of the signal is used to detect location. Spatial resolution of 0.5 mm and accuracy better than 5 mm were achieved. An impact generated-AE signal was also detected with a fiber optic Michelson interferometer [122]. Impact damage growth was measured from real-time AE events.

## 4.6 FATIGUE DAMAGE DETECTION

Although the detection of fatigue damage is important in certain applications, there have been relatively few publications in this area. This is mainly due to the fact that the fatigue test (a) is difficult to conduct, (b) requires the sensor to outlive the test specimen and (c) requires the integrity of the interfacial bond strength to be maintained for the duration of the experiment.

Waite [123] reported the use of optical fibers with the polymer buffer stripped off and treated with silane solution to enhance the bonding. The attenuation characteristics of the signal were correlated with fatigue damage, and early fatigue damage was successfully detected in glass-fiber-reinforced plastic composite specimens. The transit-time strain measurement method based on optical time-domain reflectometry (OTDR) was used for static and dynamic strain measurements in quasi-isotropic E-glass/epoxy composite specimens [124]. Polyimide-coated optical fibers (approximately 30 cm long and 100/140 μm core/cladding diameter) were embedded in the middle ply of the composite. Two reflectors, 30 cm apart (1 cm long and 1.0 mm in diameter), were connected to the embedded fibers at the end of the specimen. Strain was measured over the distance between the two reflectors. For static tensile tests, a maximum of 3% strain was measured by the sensor, which was limited by the specimen. Optical fibers embedded between two 90° composite plies demonstrated a higher chance of damage during testing than those embedded between 0° plies. Tension–tension fatigue testing was conducted by cyclically loading the specimen at a stress ratio of $R = 0.1$ and an average frequency of 1 Hz. Excellent correlation was obtained between the fiber sensor and resistance strain gauges. Vries *et al.* [62] demonstrated fatigue testing using embedded single-mode fiber EFPI sensors in concrete. Two sensors were embedded in a rebar-reinforced concrete block which was subjected to cyclic loading of 760 lb at 8.4 Hz. Both the sensors survived 100 000 cycles.

The work of Badcock and Fernando [24] on the deployment of a multi-mode EFPI sensor design was previously discussed in association with an observed drift in the signal as a function of fatigue cycles. Also discussed previously was the work of Levin and Nilsson [125] on issues relating to the reliability of embedded FBG and EFPI sensors when subjected to quasi-static and fatigue loading.

Destructive tests on FBGs embedded into glass-fiber-reinforced polyester were carried out by Simonsen *et al.* [74]. A maximum strain of 1.6% was measured. Liu *et al.* [126] embedded a FBG sensor in a cross-ply carbon-fiber-reinforced epoxy composite and fatigue-tested the specimen at a peak stress of 210 MPa and a stress ratio of −0.5. The cyclic frequency varied from 0.1 Hz to 6 Hz during the test. The sensor was found to operate satisfactorily up to 700 000 cycles when the test was terminated.

The effect of embedded OFS on the static and fatigue interlaminar shear strength of a glass-reinforced polyester composite was evaluated by Guemes and Menendez [127]. No significant differences were found in the static and fatigue strength for the specimens with and without embedded optical fibers. However, when a large volume of optical fibers was embedded in the composites, significant strength reduction was observed.

## 4.7 SIMULTANEOUS STRAIN AND TEMPERATURE MEASUREMENTS IN COMPOSITES

During the processing of AFRCs, the parameters of interest include the temperature, the strain, and the rates of reactions of the chemical species in the resin system. With reference to temperature measurements, conventional thermocouples can only be used near the edges of the composite. This is obviously an issue with large and thick AFRC structures where it is essential to ensure a uniform temperature profile to control the reaction kinetics of the resin system during processing.

With reference to on-line health monitoring, the main parameters of interest are strain, temperature and the applied stress. Most aerospace structures are subjected to both static and dynamic loading. Furthermore, the sensor system must be capable of measuring both tensile and compressive strain. A key feature of AFRCs when subjected to dynamic loading is the associated increase in the temperature in the composite [128]. The origin of the temperature rise can be attributed to friction between the reinforcing fibers and the resin matrix at points of fiber fractures and debonds, interlaminar friction at delaminations and hysteretic heating. The magnitude of the temperature rise will depend on the nature of the reinforcing fiber and the matrix, the interfacial bond strength, the rate and frequency of loading, the thermal conductivity of the composite and the external temperature. For example, surface temperature increases of 12°C and 50°C have been recorded for carbon- and polyaramid-fiber reinforced composites, respectively [128]. Therefore it is essential that techniques be developed to decouple strain from temperature.

The fundamental requirements for discrimination of strain and temperature are (a) using two sensors which have different responses to strain and temperature, or (b) using one sensor which can provide two optical parameters having different strain and temperature responses. A variety of sensor combinations have been tried in order to decouple strain from temperature, including:

1. two FBGs at different wavelengths [129];
2. using the fundamental and the second-order Bragg grating wavelengths [130];
3. polarization rocking filters and an FBG [131];
4. an FBG with two different cladding diameters [132];
5. a combination of an in-line Fabry–Perot etalon (ILFPE) with an intrinsic Fabry–Perot (IFP) interferometer or an FBG [133, 134];
6. an FBG and a low-finesse Fabry–Perot cavity (temperature and displacement measurements) [135];
7. a combination of an LPG with an FBG [136] or two resonant bands of an LPG [137];
8. a combination of interferometric and polarimetric techniques [138];
9. a combination of polarimetric and dual-mode interferometers [139];
10. complete low-coherence interferometry, i.e. group delay and dispersion, or mean and differential group delay [14];
11. a combination of an FBG and a stimulated Brillouin scattering effect [141];
12. FBG and EFPI sensors for *in situ* strain and temperature sensing in composites [142, 143].

### 4.7.1 FBGs at two different wavelengths

It is well known that the FBG reflective wavelength is a function of strain (axial strain and lateral strain) and temperature. When an FBG is only subject to axial strain and located in air or in a material which has the same thermal expansion as the quartz glass, the fractional wavelength shift can be expressed by

$$\frac{\Delta\lambda}{\lambda_b} = (1 - P_e)\varepsilon + (\alpha + \xi)\Delta T, \tag{4.7}$$

where $\lambda_b$ is the wavelength under strain-free conditions at room temperature, $P_e$ is the photoelastic constant, which is a function of the Poisson ratio, the photoelastic coefficients and the effective refractive index of the fibre core, $\alpha$ is the temperature coefficient of expansion and $\xi$ is the thermo-optic coefficient of the fiber. Since the photoelastic and thermo-optic coefficients are wavelength-dependent, the strain and temperature coefficients are also wavelength-dependent. Based on this principle, Xu *et al.* [129] have written two FBGs at different spatial periods superposed in the same section of a fiber. The differences in strain and temperature coefficients between 849 nm and 1298 nm were approximately 6.3% and 9.5%, respectively.

The dual-wavelength approach provides a simple method for decoupling strain and temperature. However, because the range of wavelengths within which the fiber supports a single guide mode is fairly limited, the difference of the strain and temperature coefficients for the fractional wavelength shift at two different wavelengths were very small, which consequently limits the detection sensitivity. Apart from superposing two different gratings, another way to generate two wavelengths is to exploit the second-order diffraction of the FBG.

### 4.7.2 Using the fundamental and second-order Bragg grating wavelengths

Second-order Bragg grating reflection results from the nonlinear process of grating recording, where continued exposure of UV light results in saturation of the index perturbation. The strength of the second-order and higher-order reflections is determined by the magnitude of their respective Fourier coefficients describing the index perturbation. The second-order wavelength is around half the value of the primary FBG wavelength, but slightly offset due to material dispersion. Xie *et al.* [144] observed the second-order diffraction from FBGs written within germanosilicate fibers at approximately 734 nm with a primary wavelength of 1451 nm. Kalli *et al.* [145] observed the second-order reflection at around 789 nm with a reflectivity of approximately 0.1%. The strain and temperature coefficients have been measured. In principle, the second-order reflection combined with the primary reflection can be used for simultaneous axial strain and temperature decoupling [130].

### 4.7.3 Photogenerated polarization couplers and combinations with low-order modes in an elliptically cored optical fiber

In fiber-photogenerated polarization couplers or rocking filters have been fabricated in elliptical core fibers using 266 nm and 532 nm sources [146]. A number of configurations have been demonstrated for strain and temperature sensing using these rocking filters. The resonant condition for a polarization rocking filter is given by

$$\lambda = \Delta n \Lambda, \tag{4.8}$$

where $\lambda$ is the resonant coupling wavelength, $\Delta n$ is the modal birefringence and $\Lambda$ is the grating period, which is equal to the polarization beat-length. Because both $\Delta n$ and $\Lambda$ vary with strain and temperature, the resonant wavelength can be used for strain and temperature sensing.

A pair of identical polarization filters 'photo-printed' in series separated by distance $L$ in one elliptical core fiber effectively becomes a compact Mach–Zehnder interferometer. Hence the resonant wavelength shift can be transferred to phase measurement. A pair of rocking filters, both with resonant wavelengths of 786.6 nm, were written externally with 266 nm light on an elliptical core D-type Andrew fiber which had a beat-length of 4.35 mm at 780 nm. The rocking filters were separated by a distance of 32 cm, each having a length of 22.44 cm and containing 52 coupling points. The coupling efficiency of the filters was 7.5%. The Mach–Zehnder interferometer was excited by one of the polarization modes and monitored in the orthogonal mode. The sensor had a temperature sensitivity of 9.1 mrad $^{\circ}C^{-1}$ $cm^{-1}$.

A simpler configuration based on a polarization rocking filter written in one elliptical core fiber which supports two pairs of low-order modes was demonstrated for strain and temperature sensing. Two different resonant wavelengths were generated corresponding to each pair of modes. Fractional wavelength shifts to strain and temperature at these two wavelengths were different; hence strain and temperature could be resolved simultaneously. Strain resolution of 40 $\mu\varepsilon$ and 0.25°C temperature resolution were reported [147].

### 4.7.4 FBGs with two different wavelengths and cladding diameters

Two FBGs of slightly different wavelengths written on two fibers of different cladding diameters and different materials spliced in series were demonstrated for simultaneous strain and temperature recovery. The combination was based on the following principles: (i) subject to the same axial stress, different cladding diameters result in different strains, and hence a different wavelength shift; (ii) FBGs at similar wavelengths but written on optical fibers made of different materials exhibit different strain and temperature coefficients. Therefore, when two such FBGs are combined, strain and temperature can be resolved. James *et al.* [132] demonstrated this sensing concept using FBGs at 826 and 837 nm written in 80 μm Corning fiber and 120 μm Spectran fiber, respectively. Their reflectivities were 60% and 75%, respectively. One edge-emitting LED (ELED) source with a center wavelength of 830 nm was used to address both FBGs. The strain and temperature coefficients of these two sensors were 6% and 20%, respectively. The advantage of this scheme is that the wavelength difference between the two FBGs is small, and hence they can be conveniently addressed with one light source. The disadvantage is that the mismatch in cladding diameter can introduce significant loss. The small difference between their strain coefficients will also limit the resolving power.

### 4.7.5 Combination of an in-line Fabry–Perot etalon (ILFPE) with an intrinsic Fabry–Perot (IFP) interferometer or an FBG

In-line Fabry–Perot etalon (ILFPE) strain sensors have similar advantages to intrinsic Fabry–Perot (IFP) sensors in terms of their small size, but with the additional advantage of minimal sensitivity to temperature and lateral strain. A combination of an ILFPE and an FBG or an IFP sensor will allow simultaneous strain and temperature sensing. Singh and Sirkis [133] demonstrated such a technique. For the ILFPE and IFP combination, an SLD light source and two path-matching interferometers were used for signal processing. For the ILFPE and FBG combination, an optical spectrum analyzer was used for FBG signal processing. Recent progress includes the development of interferometric signal processing for both the ILFPE and the FBG sensor [134], allowing simultaneous strain and temperature decoupling with resolutions of 5 $\mu\varepsilon$ and 0.5°C.

### 4.7.6 Combination of an FBG and a low-finesse Fabry–Perot cavity (temperature and displacement measurements)

An FBG and a low-finesse extrinsic FP sensor connected in tandem was reported for displacement and temperature sensing [135]. An FBG was used as a temperature sensor, while a low-finesse extrinsic Fabry–Perot sensor consisting of a cleaved fiber end and a reflective mirror was sensitive to both the displacement and the temperature. An SLD with a peak wavelength of 831 nm and bandwidth of 25 nm was used as the light source to illuminate both sensors. A fiber Mach–Zehnder interferometer with a mechanically scanned path-length was used to measure the absolute displacement. A matching FBG was actively tuned via stretching to measure the temperature-induced wavelength shift in the sensing grating.

The sensing concept was successfully demonstrated. Temperature and displacement were measured with an accuracy of approximately 6% and 3%, respectively. The application of this system is limited by the signal processing methods. The receiving Mach–Zehnder and matching FBG are prone to ambient temperature influences. The reflectivity of the referencing FBG is a critical factor in cross-talk between these two sensors.

### 4.7.7 Combination of a long-period fiber grating (LPG) with an FBG or two resonant bands of an LPG

As discussed earlier, an LPG has a few absorption bands due to the power coupling of the fundamental guide mode to the discrete cladding modes. The wavelength shifts at different absorption bands exhibit different sensitivities to strain, temperature and refractive index. This differential sensitivity allows the decoupling of strain and temperature using one LPG [137]. Strain and temperature were measured with resolutions of 15 $\mu\varepsilon$ and 2°C, and were achieved with an LPG written on a standard Corning 1310-nm fiber with a period of 320 μm. LPGs generally have a much higher temperature sensitivity than FBGs, but a comparable strain sensitivity. Hence the combination of an LPG and an FBG is capable of strain and temperature discrimination. Patrick *et al.* [136] reported a scheme using an LPG and two FBGs in series to resolve strain and temperature. The LPG has a central absorption wavelength of 1306 nm, the FBGs have reflection wavelengths of 1293 and 1321 nm. The FBG wavelengths were measured with an optical spectrum analyzer and the LPG wavelength was measured from the relative reflection intensities of the FBGs. Measurement errors of $\pm 5\,\mu\varepsilon$ and $\pm 0.5$°C were achieved.

### 4.7.8 Combination of interferometric and polarimetric techniques

The strain and temperature applied to a section of polarization fiber can be recovered by configuring the hi-bi fiber as a fiber Fabry–Perot interferometer. Two orthogonal polarization modes were used to act as two interferometers. Since the strain and temperature sensitivities are different for the two modes, the phases from these two interferometers are different. Once the strain and temperature coefficients for the hi-bi fiber are determined, two simultaneous and independent equations can be set up. Therefore, strain and temperature can be recovered. Farahi *et al.* [138] demonstrated the feasibility using a 7-cm-long hi-bi fiber which was subject to loadings of strain and temperature. Strain and temperature cross-sensitivity was studied by measuring the thermal coefficient at different strain levels.

### 4.7.9 Combination of polarimetric and dual-mode interferometers

A polarization preserving fiber with a single-mode cut-off wavelength ($\lambda_c$) can propagate two modes, $LP_{01}$ and $LP_{11}$, when operated at a wavelength shorter than $\lambda_c$, and a single mode when operated at a wavelength longer than $\lambda_c$. Vengsarkar *et al.* [139] developed a two-wavelength system to multiplex a polarimetric sensor at a longer wavelength and a modal interferometric sensor at a shorter wavelength. The system was used for simultaneous strain and temperature sensing using an 0.55-m-long elliptical core fiber. Resolutions of 10 $\mu\varepsilon$ and 5°C were achieved for strain and temperature measurements, respectively.

A modification of the above scheme was made by Wang *et al.* [148]. The dual-mode interferometer, based on an elliptical-core fiber, operated at one wavelength (632.8 nm, HeNe laser). A multimode fiber was used as a spatial filter to interrogate the phase of the modal interference signal, which can be correlated to axial strain in the elliptical-core fiber. Visibility modulation of the modal interferometer was introduced using an index-matching liquid between the end of the elliptical-core fiber and the multimode fiber, or by mounting the multimode fiber on a bi-metallic strip. In the first case, the refractive index of the liquid was a function of temperature, and hence the insertion loss and consequently the visibility of a dynamic modal interference signal was modified by temperature. In the second case, the position of the multimode fiber was modulated by temperature. Thus temperature can be modulated by measuring the visibility of the modal interference signal, and strain can be measured from the quasi-static modal interference signal. Strain and temperature measurements were achieved with resolutions of 0.1 $\mu\varepsilon$ and $\pm 1\,^{\circ}C$, respectively. The advantage of this scheme is its simplicity. The disadvantages are that strain and temperature cannot be measured simultaneously.

### 4.7.10 Complete low-coherence interferometry: group delay and dispersion or mean and differential group delay

White light interferometry normally involves a sensing interferometer and a reference interferometer which has a temporal or spatial scan unit. Absolute measurements (displacement, strain or temperature) can be made by locating the envelope of the interference fringes. An extension of low-coherence interferometry has been developed by Flavin *et al.* [149], who interpreted the complete interferogram, including both the envelope and the phase, across the entire source spectrum using dispersive Fourier transform spectroscopy (DFTS). Both the group delay and the dispersion were recovered using this method. The technique was applied to simultaneous recovery of strain and temperature from the modulation of group delay and first-order dispersion. The cross-term was found to be negligible and the transformation was very robust. Strain and temperature were recovered with accuracies of 5 $\mu\varepsilon$ and 0.35 K, respectively [150].

An analogous approach has been developed by Luke *et al.* [140], where a section of hi-bi fiber was used to fabricate a dual-mode intrinsic FP sensor. The average and the difference of their FP cavity lengths, termed mean and difference group delay, were shown to have different strain and temperature responses [151]. A low-coherence interferometric system coupled with DFTS data processing was used to recover both the mean and the differential group delay. Strain and temperature were measured for the hi-bi fiber embedded in carbon-fiber-reinforced composite. Typical measurement errors were 7 $\mu\varepsilon$ and 0.7 K, respectively. This source spectrum is much shorter than that required for the dispersion-based system. The DFTS method was found to be more accurate than locating the peaks of the interferograms.

### 4.7.11 Combination of an FBG and a stimulated Brillouin scattering effect

A combination of a truly distributed temperature-sensing system based on stimulated Brillouin scattering and a wavelength-encoded FBG sensor was reported by Davis and Kersey [141] for resolving temperature and strain. The Brillouin gain process involves a three-wave interaction between the pump signal, the continuous wave (CW) probe signal and an acoustic wave. The Brillouin gain process occurs when the difference between the pump and probe frequencies equals the Brillouin scattering frequency shift $\nu_B$. $\nu_B$ is a function of the refractive index of the fiber and the speed of the acoustic signal, which vary with temperature and strain. On the other hand, the FBG wavelength varies with temperature and strain via the refractive index of the fiber and the spatial period of the grating. Therefore, FBG and Brillouin scattering frequency shifts are independent functions of strain and temperature. The strain and temperature responses of the Brillouin scattering frequency shift when the pump wavelength was 1.3 μm were 1.21 MHz °C$^{-1}$ and 55 KHz με$^{-1}$, respectively. Strain and temperature were obtained with resolutions of 10 με and 1°C, respectively.

### 4.7.12 Combination of FBG and EFPI sensors for *in situ* strain and temperature sensing

The use of an FBG and an EFPI sensor in series to facilitate *in situ* strain and temperature measurements in carbon-fiber-reinforced epoxy composites [142] has been described previously. The FBG/EFPI combination exploited the temperature-insensitivity of the EFPI sensor. An EFPI strain sensor design with an integrated FBG in a strain-free condition [143] is shown in Fig. 4.11. Because the EFPI and the FBG have independent responses to strain and temperature, this sensor design offers a significant improvement on the reliability of decoupling strain from temperature.

The sensor shown in Fig. 4.11, consists of a single-mode lead-in/lead-out optical fiber and a multimode fiber reflector. A 5-mm-long FBG with a peak

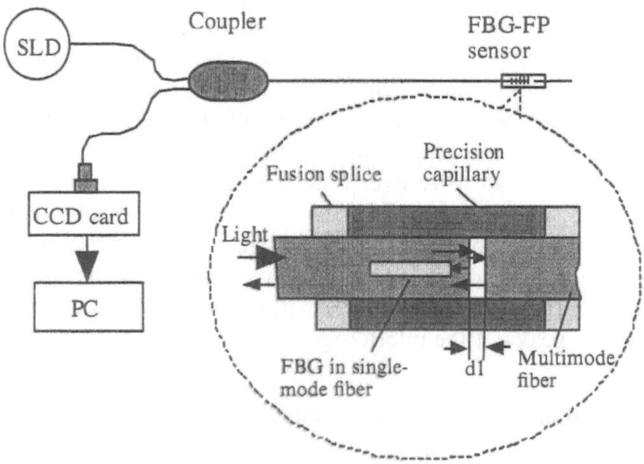


**Fig. 4.11** Schematic illustration of an *in situ* strain- and temperature-sensing system using a combined FBG and EFPI sensor design.

wavelength of 851.7 nm was written on the single-mode fiber and then cleaved 3.5 mm away from the center of the FBG. The length of the EFPI capillary (128/300 μm) housing was 20 mm. Both the single-mode and the multimode fibers were fusion-spliced onto the capillary. The fusion point on the single-mode fiber was 8 mm from the center of the FBG. This sensor design essentially consists of an FBG temperature sensor in a strain-free condition which was integrated within an extrinsic Fabry–Perot strain sensor.

A super luminescent diode (Hamamatsu L3302) was used to illuminate the sensors via a single-mode coupler. Here a narrow band of the optical power was reflected from the FBG, and the transmitted light was used to illuminate the EFPI sensor. The reflected light from both sensors was detected using a charge coupled device (CCD) spectrometer (Ocean Optics S1000). As will be seen later, the FBG had a very narrow bandwidth with a high reflectivity, whereas the EFPI, being a low-finesse interferometer, reflected only a fraction of the total optical power. Hence demultiplexing the FBG and the EFPI sensor was readily achievable.

The reflection spectrum of the EFPI sensor is modulated by its cavity length ($d$). Hence, $d$ can be measured via the free spectral range. When the sensor is embedded in, or surface mounted onto a composite specimen, the EFPI cavity length is primarily a function of the applied strain ($\varepsilon$), but it is also influenced by the orientation of the sensor in relation to the loading axis, and the nature of bonding, as well as the thermal coefficient of the composite and the adhesive.

The FBG was made from a standard 820-nm single-mode optical fiber which was photosensitized by hydrogen loading prior to UV exposure. The nominal Bragg grating wavelength was 851.7 nm, with a 3 dB bandwidth of 0.2 nm and a reflectivity of approximately 40%. The unknown vector $(\varepsilon, T)^{-1}$ can be determined by

$$\begin{bmatrix} \varepsilon \\ T \end{bmatrix} = \begin{bmatrix} P_{11} & P_{12} \\ P_{21} & P_{22} \end{bmatrix} \begin{bmatrix} \Delta d \\ \Delta \lambda \end{bmatrix} = \mathbf{P} \begin{bmatrix} \Delta d \\ \Delta \lambda \end{bmatrix} \tag{4.9}$$

where $\mathbf{P}$ is the matrix which transfers the EFPI cavity length and the FBG wavelength into strain and temperature. In these experiments, a PLS2 partial least-squares statistical calibration method [152] was used to correlate the applied strain and the temperature to the changes in the EFPI cavity length and the FBG wavelength. The values of the components in the $\mathbf{P}$ matrix were estimated using this method from a population of tests across a range of temperatures and applied strains.

The thermal response of the sensor in air was characterized before embedding the sensor in the composite. The sensor was conditioned in a thermal chamber where the temperature was controlled within ±1°C in the range 20–130°C. The temperature was increased from 20°C to 130°C in steps of 10°C. Measurements were carried out after the temperature had stabilized for a period of 15 min. Analysis of the results indicated that the FBG sensor had a wavelength/temperature coefficient of 5.6 ± 0.2 pm per °C. The mean value of the measured EFPI cavity length was 45.5 ± 0.15 μm.

The sensor was embedded between the 4th and 5th ply of a 16-ply carbon-fiber-reinforced composite sample with a ply lay-up sequence of $(0, 90_2, 0_2, 90, 0, 90)_s$. The sample was end-tabbed to protect the composite during mechanical loading. After embedment, the EFPI sensor was found to have a mean cavity length of 42.5 μm at room temperature, and a gauge length of 16.3 mm. The reduction in the EFPI cavity length has been attributed to residual fabrication strain (−185 με). The FBG sensor exhibited a negligible response to applied strain (the standard deviation of the FBG wavelength was less than 13 pm for a strain change in the range 0–1200 με).

An Instron model 1195 tensile testing machine equipped with a temperature-controlled environmental chamber was used for these trials. The specimen was loaded from the stress-free state up to 64 MPa in 8-MPa increments. This test was repeated a number of times to assess the reproducibility of the data. In addition to this, the tests were repeated as a function of temperature from ambient up to a maximum of 60°C. A K-type thermocouple was used to measure the temperature within the environmental chamber, and it also served as a reference source for the temperature. A surface-mounted foil strain gauge (gauge factor 2.07) was used as the reference source for the applied strain. A second strain gauge was surface-mounted on an unloaded specimen which was also located in the environmental chamber. The second foil gauge was used to compensate for the effect of temperature on the response of the strain gauge.

Typical responses of the sensor to applied strain and temperature are shown in Fig. 4.12. The temperature change caused shifts in the FBG peak wavelength, and

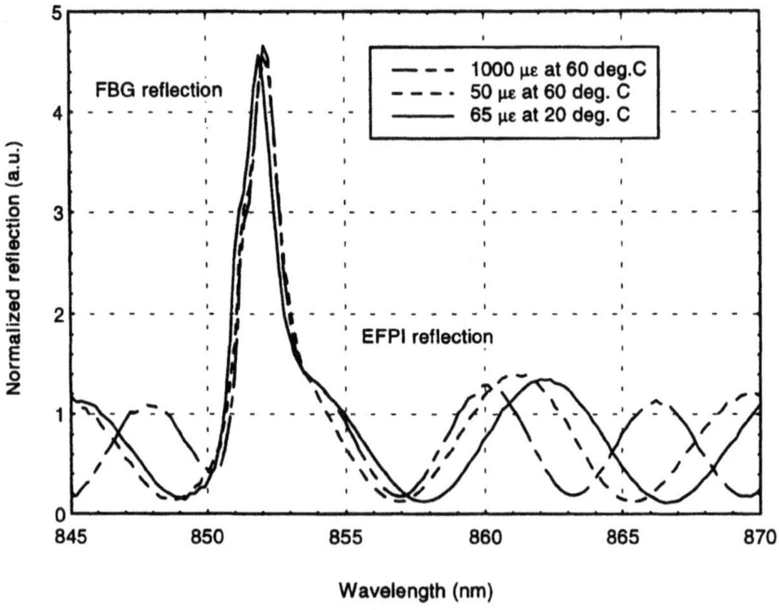


**Fig. 4.12** Individual responses of EFPI and FBG sensors.

the applied strain caused a change in the period of the interference fringes. These reflection spectra were subjected to a normalization procedure with respect to the emission spectrum of the light source. This in effect enhanced the visibility of the interference fringes.

Ninety sets of test data were randomly selected across the temperature range between 30 and 70°C, with a corresponding strain range between 0 and 1200 $\mu\varepsilon$. A PLS2 partial least-squares algorithm with random cross-validation was used to characterize the relative response of the sensors. The conversion matrix for $P$ was obtained as follows:

$$\begin{bmatrix} \varepsilon \\ T \end{bmatrix} = \mathbf{P} \begin{bmatrix} \Delta d \\ \Delta\lambda \end{bmatrix} = \begin{bmatrix} 63.22 & -139.37 \\ 0.145 & 163.28 \end{bmatrix} \begin{bmatrix} (d - 42.5) \\ (\lambda - 851.75) \end{bmatrix} \tag{4.10}$$

The EFPI gauge length and the FBG thermal coefficient deduced from $P_{11}$ and $P_{22}$ were 15.8 nm and 6.12 pm per °C, respectively, whereas the pre-calibrated EFPI gauge length at room temperature was 16.3 mm, and the thermal coefficient for the FBG prior to embedding was $5.6 \pm 0.2$ pm per °C. In principle, $P_{12}$ and $P_{21}$ should be zero; the statistically derived value of $-139.37\,\mu\varepsilon\,\text{nm}^{-1}$ and $0.145°\text{C}\,\mu\text{m}^{-1}$ were not significant. $P_{12}$ was mainly attributed to the shear strain applied to the specimen during tensile tests. $P_{21}$ was mainly attributed to the thermal apparent strain applied to the EFPI. Cross-sensitivity between strain and temperature also contributed to these two terms.

Figures 4.13 and 4.14 illustrate the observed relationship between the optically measured strain and temperature when compared with the reference data from

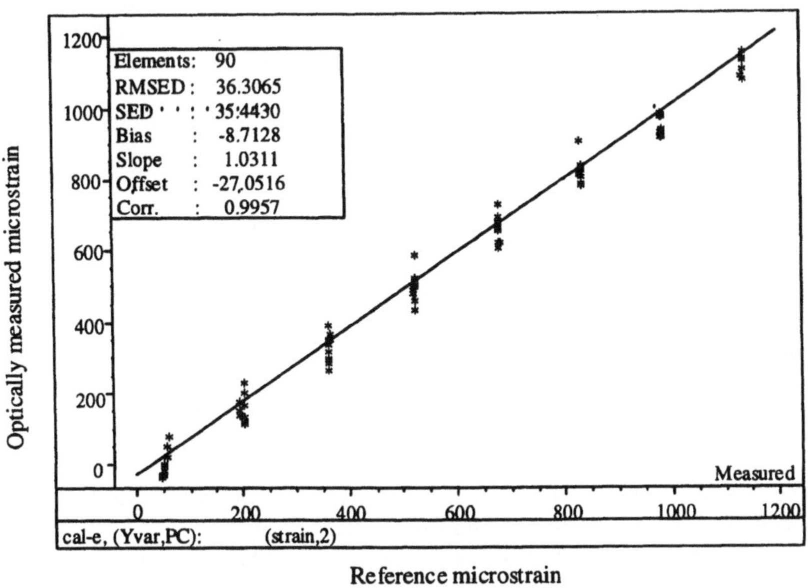


**Fig. 4.13** Optically measured strain *vs.* reference data from a surface-mounted strain gauge.

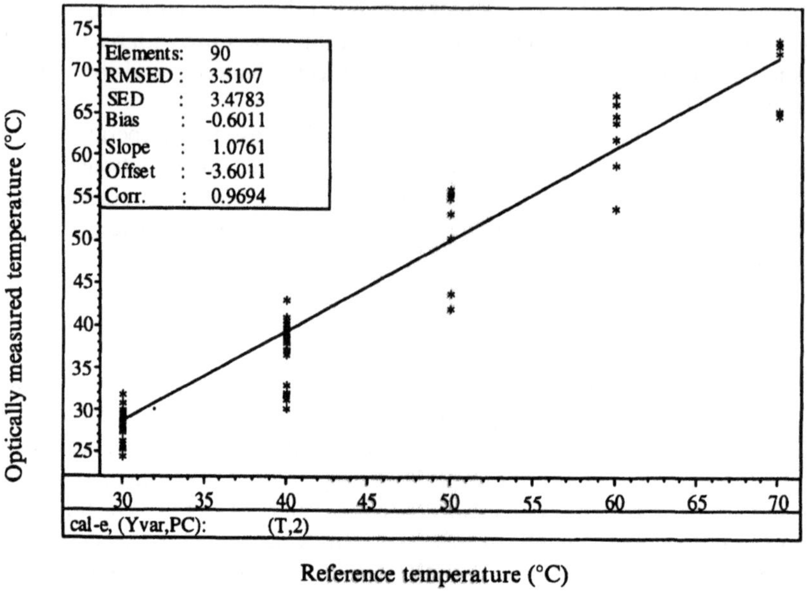


**Fig. 4.14** Optically measured temperature *vs.* data from a thermocouple.

the thermocouple and strain gauge, respectively. The correlation between the measured strain and the reference (strain gauge) data was 99.6%. The corresponding standard deviation was 36 $\mu\varepsilon$ for a mean strain of 605 $\mu\varepsilon$. The temperature measurement had a correlation of 98.7% with the reference (thermocouple) data, along with a standard deviation of 3.5°C for a mean temperature of 34.5°C.

## 4.8 SUMMARY

This chapter has reviewed the application of optical-fiber sensors in AFRCs for *in situ* health monitoring. The areas considered were impact and fatigue damage detection, along with strain and temperature monitoring. Both intensity and interferometric types of sensors were considered.

On the basis of the information presented in this and the previous two chapters, it can be seen that a number of areas still remain to be addressed, including:

1. low-cost sensor systems for on-line process monitoring and *in situ* health monitoring of AFRCs;
2. the design of robust sensor systems and associated instrumentation;
3. the design and evaluation of multifunctional sensors;
4. the development of fast sensor response times and high sensitivity;
5. establishing the long-term reliability of embedded sensors;
6. the long-term structural integrity of AFRC structures with embedded sensors;

7. the long-term durability of the interfacial bond strength between the matrix and sensors under specified loading regimes and environments;
8. the interpretation and correlation of the sensor signal to the structural integrity of the composite;
9. detailed studies on the above-mentioned issues.

## ACKNOWLEDGMENTS

The authors wish to thank Prof. B. Ralph (Brunel University), D. Brooks (Brunel University), R. Badcock (DERA), A. Martin (BAe), S. Hayes (Sheffield University), C. Doyle (Brunel University), Prof. P. Curtis (DERA), Prof. M. Hinton (DERA), Dr. D. Pullen (DERA), A. Howard (DERA), B. Ridoutt (Cape Durasteel) and D. Coward (RM Consulting) for supporting the Sensors Research Group at Brunel University. The authors also wish to thank Prof. I. Bennion (Aston University), D. A. Jackson (University of Kent), K. T. V. Grattan (City University), the EPSRC and the Royal Society for assistance and support.

## REFERENCES

1. Brooks, D. (1997) Unpublished results. DERA, Farnborough, Second Progress Report, Brunel University, UK, August.
2. Reifsnider, K. L. (ed.) (1991) *Fatigue of Composite Materials*, Vol. 4, Composite Materials Series, Elsevier, Amsterdam.
3. Stinchcomb, W. W. (1980) *Mechanics of Nondestructive Testing*, Plenum Press, New York.
4. Reinhart, T. J. (ed.) (1987) *Composites*, Vol. 1, *Engineering Materials Handbook*, ASM International, Ohio.
5. Purslow, D. (1988) Fractography of fibre-reinforced thermoplastics, Part 3. *Composites*, **19**(5), 358–366.
6. Fernando, G. F., Dixon, R. F., Adam, T., Reiter, H. and Harris, B. (1988) Fatigue behaviour of hybrid composites. 1. Carbon/Kevlar hybrids. *Journal of Materials Science*, **23**, 3732–3743.
7. Curtis, P. T., Gates, J. and Margerison, G. G. (1993) The selection of cyclic load frequency for the fatigue testing of fibre reinforced polymeric composites. RAE Technical Report 93017, Farnborough, Guildford.
8. Al-Khodairi, F. A. (1996) Static and dynamic properties of unidirectional hybrid resin and hybrid fibre composites. PhD Thesis, Brunel University.
9. Guild, F. J., Phillips, M. G. and Harris, B. (1980) Acoustic emission studies of damage in GRP. *NDT International*, **13**(5), 209–218.
10. Maslouhi, A. Proulx, D., Roy, C., Liu, K., McEwen, K., Measures, R. M. and Zimcik, D. G. (1991) Use of optical fibre sensors for acoustic emission detection within composite materials. *International SAMPE Symposium and Exhibition*, **36**, 259–271.
11. Tapanes, E. (1991) Real-time structural integrity monitoring using a passive quadrature demodulation, localised Michelson optical fibre interferometer capable of simultaneous strain and acoustic emission sensing. *Fibre Optic Smart Structures and Skin, IV, SPIE*, **1588**, 356–367.

12. Baillie, P. W. R., Hale, K., Fernando, G. F. and Jones, B. E. (1995) Optical fibre sensing of acoustic emission in fibre reinforced composites. *Proc. Sensors and their Applications VII*, IoP, Dublin.
13. Davila, A., Kerr, D. and Kaufmann, G. H. (1996) Fast electro-optical system for pulsed ESPI carrier fringe generation. *Optics Communications*, **123**, 457–464.
14. Chan, Y. W. (1994) Measurement of in-plane residual displacement due to interference fitting by TV–holographic interferometry. *Strain*, **August**, 91–94.
15. Steinchen, W., Schuth, M. and Yang, L. X. (1994) Strain measured on plane and curved surfaces by means of the shearographic methods. Part 1. *Strain*, **August**, 105–108.
16. Petzing, J. N. and Tyrer, J. R. (1996) Speckle shearing interferometry applied to the measurement of sonar transducer mode shapes. *Proc. Applied Optics and Optoelectronics Conf.*, IOP, Reading, pp. 205–210.
17. Hale, K. F. (1985) Using optical fibres to detect cracks. *Process Engineering*, **66**(8), 27.
18. Hayes, S., Liu, T., Brooks, D., Monteith, S., Ralph, B., Vickers, S. and Fernando, G. F. (1997) In situ self-sensing fibre-reinforced composites. *Journal of Smart Structures and Materials*, **6**, 432–440.
19. Martin, A. R., Fernando, G. F. and Hale, K. F. (1997) Impact damage detection in filament wound tubes using embedded optical fibre sensors. *Journal of Smart Materials and Structures*, **6**, 470–476.
20. Badcock, R. A. (1997) PhD Thesis, Brunel University.
21. Fernando, G. F. and Khan, N. (1997) Unpublished results. Brunel University.
22. Dry, C. M. (1992) Smart materials which sense, activate and repair damage; hollow porous fibres in composites release chemicals from fibres for self-healing, damage prevention, and/or dynamic control. *Proc. 1st European Conf. on Smart Structures and Materials*, pp. 367–371.
23. Vaziri, M. and Chen, C. L. (1992) Etched fibre as strain gauges. *Journal of Lightwave Technology*, **10**(6), 836–841.
24. Badcock, R. and Fernando, G. F. (1995) An intensity-based optical fibre sensor for fatigue damage detection in advanced fibre-reinforced composites. *Journal of Smart Materials and Structures*, **4**, 223–230.
25. Martin, A., Badcock, R., Nightingale, C. and Fernando, G. F. (1997) A novel optical fibre-based strain sensor. *IEEE Photon Technology Letter*, **9**(7), 982–984.
26. Spillman, W. B. Jr. and Lord, J. R. (1987) Self-referencing multiplexing technique for intensity-modulating sensors, *SPIE*, **718**, 182–191.
27. Berthold, J. W. (1995) Historical review of microbend sensors. *Journal of Lightwave Technology*, **13**(7), 1193.
28. Lammerink, T. S. J. and Fluitman, J. H. J. (1984) Measuring method for optical fibre sensors. *Journal of Physics E: Scientific Instruments*, **17**, 1127–1129.
29. Dakin, J. P., Wade, C. A. and Withers, P. B. (1988) An optical fibre sensor for the measurement of pressure. *Fibre and Integrated Optics*, **7**, 35–46.
30. Johnston, S. F. (1992) Gas monitors employing infrared LEDs. *Measurement Science and Technology*, **3**, 191.
31. Senior, J. M., Murtaza, G., Stirling, A. I. and Wainwright, G. H. (1992) Single LED-based dual wavelength referenced optical fibre sensor system using intensity modulation. *Optics and Laser Technology*, **24**, 187.
32. Wang, A., Miller, M. S., Plante, A. J., Sunther, M. F., Murphy, K. A. and Claus, R. O. (1996) Split-spectrum intensity-based optical fibre sensors for measurement of microdisplacement, strain and pressure. *Applied Optics*, **35**, 2995–2601.

33. Hogg, W. D., Turner, R. D. and Measures, R. M. (1989) Polarimetric fibre optic structural strain sensor characterisation. *SPIE*, **1170**, *Fibre Optic Structures and Skins II*, pp. 542–549.

34. Zheng, G., Campbell, M. C., Wallace, P. A. and Holmes-Smith, A. S. (1996) Configurations of remote birefringent fibre strain sensors using a frequency modulation continuous wave technique. *Proc. Appl. Opt. Div. Conf.* Grattan, K. T. V. (ed.) Reading, pp. 380–385.

35. Meltz, G. and Dunphy, J. R. (1985) Fibre optic sensor for the non-destructive evaluation of composite materials. *Proc. SPIE Symposium on Fibre Optics and Laser Sensors*, Bellingham, WA, *SPIE*, **566**, 243–248.

36. Waite, S. R., Tatam, R. P. and Jackson, D. A. (1988) Use of optical fibre for damage and strain detection in composite materials. *Composites*, **19**(6), 435–442.

37. Layton, M. R. and Bucaro, J. A. (1979) Optical fibre acoustic sensor utilizing mode–mode interference. *Applied Optics*, **18**(5), 666–670.

38. Murphy, K. A., Miller, M. S., Vengsarkar, A. M. and Claus, R. O. (1990) Elliptical-core two-mode optical fibre sensor implementation methods. *Journal of Lightwave Technology*, **8**(11), 1688–1696.

39. Measures, R. M. (1993) Fibre optic sensing for composite structures. *Composites Engineering*, **3**(7–8), 715–750.

40. Bennett, K. D. and O'Brien, N. F. (1992) Optimal detection schemes for dual mode fibre sensor signals. *Proc. Fibre Optic Sensor-Based Smart Materials and Structures Workshop*, Blacksburg, VA, pp. 61–66.

41. Jackson, D. A. and Jones, J. D. C. (1986) Fibre optic sensors. *Optica Acta*, **33**, 1469–1503.

42. Valis, T., Tapnese, E. and Measures, R. M. (1989) Localised fibre optic strain sensors embedded in composite materials. *SPIE*, **1170**, 495–503.

43. Liu, K., Ferguson, S. M. and Measures, R. M. (1989) Damage detection in composites with embedded fibre optic interferometric sensors. *SPIE*, **1170**, 205–210.

44. Liu, K., Ferguson, S. M., McEwen, K., Tapanese, E. and Measures, R. M. (1990) Acoustic emission detection for composites using embedded ordinary single-mode fibre optic interferometric sensors. *SPIE*, **1370**, 316–323.

45. Zheng, S. X., McBride, R., Barton, J. S., Jones, J. D. C., Hale, K. F. and Jones, B. E. (1992) Intrinsic optical fibre sensor for monitoring acoustic emission. *Sensors and Actuators. A. Physical*, **31**(1–3) Pt. 3, 110–114.

46. Lee, C. E., Gibler, W. N., Atkins, R. A. and Taylor, H. F. (1992) In-line fibre Fabry–Perot interferometer with high reflectance internal mirrors. *Journal of Lightwave Technology*, **10**(10), 1376–1379.

47. Valis, T., Hogg, D. and Measures, R. M. (1990) Composite material embedded fibre optic Fabry–Perot strain rosette. *Spie*, **1370**, 154–161.

48. Friebele, E. J., Putnam, M. A., Kersey, A. D., Greenblatt, A. S., Ruthven, G. P., Kim, M. H. and Gottschalck, K. S. (1997) Ultrahigh sensitivity strain sensing using fibre cavity etalon. *SPIE*, **3042**, 100–110.

49. Stone, J. and Marcuse, D. (1986) Ultrahigh finesse fibre Fabry–Perot interferometers. *Journal of Lightwave Technology*, **LT-4**, 382–385.

50. Kidd, S. R., Sinha, P. G., Barton, J. S. and Jones, J. D. C. (1993) Wind tunnel evaluation of novel interferometric optical fibre heat transfer gauges. *Measurement Science and Technology*, **4**, 362–368.

51. Dorighi, J. F., Krishnaswamy, S. and Achenbach, J. D. (1995) Stabilisation of an embedded fibre optic Fabry–Perot sensor for ultrasound detection. *IEEE Transactions on Ultrasonics, and Frequency Control*, **42**(5), 820–824.

52. Atkins, R. A., Gardner, J. H., Gibler, W. N., Lee, C. E., Oakland, M. D., Spears, M. O., Swenson, V. P., Taylor, H. F., McCoy, J. and Beshouri, G. (1994) Fibre-optic pressure sensor for internal combustion engines. *Applied Optics*, **33**, 1315–1320.
53. Sirkis, J., Berkoff, T. A., Jones, R. T., Singh, H., Kersey, A. D., Friebele, E. J. and Putman, M. A. (1995) In-line fibre etalon (ILFE) fibre-optic strain sensors. *Journal of Lightwave Technology*, **13**(7), 1256–1263.
54. Bhatia, V., Murphy, K. A., Clause, R. O., Jones, M. E., Grace, J. L., Tran, T. A. and Greene, J. A. (1995) Multiple strain state measurements using conventional and absolute optical fibre-based extrinsic Fabry–Perot interferometric strain sensors. *Journal of Smart Materials and Structures*, **4**, 240–245.
55. Bellivelli, C. and Duplain, G. (1993) White-light interferometric multimode fibre-optic strain sensor. *Optics Letters*, **18**(1), 78–80.
56. Liu, T. D., Brooks, A., Martin, R., Badcock, R. B. and Fernando, G. F. (1997) A multi-mode extrinsic Fabry–Perot interferometric strain sensor. *Journal of Smart Materials and Structures*, **6**, 464–469.
57. Narendran, N. and Weiss, J. (1996) Fibre-optic strain and temperature sensors for power plant applications. *SPIE*, **2594**, 149–158.
58. Bhatia, V., Sen, M. B., Murphy, K. A. and Claus, R. O. (1996) White light interferometry for highly sensitive strain and temperature measurements. *Proc. 11th Int. Conf. on Optical Fibre Sensors*, Saporro, Japan, pp. 308–311.
59. Gunderson, L. C. (1990) Fibre optic sensor applications using Fabry–Perot interferometry. *SPIE*, **1267**, *Fibre Optic Sensors IV*, pp. 194–204.
60. Beard, P. C. and Mills, T. N. (1997) Miniature optical fibre ultrasonic hydrophone using a Fabry–Perot polymer film interferometer. *Electronics Letters*, **33**(9), 801–803.
61. Lawrence, C. M., Nelson, D. V., Spingarn, J. R. and Bennet, T. E. (1996) Measurement of process-induced strain in composite materials using embedded fibre optic sensors. *SPIE*, **2718**, 60–68.
62. Vries, M. D., Nasta, M., Bhatia, V., Tran, T., Greene, J. and Claus, R. O. (1995) Performance of embedded short-gauge-length optical fibre sensors in a fatigue-loaded reinforced concrete specimen. *Journal of Smart Materials and Structures*, **4**, A107–113.
63. Liu, T., Brooks, D., Martin, A., Badcock, R. and Fernando, G. F. (1996) Design, fabrication and evaluation of an optical fibre sensor for tensile and compressive strain measurements via the use of white light interferometry. *SPIE*, **2718**, 408–416.
64. Hill, K. O., Fujii, Y., Johnson, D. C. and Kawasaki, B. S. (1978) Photosensitivity in optical fibre waveguides: Application to reflection filter fabrication. *Applied Physics Letters*, **32**, 647.
65. Hand, D. P. and Russell, R. St. J. (1990) Photoinduced refractive index changes in germanosilicate fibres. *Optics Letters*, **15**, 102.
66. Lemaire, P. L., Atkins, R. M., Mizrahi, V. and Reed, W. A. (1993) High-pressure $H_2$ loading as a technique for achieving ultrahigh UV photosensitivity and thermal sensitivity in $GeO_2$-doped optical fibres. *Electronics Letters*, **29**, 1191.
67. Askins, C. G., Tsai, T.-E., Williams, G. M., Putnam, M. A., Bashkansky, M. and Friebele, E. J. (1992) *Fibre Optic Sensor-Based Smart Materials and Structures*. R. O. Claus (ed.), IOP Publishing, Bristol, Philadelphia, PA, pp. 85–88.
68. Dianov, E. M., Starodubov, D. S. and Frolov, A. A. (1996) UV argon laser induced luminescence changes in germanosilicate fibre preforms. *Electronics Letters*, **32**(3), 246–247.
69. Starodubov, D. S., Grubsky, V., Feinberg, J. and Erdogan, T. (1997) Near-UV fabrication of ultrastrong Bragg gratings in hydrogen-loaded germanosilicate fibres. *CLEO'97 Tech. Dig.*, Post-deadline paper CPD-24.

70. Starodubov, D. S., Grubsky, V. and Feinberg, J. (1997) Efficient Bragg grating fabrication in a fibre through its polymer jacket using near-UV light. *Electronics Letters*, **33**, 1131–1332.

71. Meltz, G., Morey, W. W. and Glenn, W. H. (1989) Formation of Bragg gratings in optical fibre by a transverse holographic method. *Optics Letters*, **14**, 823.

72. Hill, K. O., Malo, B., Bilodeau, F., Johnson, D. C. and Albert, J. (1993) Bragg grating fabricated in monomode photosensitive optical fibre by UV exposure through a phase mask. *Applied Physics Letters*, **62**, 1035–1037.

73. Morey, W. W., Meltz, G. and Glenn, W. H. (1989) Fibre optic Bragg grating sensors. *SPIE*, **1169**, *Fibre Optic and Laser Sensors VII*, pp. 98–107.

74. Simonsen, H. D., Paetsch, R. and Dunphy, J. R. (1992) Fibre Bragg grating sensor demonstration in a glass-fibre reinforced polyester composite. *Proc. 1st European Conference on Smart Structures and Materials*, Glasgow, pp. 73-76.

75. Davis, M. A., Bellemore, D. G., Kersey, A. D., Putam, M. A., Friebele, E. J., Idriss, R. L. and Kodindouma, M. B. (1996) High-sensor-count Bragg grating instrumentation system for large-scale structural monitoring applications. *SPIE*, **2718**, 303–309.

76. Kersey, A. D., Berkoff, T. A. and Morey, W. W. (1992) Fibre-grating based strain sensor with phase sensitive detection. *Proc. 1st European Conf. on Smart Structures and Materials*, Glasgow, pp. 61–67.

77. Vengsarkar, A. M., Lemaire, P. J., Judkins, J. B., Bhatia, V., Sipe, J. E. and Ergodan, T. E. (1995) Long-period fibre gratings as band-rejection filters. *Proc. Conf. on Optical Fibre Communications*, Post-deadline Paper PD4-2.

78. Bhatia, V., D'Alberto, T., Murphy, K. A. and Claus, R. O. (1996) Comparison of optical fibre long-period and Bragg grating sensors. *SPIE*, **2718**, 110–121.

79. Johnson, M. and Ulrich, R. (1978) Fibre-optical strain gauges. *Electronics Letters*, **14**(14), 432.

80. Kashyap, R. and Reeve, M. H. (1980) Single-ended fibre strain and length measurement in frequency domain. *Electronics Letters*, **16**(18), 2867–2869.

81. Zimmermann, B. D., Thomas, D. D. and Claus, R. O. (1990) Recent progress in high resolution optical fibre time domain methods and its impact on dynamic environment sensing applications. *SPIE*, **1370**, *Fibre Optic Smart Structures and Skins III*, San Jose, CA, pp. 197–204.

82. Arjyal, B., Paipetis, A. and Galliotis, C. (1996) Stress/strain measurements in advanced composites using remote laser Raman microscopy. *Nondestructive Testing and Evaluation*, **12**, 355–366.

83. Galiotis, C. (1993) Study of mechanisms of stress transfer in continuous and discontinuous-fiber model composites by laser Raman spectroscopy. *Composite Science & Technology*, **48**(1–4), 15–28.

84. Inaudi, D., Elamari, A., Pflug, L., Gisin, N., Breguet, J. and Vurpillot, S. (1994) Low-coherence deformation sensors for the monitoring of civil engineering structures. *Sensors and Actuators A*, **44**, 125–130.

85. Fan, N. Y., Huang, S. Y., and Measures, R. M. (1997) Localised long-gauge-length fibre optic sensor demodulated with wavelength tuning technique. *SPIE*, **3042**, 366–371.

86. Sirkis, J. S. and Lo, Y. L. (1994) Simultaneous measurement of two strain components using 3X3 and 2X2 coupler-based passive demodulation of optical fibre sensors. *Journal of Lightwave Technology*, **12**(12), 2153–2161.

87. Carman, G. P., Lesco, J. J., Case, S. W., Fos, B. and Claus, R. O. (1992) Development of an embedded Fabry–Perot fibre optic strain rosette sensor (FP–FOSRS). *Proc. Fibre Optic Sensor-Based Smart Materials and Structures Workshop*, Blacksburg, VA, pp. 45–50.

88. Udd, E., Nelson, D. V., Lawrence, C. M. and Ferguson, B. A. (1996) Three-axis strain and temperature fibre optic grating sensor. *SPIE*, **2718**, 104–109.

89. Udd, E., Lawrence, C. M. and Nelson, D. V. (1997) Development of a three-axis strain and temperature fibre optic grating sensor. *SPIE*, **3042**, 229–236.

90. Healy, P. (1986) Instrumentation principles for optical time domain reflectometry. *Journal of Physics, E: Science Institute*, **19**, 334–341.

91. Dakin, J. P. (1989) Distributed optical fibre sensor systems. *Optical Fibre Sensors: Systems and Applications*, Vol. 2, B. Culshaw and J. Dakin (eds), Artech House, London, pp. 575–598.

92. Rogers, A. J. (1980) Polarisation optical time domain reflectometry. *Electronics Letters*, **16**, 489–490.

93. Horiguchi, T., Kurashima, T. and Tateda, M. (1989) Tensile strain dependence of Brillouin frequency dependency shift in silica optical fibres. *IEEE Photon Technology Letters*, **1**, 107–108.

94. Horiguchi, T., Kurashima, T. and Tateda, M. (1990) A technique to measure distributed strain in optical fibers. *IEEE Photon Technology Letters*, **PTL-2**, 352–354.

95. Kurashima, T., Horiguchi, T., Yoshizawa, N., Tada, H. and Tateda, M. (1991) Measurement of distributed strain due to laying and recovery of submarine optical fibre cable. *Applied Optics*, **30**(3), 334–337.

96. Bao, X., Webb, D. J. and Jackson, D. A. (1995) Recent progress in distributed fibre optic sensors based upon Brillouin scattering. *SPIE*, **2507**, *Proc. Distributed and Multiplexed Fibre Optic Sensors V*, Munich, pp. 175–185.

97. Parker, T. R., Farhadiroushan, M., Handerek, V. A. and Rogers, A. J. (1997) A fully distributed simultaneous strain and temperature sensor using spontaneous Brillouin backscatter. *IEEE Photon Technology Letters*, **9**(7), 979–981.

98. Kersey, A. D. (1995) Fibre optic sensor multiplexing techniques, in *Fibre Optic Smart Structures*, E. Udd (ed.) Wiley, New York.

99. Koch, A. and Ulrich, R. (1990) Displacement sensor with electronically scanned white-light interferometer. *Fibre Optic Sensors IV*, R. T. Kersten (ed.), *Proc. Soc. Photo-Optical Instrumentation Engineers*, **1267**, 126–133.

100. Davis, M. A., Bellemore, D. G., Putam, M. A. and Kersey, A. D. (1996) A 60 element fibre Bragg grating sensor system. *Proc. 11th Int. Conf. on Optical Fibre Sensors*, Sapporo, Japan, IEICE and IEEJ, pp. 100–103.

101. Jones, R. T., Berkoff, T. A., Bellemore, D. G., Early, D. A., Sirkis, J. S., Putnam, M. A., Friebele, E. J. and Kersey, A. D. (1996) Cantilever-plate-deformation monitoring using wavelength-division multiplexed fibre Bragg grating sensors, *SPIE*, **2718**, 258–268.

102. Kersey, A. D., Davis, M. A., Berkoff, T. A., Dandridge, A., Jones, R. T., Tsai, T., Cogdell, G., Wang, G., Havsgaard, G. B., Pran, K. and Knudsen, S. (1997) Transient load monitoring on a composite hull ship using distributed fibre optic Bragg grating sensors. *SPIE*, **3042**, 421–430.

103. Rao, Y. J., Lobo-Ribeiro, A. B., Jackson, D. A., Zhang, L. and Bennion, I. (1995) Combined spatial- and time-division-multiplexing scheme for fibre grating sensors with drift-compensated phase-sensitive detection. *Optics Letters*, **20**(20), 2149–2151.

104. Sirkis, J. S. (1993) Unified approach to phase–strain–temperature models for smart structure interferometric optical fibre sensors. Part 1. Development. *Optical Engineering*, **32**(4), 752–761.

105. Sirkis, J. (1993) Unified approach to phase–strain–temperature models for smart structure interferometric optical fibre sensors. Part 2. Applications. *Optical Engineering*, **32**(4), 762–773.

106. Hannah, J. and Hiller, M. J. (1978) *Applied Mechanics*. Pitman, London.

107. Doyle, C. and Fernando, G. F. (1997) Condition monitoring engineering materials with an optical fibre vibration sensor system. *SPIE*, **3042**, 310–318.

108. Spillman, W. B. Jr., Kiln, B. R., Maurice, L. B. and Fuhr, P. L. (1989) Statistical-mode sensor for fibre optic vibration sensing uses. *Applied Optics*, **28**(15), 3166–3176.

109. Murphy, K. A., Fogg, B. R. and Claus, R. O. (1992) Spatially weighted fibre optic sensors for smart materials and structures. Active materials and adaptive structures. *Proc. ADPA/AIAA/ASME/SPIE*, Alexandria, VA, G. J. Knowles (ed.), IOP Publishing, Bristol, Philadelphia, PA, pp. 43–46.

110. Murphy, K. A., Miller, W. V. III, Vensarkar, A. M. and Claus, R. O. (1989) Surface acoustic wave detection using Michelson-type fibre optic interferometer. *Proc. Fibre Optic Smart Structures and Skins II, SPIE*, **1170**, 558–565.

111. Measures, R. (1991) The detection of damage and the measurement of strain within composites by means of embedded optical fibre sensors. *Review of Progress in Quantitative Nondestructive Evaluation*, **10B**, 1247–1258.

112. Doyle, C., Hayes, S., Martin, A., Liu, T., Wu, M. and Fernando, G. (1996) Intensity-based optical sensors for composite health monitoring. *Fibre Optic Sensors V, SPIE*, **2895**, 288–299.

113. Martin, A. R. (1997) PhD Thesis, Material Engineering, Brunel University.

114. Chang, C. C. and Sirkis, J. S. (1995) Optical fibre sensors embedded in composite panels for impact damage detection. *SPIE*, **2444**, 502–513.

115. Jones, T. R., Sirkis, J. S., Friebele, E. J. and Kersey, A. D. (1995) Location and magnitude of impact detection in composite plates using neural networks. *SPIE*, **2444**, 469–480.

116. Spillman, W. B. and Fuhr, P. (1990) Impact detection system for smart skin applications. *SPIE*, **1370**, 308–315.

117. Fuhr, P. (1994) Single-fibre simultaneous vibration sensing and impact detection for large space structures. *Journal of Smart Materials and Structures*, **3**, 124–128.

118. Fan, N. Y., Huang, S. Y., Alavie, A. T. and Measures, R. M. (1995) Rare earth dope fibre for structural damage assessment. *Journal of Smart Materials and Structures*, **4**, 179–185.

119. Noharret, B., Chazelas, J., Bonniau, P., Lecuellet, J. and Turpin, M. (1995) Impact detection on airborne multi-layered structures. *SPIE*, **2444**, 460–468.

120. Bonniau, P., Chazelas, J., Lecuellet, J., Gendue, F., Turpin, M., Peasant, J. P. Le and Bressignon, M. (1991) Damage detection in woven composite materials using embedded fibre optics. *SPIE*, **1588**, *Fibre Optic Smart Structures and Skins IV*, pp. 52–63.

121. Greene, J. A., Tran, T. A., Bhatia, V., Gunther, M. F., Wang, A., Murphy, K. A. and Claus, R. O. (1995) Optical fibre sensing technique for impact detection and location in composites and metal specimens. *Journal of Smart Materials and Structures*, **4**, 93–99.

122. Liu, K., Ferguson, S. M., McEwen, K., Tapanese, E. and Measures, R. M. (1990) Acoustic emission detection for composite using embedded ordinary single-mode fibre optic interferometric sensors. *SPIE*, **1370**, *Fibre Optic Smart Structures and Skins III*, pp. 316–323.

123. Waite, S. R. (1990) Use of embedded optical fibre for early fatigue damage detection in composite materials. *Composites*, **21**(2), 148–153.

124. Yaniv, G., Zimmermann, B. and Lou, K. A. (1993) Development of an optical fibre time domain sensor for monitoring static and fatigue strains in composite laminates. *SPIE*, **1918**, 377–387.

125. Levin, K. and Nilsson, S. (1996) Examination of reliability of fibre optic sensors embedded in carbon/epoxy composites. *Proc. Third ICIM/ECSSM*, Lyon, pp. 222–229.

126. Liu, T., Fernando, G. F., Rao, Y. J., Jackson, D. A., Zhang, L. and Bennion, I. (1996) *In-situ* strain measurements in composites during fatigue testing using optical fibre Bragg gratings and a portable CCD detection system. *SPIE*, **2895**, 249–257.

127. Guemes, A. and Menendez, J. M. (1996) Fatigue strength of glass reinforced polyester (GRP) laminates with embedded optical fibres. *SPIE*, **2779**, 217–221.

128. Fernando, G. F. (1989) PhD Thesis, University of Bath.

129. Xu, M. G., Archambault, J.-L., Reekie, L. and Dakin, J. P. (1994) Discrimination between strain and temperature effects using dual-wavelength fibre grating sensors. *Electronics Letters*, **30**, 1085–1087.

130. Brady, G. P., Kalli, K., Webb, D. J., Jackson, D. A., Zhang, D. A. and Bennion, I. (1994) Recent developments in optical fibre Bragg gratings. *SPIE*, **2839**, 8–19.

131. Kanellopoulos, S. E., Handerek, V. A. and Rogers, A. J. (1995) Simultaneous strain and temperature sensing with photogenerated in-fibre gratings. *Optics Letters*, **20**(3), 333–335.

132. James, S. W., Dockney, M. L. and Tatam, R. P. (1996) Independent measurement of temperature and strain using in fibre Bragg grating sensors. *Proc. 11th Int. Conf. Optical Fibre Sensors*, Sapporo, Japan, FR3-3, pp. 10–13.

133. Singh, H. and Sirkis, J. (1996) Simultaneous measurement of strain and temperature using optical fibre sensors: two novel configurations. *Proc. 11th Int. Conf. Optical Fibre Sensors*, Sapporo, Japan, pp. 108–111.

134. Jin, X. D. and Sirkis, J. (1997) Optical fibre sensors for simultaneous measurement of strain and temperature. *SPIE*, **3042**, 120–127.

135. Ferreira, L. A., Lobo-Ribeiro, A. B., Santos, J. L. and Farahi, F. (1996) Simultaneous measurement of displacement and temperature using a low finesse cavity and a fibre Bragg grating. *IEEE Photon Technology Letters*, **8**(11), 1519–1521.

136. Patrick, H., Williams, G. M., Kersey, A. D., Pedrazzani, J. R. and Vengsarkar, A. M. (1996) Hybrid fibre Bragg grating/long period fibre grating sensor for strain/temperature discrimination. *IEEE Photon Technology Letters*, **8**(9), 1223–1225.

137. Bhatia, V., Murphy, K. A. and Claus, R. O. (1996) Simultaneous measurement systems employing long period grating sensors. *Proc. 11th. Int. Conf. Optical Fibre Sensors*, Sapporo, Japan, pp. 702–705.

138. Farahi, F., Webb, D. J., Jones, J. D. C. and Jackson, D. A. (1990) Simultaneous measurement of temperature and strain: cross-sensitivity considerations. *Journal of Lightwave Technology*, **8**, 138–142.

139. Vengsarkar, A. M., Michie, W. C., Jaankovic, L., Culshaw, B. and Claus, R. O. (1994) Fibre-optic dual-technique sensor for simultaneous measurement of strain and temperature. *Journal of Lightwave Technology*, **12**, 170–177.

140. Luke, D. G., McBride, R., Lloyd, P., Burnett, J. G., Greenaway, A. H. and Jones, J. D. C. (1996) Strain and temperature measurements in composite-embedded highly birefringent optical fibre using mean and differential group delay. *Proc. 11th Int. Conf. Optical Fibre Sensors*, Sapporo, Japan, pp. 200–203.

141. Davis, M. A. and Kersey, A. D. (1996) Separating the temperature and strain effects on fibre Bragg grating sensors using stimulated Brillouin scattering. *SPIE*, **2718**, 270–278.

142. Liu, T., Fernando, G. F., Rao, Y. J., Jackson, D. A., Zhang, L. and Bennion, I. (1997) Simultaneous strain and temperature measurements in composite using a multiplexed fibre Bragg grating sensor and an extrinsic Fabry–Perot sensor. *SPIE*, **3042**, 203–212.

143. Liu, T., Fernando, G. F., Zhang, L., Bennion, I., Rao, Y. and Jackson, D. A. (1997) Simultaneous strain and temperature measurement. *Proc. 12th Int. Conf. on Optical Fibre Sensors*, Washington, DC, pp. 40–43.

144. Xie, W. X., Douay, M., Bernage, P., Niay, P., Bayon, J. F. and Georges, T. (1993) Second order diffraction efficiency of Bragg gratings written within germanosilicate fibres. *Optics Communications*, **10**(1/2), 85–91.

145. Kalli, K. *et al.* (1994) Possible approach for the simultaneous measurement of temperature and via first and second order diffraction from Bragg grating sensors. *OFS'94*, Glasgow, post-deadline paper.

146. Kanellopoulos, S. E., Handerek, V. A. and Rogers, A. J. (1992) Photoinduced polarisation couplers in elliptical core optical fibres written using 532 nm and 266 nm sources. *Electronics Letters*, **28**(6), 1558–1560.

147. Kanellopoulos, S. E., Handerek, V. A. and Rogers, A. J. (1994) Simultaneous strain and temperature sensing employing a photogenerated polarisation coupler and low-order modes in an elliptically cored optical fibre. *Electronics Letters*, **30**(21), 1786–1787.

148. Wang, A., Wang, G. Z. and Claus, R. O. (1995) Two-mode elliptical core fibre sensors for strain and temperature measurement. *Journal of Smart Materials and Structures*, **4**, 42–49.

149. Flavin, D. A., McBridge, R., Jones, J. D. C., Burnett, J. G. and Greenaway, A. H. (1994) Combined temperature and strain measurement with a dispersive Fourier-transform spectrometer. *Optics Letters*, **19**(24), 2167–2169.

150. Flavin, D. A., McBride, R. and Jones, J. D. C. (1995) Interferometric fibre-optic sensing based on the modulation of group delay and first order dispersion: application to strain–temperature measurand. *Journal of Lightwave Technology*, **13**(7), 1314–1323.

151. Zhang, F. and Lit, J. W. Y. (1993) Temperature and strain sensitivity measurements of high-birefringent polarisation-maintaining fibres. *Applied Optics*, **32**(13), 2213–2218.

152. Martens, H. and Naes, T. (1989) *Multivariate Calibration*, Wiley, Chichester.

# 5

# Mathematical techniques in fiber optic sensor applications

K. T. V. Grattan and B. M. A. Rahman

## 5.1 INTRODUCTION

There is a continuing need for high quality and repeatable measurement techniques for application in a wide variety of industries, and with that the requirement for reliable and trustworthy instrumentation. Fiber optics have a significant part to play in achieving this goal and could result in an opportunity for cost savings, which increases the competitive potential, with benefits being passed on to the customer. Further, the meeting of increased legislative requirements world-wide are often seen to be driving the field, especially in safety and environmental matters. The technology has developed rapidly in recent years, and is now being used by physicists, chemists, engineers and biologists, as well as those comparatively untrained in the fundamentals of the subject. With the wide variety of techniques (which include optical fiber systems) available to these investigators, it is essential that there is a clear understanding of the appropriate guidelines for the selection and use of any particular method of instrumentation, and its limitations as well as its desirable features. Also, the degree to which it is 'user friendly' can considerably influence its acceptability for any specific application.

All this imposes considerable pressure on the instrument designer, ahead of that on the instrument user. The selection of a particular class of technique is made more difficult by the variety of methods from which to choose, and having chosen one approach, on the basis, perhaps, of compatibility with other systems or of cost, there is then the need to select the most appropriate instrumental method. A full understanding of the implications of that choice may only come about after detailed investigation, which may be either experimental or theoretical. Thus the prime value of the role of modeling in instrumental selection lies in the ability to make an assessment of the technique, and its suitability for the purpose, in terms of the important systems parameters. There is often a considerable cost saving involved when effective modeling is carried out in advance of the selection of the best experimental technique. In the design of new systems, often involving components developed for other purposes (such as telecommunications), the advantage of

*Optical Fiber Sensor Technology*, Vol. 3. Edited by K. T. V. Grattan and B. T. Meggitt.
Published in 1998 by Kluwer Academic Publishers, London. ISBN 978-1-4419-4736-9

high-quality modeling is especially clear, and the potential performance, and in particular any negative features of the instrument characteristics, can often be predicted before expensive construction work is undertaken. The definition and role of these more general concepts is discussed in the work of Finkelstein and Grattan [1].

This chapter will focus on the impact of mathematical techniques and tools on fiber optic measurements and sensor and instrumentation development. The growth of many aspects of fiber optic sensor systems has often been largely empirical, and based on 'good practice' and the experience and knowledge of the systems designer. Whilst this may be satisfactory, most experimentalists are familiar with the need for, and value of, confirmatory theoretical underpinning of the methods and technology of the fiber optic systems they design or employ – in that way the limitations and restrictions of such systems do not represent something unforeseen, even though the agreement achieved between theory and experiment may be limited, or in some cases approximate. Thus, analytical solutions of equations governing the performance of an optical instrumentation system can be a useful guide, and a few of the many examples where these methods have been valuable are considered. More fundamental, however, is the use of advanced numerical methods to understand the performance of optical systems, be they fiber-based, open air-path or integrated optics. In particular, complex numerical modeling can yield information which is unavailable by simple analytical methods, where solutions to governing equations are only an unsatisfactory approximation, and where this more sophisticated approach can aid the systems designer. Further, in the treatment of the output of a fiber optic instrument, matters of calibration and response are particularly important, and novel neural network methods are being applied to optical systems to interpret such output, and to find a balance between the use of less complex instrumentation and advanced signal processing where such tools are available. Finally, the 'intelligent instrument' and the smart sensor are terms familiar to designers and users of electronic instrumentation where an intelligent system is incorporated into the sensor head itself. The impact of this, and with it the necessary mathematical processing to achieve effective intelligence in a fiber optic system, is considered briefly. In this way fiber optic instrumentation systems can make full use of the advances in electronic processing technology to advance their utility and widen their applicability into the future.

## 5.2 MODELING TECHNIQUES

As photonic technology has reached maturity, the lightwave devices themselves, and their associated systems have become more complex. The optimization of the efficiency of such optical devices often requires a knowledge of the methods of wave propagation and related field distributions, and their dependence on system fabrication parameters. There is also a significant need for the use of methods, now usually numerical, to simulate complex structures such as those used in multilayer diode lasers or integrated optics in order to optimize existing designs and evaluate novel devices, either prior to, or as a substitute for, fabrication and experimental testing, and to facilitate optimal optical coupling, e.g. to fiber optics. This has been a stimulus for the high degree of interest in the development and use of effective theoretical methods, and in particular computer-aided design techniques.

Modern optical sensor and communications systems consist of many different types of lumped or distributed optical devices. The effective use of such systems requires a clear understanding of the fundamental physics, and thus the operating principles, of their individual components. Optical waveguides are the building blocks of all guided wave optical systems, including those using fiber optics. Unfortunately, analytical techniques are not adequate to model lightwave devices without significant approximations. Only the simplest planar waveguides can be analyzed by solving their corresponding transcendental equations, which are developed by applying field continuity conditions at the interface boundaries. Even then, it is not possible to find a closed-form solution of such transcendental equations, and iterative procedures must be used to find the zeros of such equations. From the boundary conditions for multilayered planar structures, a matrix equation can be generated, and similarly the roots of the determinants can be found iteratively, to yield the propagation constants of such planar structures. However, in all practical optical waveguides the optical power is confined in both the transverse directions, and for such a waveguide, an analytical solution is not possible. The standard optical fiber with circular symmetry is one of the simplest of such waveguides, and is used extensively in optical communications and sensing. Although these types of optical fiber have been investigated extensively and their solutions can be found in almost every textbook on this subject, it is not possible to find an exact analytical solution for even the simplest arrangement, i.e. one with perfect circular symmetry, a step-index difference between core and cladding, and infinite cladding. Currently, as well as step-index single-mode fibers, graded index, multilayered cladded, noncircular, stressed or doped exotic fibers, and a wide range of integrated optic waveguides are used in photonic integrated circuits (PICs), which have significant potential in both optical communications and sensor systems. For more practical fibers, such as those with high levels of doping of the silica host in a graded index core structure, the techniques required become more demanding, and more exotic fibers such as high birefringence (hi-bi) fibres can only be analyzed effectively using advanced numerical techniques. Among the many methods used so far in such analysis, those employing the staircase approximation, direct numerical integration, point matching [2], the Wentzel–Kramers–Brillouin (WKB) method, the effective index method [3], the finite difference method [4] and the finite element method [5] may be used.

## 5.3 ANALYTICAL METHODS IN OPTICAL FIBER SENSORS

A glance at any text book on the development of optical fiber sensor systems will show the value (and limitations) of analytical methods. Even when supported by the use of computer methods and such equation-solving tools as are available, and aided by an average personal computer, the method still relies upon the development of an equation, or set of equations, to govern and thus represent a particular situation. The result of a simulation for the variation of one parameter with another, especially the output parameter in the process of calibration of the overall optical instrument, is vital to the successful functioning of the system and the development of satisfactory agreement with the results of experiment. Such a process is regularly engaged in by the average experimenter, most usually to show that the results obtained have a degree of underpinning support in the basic physics governing the situation, and thereby that the results of the experimentation are trustworthy. As all experimenters know mistakes can be made, at any one of several stages. In particular, the process of approximation that often accompanies the development of an analytical approach is fraught with danger. However, it is a vital and basic tool, and usually only when it fails is there the need to apply more complex numerical methods which may be less transparent to errors either in design or execution. The examples in fiber optic instrumentation given here are merely illustrative of the value of these analytical methods, and they are widely seen in the work reported in other chapters of this book. Two simple illustrations are given to show the value of the simple technique.

### 5.3.1 Interferometry

In interferometry, a knowledge of the anticipated performance of a white light interferometer can yield vital information about the way it operates under a wide variety of conditions. In the work of Wang *et al.* [6], for example, a simple analytical theory is described which accurately predicts the performance of a white light interferometer used for displacement sensing. This gives a theoretical description from which the predicted output visibility profile of the sensor may be obtained, and a theoretical graph of signal intensity *vs.* optical path difference yields good agreement with the results of an experiment carried out with components whose characteristics have thus been obtained matching those of the theoretical approach. The performance of multiwavelength sources in these interferometers, for example, can be predicted, and both two- and three-wavelength combination sources yield complex interference fringe patterns where the important central fringe is more readily identifiable, and the agreement obtained between the results of theory and experiment is very good, as shown in Fig. 5.1, acting as a real aid to the experimenter. Such an analysis can be extended to nonideal situations [7] where the deleterious effects of dispersion, especially in the main beam splitting optics used, can be used to show the degree to which such an effect becomes a problem for central fringe identification. In this way, with such a suite of equations governing the performance of the system, its characteristics can be optimized, e.g. the choice of wavelengths in a multiwavelength system and the avoidance of sources of dispersion error can be predicted, and a better experimental performance can be obtained, from systems designed in this way.

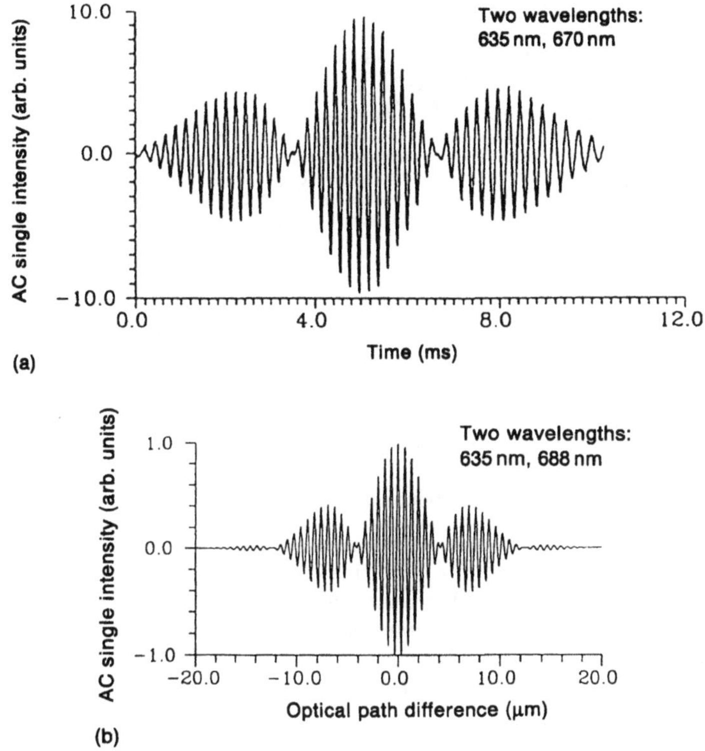


**Fig. 5.1** (a) Experimentally obtained visibility pattern for a two-wavelength combination source of coherence length approximately 20 μm. (b) Computer-generated visibility pattern for a two-wavelength combination source of coherence length approximately 20 μm.

### 5.3.2 Fiber optic 'decay-time' thermometer

A further example lies in the use of the 'decay time' of the fluorescence of a material, which changes as a function of temperature, for applications of thermometry, as discussed in detail in a previous chapter. To assess fully the potential of a material to be used in this way, there is a real need to understand the fundamental interactions in the fluorescent material in order to determine its most effective temperature range, its potential accuracy and its speed of response, for example. An illustration of such a system, developed with the aid of analytical modeling techniques, is that based on the fluorescence of chromium-doped ($Cr^{3+}$) crystals [8, 9]. Modeling of these materials has revealed that the crystal field strength has a significant influence on the fluorescence characteristics of the material, and the use of single configurational coordinate models has been employed by Zhang *et al.* [10] in this type of assessment, as shown in Fig. 5.2. Thus valuable analytical expressions for the temperature-dependencies of the

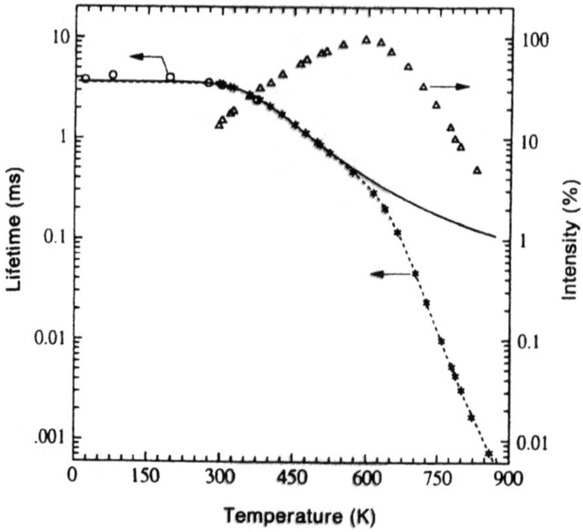


**Fig. 5.2** Temperature-dependence of ruby fluorescence. Asterisks denote the fluorescence lifetime data obtained experimentally; open circles denote early theoretical data. Solid line denotes the imperfect least-squares-based curve-fitting profile, which becomes significantly in error at $T > 500\,\mathrm{K}$; dashed line denotes the well-fitted curve, using a more accurate algorithm, with a good fit over the entire temperature region. Triangles denote the fluorescence intensity from Zhang *et al.* [10].

fluorescence lifetimes in different crystal media can be obtained, together with data on the intensity of the fluorescence signal expected from a particular input power, which is vital in determining the expected signal-to-noise ratio and the performance characteristics experienced in the experimental system. The analysis can be used to confirm which materials are best suited to certain temperature regions, and to predict how the signal strength and sensitivity of the sensor will vary as a function of temperature itself. Care must be taken to balance the approximations used in the model with the sensor range expected, as is also shown in Fig. 5.2. Here, the analytical approximation used is seen to be valid up to approximately 500 K (approximately 230°C) – beyond that it fails catastrophically. The revision of the model, as shown, gives a change in the characteristics at higher temperatures, and thus it yields full agreement with the results of experiment, not only at the higher range of temperatures but also in the lower ones. These analytical models have been very helpful in the design of a group of sensors for temperature measurement in specialist regions such as the cryogenic range (<0°C), the biomedical range (30–40°C) and for furnace and engine testing (>500°C, for example).

Although a wider range of successful examples could be given, the use of these techniques fails for optical waveguides, which are especially suited to numerical methods of analysis, such as those discussed below.

## 5.4 NUMERICAL METHODS

A wide range of methods that are not purely analytical have been discussed for applications to a range of modern waveguide devices.

Following the early work of Goell [11] and Marcatili [12], Knox and Toulios [13] introduced the effective index method, which being semi-analytic has been widely adopted as an approximate technique. Although this simple method gives 'rule-of-thumb' optical waveguide design criteria, its application is limited to the simpler waveguide geometries, and it also fails in many situations when one of the slab approximations is below the cutoff. Over the last two decades other approximate methods have been developed, such as the matrix method [14], the spectral index method [15], the method of lines [16], the finite difference method [17] and the mode matching method [18]. However, there is a significant need for the use of more powerful numerical methods to simulate complex structures in order to optimize existing designs and evaluate novel devices, either prior to, or instead of, fabrication and experimental testing. These would also be useful in providing design guidelines for experimental developments. This has been a valuable stimulus for the high degree of interest in the development and uses of accurate and powerful numerical methods and, in particular, computer-aided design techniques, as a result of readily available PCs or workstations, at reasonable cost. The development and status of various numerical methods for the analysis of optical waveguides have been discussed in a recent research monograph edited by Huang [19].

Among the different numerical approaches considered so far, the finite element method (FEM) has been established as one of the most powerful and versatile methods in many branches of engineering, and is used to model complicated structural, fluid flow, thermal and electromagnetic problems. During the last two decades this method has been used to model optical waveguides [20], directional couplers, butt-coupling situations, gratings, electro-optic modulators, second harmonic generators and semiconductor lasers, for example devices which find wide use in optical and photonic systems.

### 5.4.1 Fundamentals of the finite element method (FEM)

As the range of optical guiding structures becomes more intricate, so the need for effective computer analysis becomes greater and more demanding. The finite element method is particularly advantageous for electromagnetic field problems [21] because of its applicability to waveguides with arbitrary shape or arbitrary index profile, and even with anisotropic, active or nonlinear materials. The technique has been extensively used in a wide range of activities in the analysis of various optical waveguide structures, by many research workers [22–25].

Finite element formulations are usually established via a variational or a Galerkin approach [20]. Several different such variational formulations have been proposed for use with the finite element method. The simplest is the scalar formulation, in terms of the longitudinal field component of the transverse electric (TE) or transverse magnetic (TM) modes present in the structure. This can be used in situations where the modal fields can be described as predominantly TE or TM type. This formulation has been employed for optical waveguides with both isotropic [23] and anisotropic [26] materials. However, the single scalar formulation is inadequate (except as an approximation) for the inherently hybrid nature of the modes of anisotropic, or genuinely inhomogeneous, waveguides with two-dimensional confinement. The finite element method, in terms of both the axial components of the electric and magnetic fields, has been used for analyzing a range of optical waveguides [22]. However, a more suitable formulation has been developed by Rahman and Davies [24], which has been widely used in recent years for the analysis of a considerable range of optical waveguides.

In the finite element approximation, the entire waveguide cross section is divided into a patchwork of subregions or elements. Elements are classified as one-, two- or three-dimensional. In two dimensions, the elements are usually triangles or quadrilaterals. Triangles are commonly used, as shown in Fig. 5.3, because they are easy to adapt. Using many elements, any cross section with a complex boundary can be accurately represented. Since each of the triangles can be of arbitrary

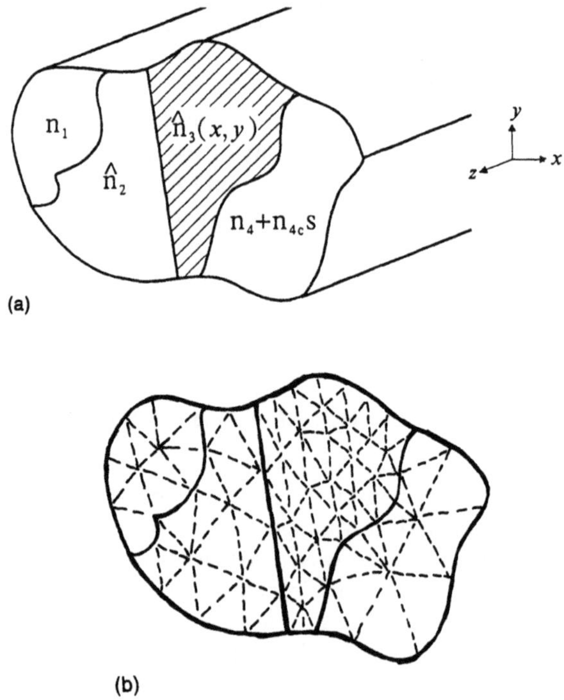


**Fig. 5.3** (a) A general arbitrary shaped optical waveguide geometry, and (b) the finite element representation of its cross-section.

refractive index value, so any complicated refractive index profile, such as diffused or graded index profiles, can also be approximated. The simplest triangular element assumes a linear interpolation between the field values at the vertices of the triangles. By expressing the fields in terms of the nodal values, and with the assumed shape functions, the resulting field components are continuous over the whole waveguide cross section.

The continuous vector field function in the problem domain may be replaced by two or three scalar sets of discrete nodal field values. The simplest triangular element assumes a linear interpolation between the field values at the corner points (vertices) of the triangle. Within each of these elements the trial function is approximated by a suitably chosen polynomial. By this procedure the transverse plane is covered with a grid of discrete nodes, which coincide with the corners of the triangles. The field function will be continuous across adjacent triangles, so that if the potential is interpreted as a third dimension, it can be viewed as a surface with many triangular facets. To be admissible functions, they must satisfy some specific conditions between the elements, where usually the continuity of the field across the boundaries is preferred.

The application of standard finite element techniques to find the stationary condition to the vector H-formulation will yield a generalized eigenvalue equation. When solving optical waveguide problems by finite elements, the key factor affecting storage requirements and computational effort is the choice of the algorithm to solve the matrix equation. In the FE formulation, the matrices are very sparse, and so the use of a sparse matrix eigenvalue equation solver based on the sub-space iteration [27] is one of the most efficient approaches.

### 5.4.2 Application of FEM to waveguide structures

The FEM has been used for the accurate characterization of lightwave propagation through a wide range of waveguides, such as the semiconductor ridge [28], the titanium-diffused $LiNbO_3$ channel [29] and the ion exchanged glass waveguide, as well as a range of optical fibers. In particular, the problems of coupling light from solid state devices, via fiber optics, is very important for photonic sensor systems, and the FEM is able to offer solutions to the characterization problem.

#### *(a) Semiconductor waveguides*

GaAs/AlGaAs or InGaAsP rib waveguides are well suited for use in semiconductor guided-wave optical devices for the integration of coherent light sources and devices. The waveguide edges are quite often not vertical, and the material shows slight anisotropy, but since this increases for finely layered structures, the injected carrier induces spatially varied refractive index profiles with local gain, and in these situations the vector finite element method is best placed to find the accurate modal solutions of such semiconductor devices, and to understand and thus define the operation of the devices. In Fig. 5.4, the results of the FEM solution

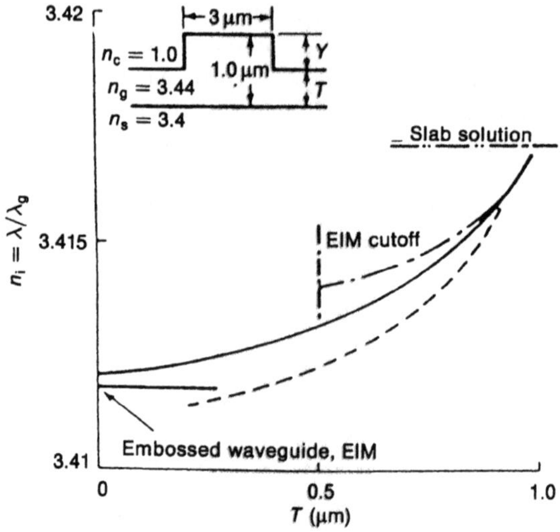


**Fig. 5.4** Variation of the normalized propagation constant for the fundamental TE mode as a function of the slab height, $T$. ——, FEM; ----, SVM; ·-·-·, EIM.

are compared with the results of Austin [30] using the effective index method (EIM) and a scalar variational method (SVM) for the dominant TE mode. The guide and substrate refractive indices are considered to be typical values of 3.44 and 3.4. In this case the rib height has been varied, keeping the rib width 3 μm. When the rib height is small, the EIM gives a reasonably good solution, as in this case the guide can be well approximated as a slab waveguide. The EIM has failed to calculate any eigenvalues when the rib height was larger than 0.5 μm, because above this height the slabs on the sides of the central rib were below the cutoff. The solution obtained by using the SVD with the Hermite–Gaussian function is also of limited accuracy, which shows the value of the FEM for such devices.

*(b) Dielectric waveguides*

As well as semiconductor waveguides, there has been considerable interest in guides formed by silver–sodium exchange, thermal back-diffusion of $K^+$ ions in soda lime glass, proton exchange in $LiNbO_3$ or $LiTaO_3$, annealed proton exchange in $LiNbO_3$ and Ti diffusion of $LiNbO_3$, which has received considerable attention. Again such devices have wide applications in photonic systems, coupled with fiber optics, for sensing or communications. Ti:$LiNbO_3$ is particularly important because of its properties of low loss, its large electro-optic, piezoelectric and elasto-optic coefficients, and its high second-order nonlinearity. However, owing to the nature of the fabrication process, all of the above devices will result in a waveguide with a diffused refractive index profile, and in all cases one which is strongly anisotropic in nature. These limitations in fabrication give

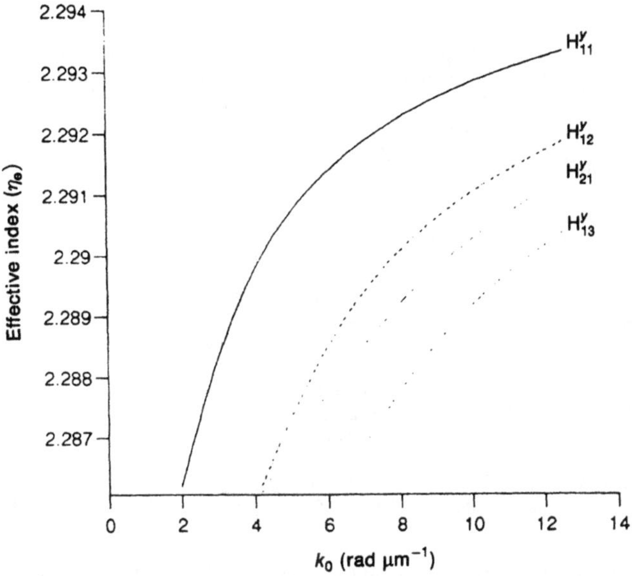


**Fig. 5.5** Variation of the modal effective indices in an anisotropic and diffused channel waveguide.

actual devices with properties which may be different from what the designer specified originally. Figure 5.5 shows the variation of the effective index ($\beta/k_0$) with the wave number ($k_0$) for the first four $H^y_{mn}$ (hybrid TE) for titanium-diffused $LiNbO_3$ waveguides. In this case, the extraordinary and ordinary refractive indices of the substrate are 2.20 and 2.286, respectively, and the maximum extraordinary and ordinary indices in the core are 2.21 and 2.296, respectively. In this example, it is assumed that the optical axis is in the $y$-direction, and that the Gaussian diffusion profiles in the $x$-direction and the exponential diffusion profile in the $y$-direction have diffusion depths of $D_x = 3.0\,\mu m$ and $D_y = 6.0\,\mu m$, respectively [29].

### *(c) Optical fiber waveguides*

Optical fiber is the most extensively used of all the potential optical waveguides, but it is well known in sensor applications that the commonly used single-mode fibers with complete circular symmetry cannot effectively maintain a state of polarization, as they are sensitive to a number of handling factors. Single-mode polarization-maintaining optical fibers, that can maintain a state of polarization over a long propagation distance, place an important role in coherent communications, as well as in sensing or signal processing applications. These fibers are fabricated by using an axially nonsymmetrical refractive index distribution, or in the case of the highly birefringent (hi-bi) fibers, by a

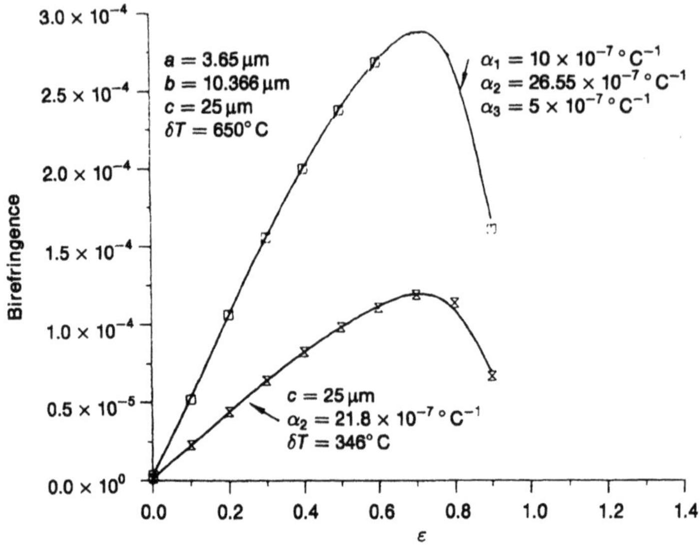


**Fig. 5.6** The birefringence dependence on the ellipticity of the optical fiber made by a preform deformation technique.

nonsymmetrical stress distribution to reduce the mode coupling between two orthogonally polarized modes. The FEM has been used to model such polarization-maintaining hi-bi fiber [31], in particular when thermally induced elasto-optic effects are being considered. The most significant characteristic of such polarization-maintaining fiber is the elliptical nature of its cladding and jacket, and hence, the independence of its birefringence property on the ellipticity of the fiber cladding and jacket ($\varepsilon$) has been studied [32]. The birefringence-dependence of the fibers on the ellipticity for a two-molar concentration of $B_2O_3$ is obtained by use of a finite element simulation, as shown in Fig. 5.6. In this simulation, the tensor stress profile in the fiber geometry, due to temperature change and the unequal thermal expansion coefficients in different regions, is first obtained by using the finite element method. Following that, the profile of the refractive index change, which is tensor in nature, is calculated using the elasto-optic effects of the materials. Finally, the propagation constants for two orthogonal polarized propagating waves are calculated by using the vector H-field FEM to find the modal birefringence.

### *(d) Nonlinear devices*

The study of the characteristics of the lightwave propagation through nonlinear optical devices has stimulated considerable interest, as these devices have potential applications in all-optical computing, signal processing, sensing and communications. The FEM has been used to obtain modal solutions for non-linear optical waveguides with one- [33] or two-dimensional [34] confinement, including the effects of material saturation. The novel features (enhanced

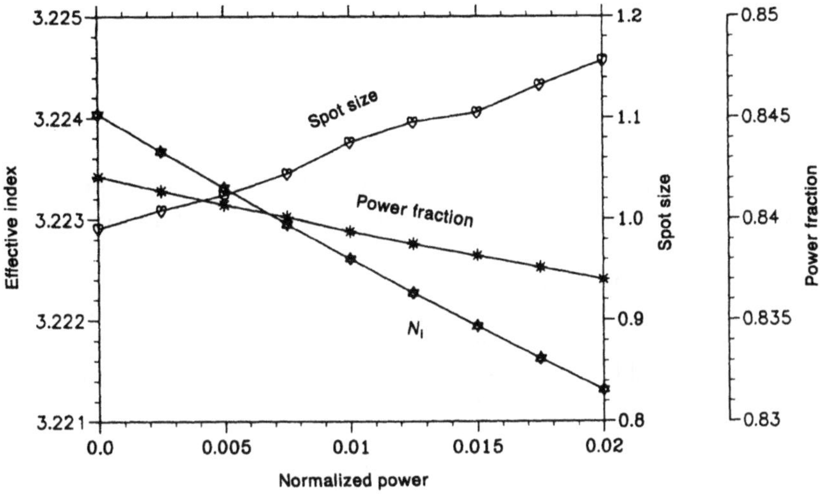


**Fig. 5.7** Variation of the optical parameters with the total power for a self-defocusing nonlinear MQW waveguide.

nonlinearity, absorption) involved with multiple quantum well (MQW) structures make them attractive for use in advanced semiconductor lasers and all-optical switching and signal processing devices, which are very important components of many optical sensor systems. Recently, the FEM has also been used to characterize lightwave propagation through optical waveguides incorporating submicron sized MQW [35] regions. Figure 5.7 shows the variation of the effective index, spot-size and power fraction with the total optical power for a MQW rib waveguide with self-defocusing media. In this case, 70 periods of the GaAsInP well and InP barrier layers with 4 nm and 10 nm thickness, respectively, for each layer, are considered. One of the key features of modern optical systems, especially multilayered sensors, is the optimum use of the power budget from convenient sources, such as these lasers. This type of modeling allows not only the optimization of the source itself, but also of the output beam characteristics to allow the optimization of the coupling to a range of different optical fibers.

### *(e) Loss/gain analysis and surface plasmon devices*

The FEM has also been used to model waveguides incorporating a thin metal layer that can support surface plasmon modes, which are particularly useful for evanescent sensors and polarization-sensitive devices, and which are highly sensitive in their operation to the beam coupling and propagation conditions in the systems. Recently, there has been some valuable modeling activity to develop vector formulations using only the transverse magnetic field components [36], as these formulations do not produce spurious modes, and can also be used to calculate modal losses. However, for practical guided-wave devices, the perturbation technique can be used [37] to calculate modal loss or gain, which is relatively simple and thus satisfactory for most situations.

## 5.5 PROPAGATION MODELS

The vector formulation described above can provide accurate **modal** solutions for any complicated **uniform** waveguides, whether linear, nonlinear, isotropic, anisotropic, homogeneous, inhomogeneous, passive or active. However, this approach cannot provide the evolution of optical waves through tapered structures. So far, the most established and widely used approach is the Fourier-based beam propagation method [38, 39]. However, recently the finite-difference-based beam propagation method (BPM) using the split-step Crank–Nicolson technique [40] has been used to characterize such nonuniform optical guided-wave devices.

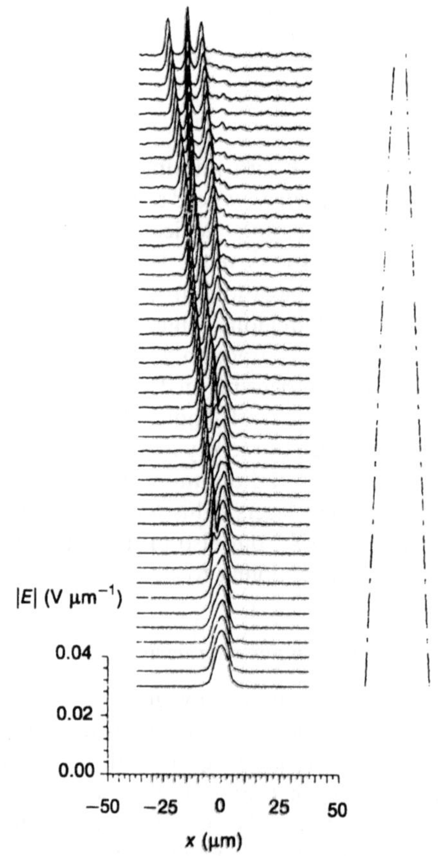


**Fig. 5.8** CW evolution of an optical soliton in a slowly tapered optical waveguide with nonlinear cladding.

Finite-element-based BPM procedures are being developed [41, 42] by discretizing the transverse domain using irregular mesh divisions to achieve a range of computational advantages. During the past few years there has been a great deal of interest in all-optical signal processing and communications with nonlinear optical devices as part of various optical systems. The use of optical solitons can potentially revolutionize optical communications by compensating the material dispersion on the 1.55 μm wavelength band for minimum fiber loss, and could have an equal effect on long-length optical sensor systems. The modal solution of a uniform nonlinear optical waveguide is not strictly valid as the superposition of the modes does not hold for such structures. Recently, this FEM–BPM method, along with the finite difference scheme for axial and time dimensions, has been developed to model optical pulse propagation through nonlinear tapered optical waveguides [41] with a finite material response time [43]. This finite-element-based propagation model can also provide information on the evolution of lightwave propagation in $z$-variant guided-wave structures such as tapers, or $x$- or $y$-junctions, using nonuniform adaptive meshes, infinite elements and flexible boundary conditions. Figure 5.8 shows an example of the evolution of solitary waves through a nonlinear slowly tapered optical waveguide using the FEM–BPM.

## 5.6 DISCONTINUITY IN OPTICAL STRUCTURES

In general, the coupling of optical power from one waveguide to another is achieved using butt-coupling, and the effect is modeled by applying the overlap integral method, which is simple to apply but provides only an approximate estimation. The discontinuity in a guided-wave structure can be characterized more accurately by applying the least squares boundary residual (LSBR) method [44], which is more rigorous. In this approach, the continuity of the tangential components of both the electric and magnetic fields are enforced in a global sense rather than at discrete points, as in the point-matching technique. It is well known that owing to a specific source of illumination at the discontinuity plane, the fundamental and also the higher order modes will be generated at this plane and they will propagate along the guides. An example of this is seen in the variation of the transmission coefficients with vertical misalignment ($\Delta y$) between two identical GaAlAs rib waveguides, as shown in Fig. 5.9.

## 5.7 DIRECTIONAL COUPLER-BASED DEVICES

Directional coupler-based power splitters and combiners are key elements of optical networks for the transmission and multiplexing of sensors. The FEM can provide accurate propagation constants and field profiles for supermodes in a coupled system, even when the individual waveguides are not identical [45], as in the case of some directional coupler-based devices, to obtain the coupling length of the device.

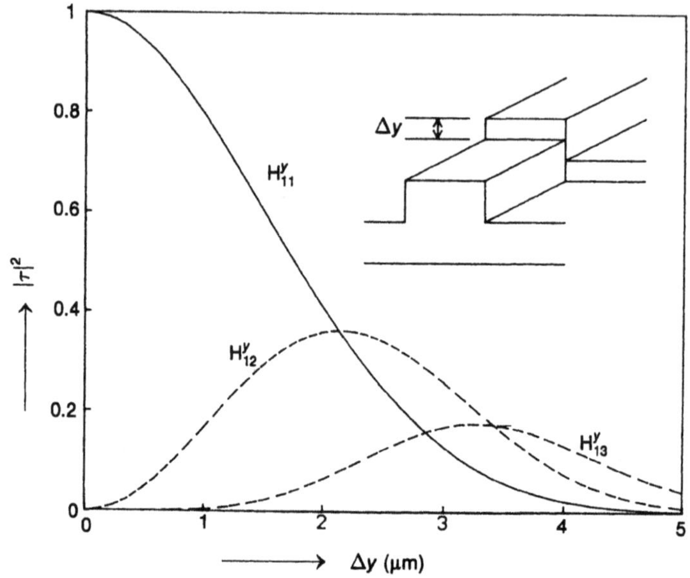


**Fig. 5.9** Variation of the transmission coefficients with vertical misalignment for a GaAlAs rib waveguide.

In the example shown in Fig. 5.10, the thicknesses of the two planar InGaAsP semiconductor waveguides are 1.6 μm and 0.53 μm, respectively. The refractive indices of the guides, cladding and separating regions vary with the operating wavelength, and are designed so that the two guides are phase-matched at a typical system wavelength of $\lambda = 1.3\,\mu m$. Apart from this phase-matching wavelength, the guides are not phase-matched at any other wavelengths. In the situation considered here, it can be shown that the coupling length is maximum when $\lambda = 1.3\,\mu m$, and reduces monotonically for both lower and higher operating wavelengths. Although the guides are nonidentical, since they are phase-matched at this wavelength, the supermode field profile is equally distributed in both waveguides. However, when the guides are not phased-matched, the power transfer between the coupled waveguides is insignificant. The power transfer efficiency between nonidentical and strongly coupled waveguides has also been studied and has applications to the power budget of a multiplexed optical sensor network. This is also a significant parameter in the design of a range of important components such as optical polarizers, modulators and filters, which are widely used in optical sensors. These optical devices can be characterized by using the FEM along with either the coupled mode approach [46] or the LSBR method [47]. Figure 5.10 shows the power transfer efficiency between the guides against the operating wavelength. The use of a nonidentical, but synchronously coupled waveguide as an optical filter is demonstrated.

Many sensor systems rely upon high-power lasers, for example as sources where external modulation of the input signal is required, for sensor identification or for

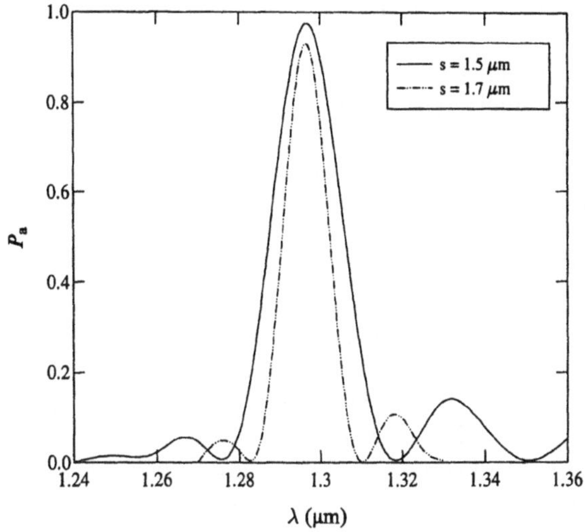


**Fig. 5.10** Variation of the power transfer efficiency with the operating wavelength for an optical filter.

signal processing. The FEM has also been used to characterize electro-optical modulators by taking into account the phase changes due to the linear electro-optical effect in $LiNbO_3$ [48] and additional current depletion in semiconductor modulators [49]. Figure 5.11 shows the reduction of the power transfer between two optical waveguides against the applied modulating/switching potential.

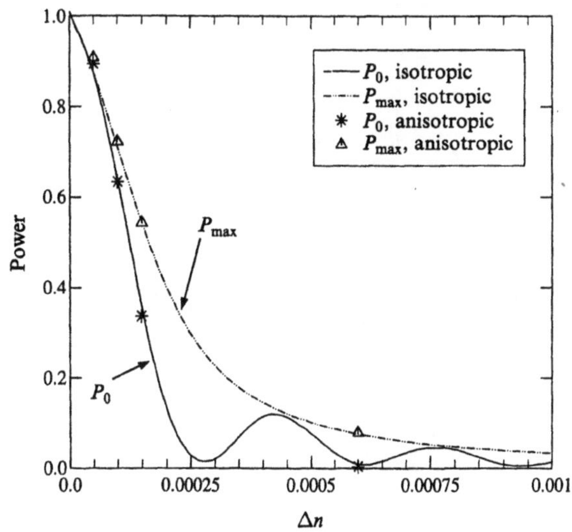


**Fig. 5.11** Variation of the power transfer efficiency with the index change in an electro-optic modulator.

## 5.8 MULTI-MODE INTERFERENCE-BASED DEVICES

In recent years, there has been a growing interest in compact photonic devices using the multimode interference (MMI) principle for signal-routing and signal-processing in communications and sensing. The operation of an MMI device is based on the self-imaging property of a multimode waveguide due to constructive interference of the waveguide modes. Their unique properties, such as compact size, low loss, stable splitting ratio, low cross-talk and imbalance, large optical bandwidth, insensitivity to polarization and ease of production have led to their rapid incorporation into more complex photonic integrated circuits. Several MMI-based devices have been reported, such as 3-dB couplers, 90° hybrids and symmetric power splitters, Mach–Zehnder switches, ring lasers and arrayed waveguide filters. The use of LSBR along the modal solution by the FEM has also been used to characterize MMI devices. Figure 5.12 shows the evolution of the optical field inside a wide MMI section. The MMI section can be used as a power divider circuit or as a mode expander [50].

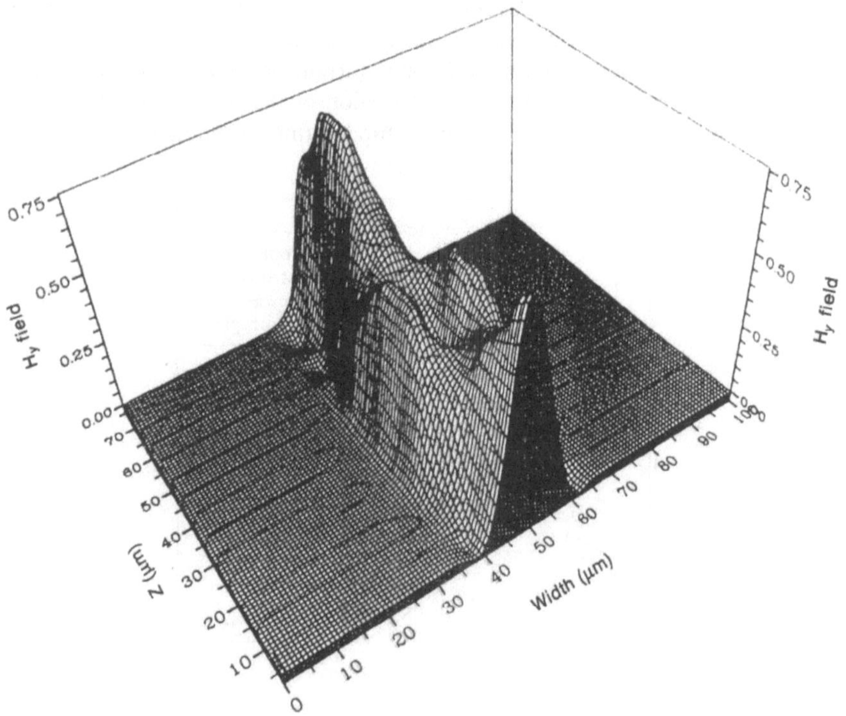


**Fig. 5.12** Evolution of the optical field along an MMI structure.

## 5.9 OTHER OPTICAL DEVICES

As well as the few examples of photonic devices and waveguides discussed above, many more important devices have also been characterized using the finite element method. Recently, Bragg devices using many periodical gratings have been applied to semiconductor and fiber lasers and also as fiber optic filters. Optical filters are important components for multi/demultiplexing of optical signals for optical fiber communications and sensing. The Bragg gratings are usually characterized by using the transfer matrix method (TMM), but a combination of the FEM and the LSBR has recently been used for more rigorous characterization. Similarly, vertical cavity surface emitting laser (VCSEL) structures can be simulated as a layered structure using the TMM, but this type of structure can be more accurately characterized by considering it as a three-dimensional resonating cavity. The finite-element-based propagation model has also been used to model Cherenkov-type second-harmonic generators [51, 52]. This method has also found applications in the self-consisting modeling of semiconductor lasers by considering their electrical, optical and thermal equations [53, 54]. Carrier profiles in a semiconductor laser produce refractive index changes with complex values and a transverse profile which depends on the fabrication parameters. The modal gain in active semiconductor laser structures can be characterized by using the vector $H_t$ formulation [36], or by using the vector H formulation with a perturbation technique [37]. Heat is generated inside an active optical device, and the performance of these devices can be improved by identifying hot-spots and designing an appropriate heat-sink to reduce the temperature in the active region. Thermal modeling of active semiconductor devices can also be characterized by using the FEM [55].

The examples given have shown that numerical methods, and especially the FEM, are powerful and versatile tools in accurate modeling of the optical, thermal, elasto-optic, electro-optic and electronic effects in passive or active guided-wave devices. Computer-aided design and evaluation of important photonic devices, and the identification of important fabrication parameters by performing numerical experiments are invaluable tools in the optimization of such advanced optical devices, and above all they reduce both the cost and the time required in the experimental implementation of new device structures.

## 5.10 NEURAL NETWORK TECHNIQUES

Neural network techniques are now widely applied for the processing of images obtained from video inspection systems, and for pattern recognition, target tracking and system identification. However, as sensor systems become more complex, or the demands upon them become more sophisticated, there is a need to apply methods to extract the maximum amount of valuable information from the sensor used, ideally configured with the minimum of complexity and cost. For example, the demands of the process industries for distributed or multiparameter measurements is a major driving force. Johnson [56] has reported that even where the huge amounts of data which modern fiber sensor arrays can give are available, three major tasks remain – the validation of the data, the extraction of useful information from the mass of data obtained, and the effective use of this information in process control. One vital aspect of the first task is instrument calibration, but the tools and techniques involved are also applicable to other tasks.

### 5.10.1 Calibration of fiber-optic sensors

The calibration of any instrument, be it fiber optic or otherwise, is often less simple than may at first appear [1]. In particular, where fiber optical techniques are used in chemical sensing, the techniques used both in the laboratory and in the field with the instrument are often the same, and contrast sharply with the 'gold standard' of instrument calibration, illustrated, for example, in the measurement of some of the base units of standards. For example, a familiar method of calibration involves taking spectroscopically based chemical sensing and 'known' reference standards and plotting one data set against another, usually employing a 'best fit' straight line or function, to give a 'calibration graph' and a unique definition of the calibration, which is often assumed to be linear. Often, however, such systems are nonlinear, with any linear fit of one data set to another being an approximation. For example, corrections must often be applied when high precision is required, e.g. in laser interferometers for displacement monitoring when issues of the temperature, pressure and humidity of the air and its effect on the refractive index are concerned.

Alternative approaches to system calibration lie in the use of neural networks (NNs). A neural network may be defined as a mathematical function mapping from one data set to another, conventionally depicted as a set of nodes (Fig. 5.13) [56]. Each node sums the values on all the previous-stage nodes connected to it and performing a non-linear function on it. In the simple three-layer

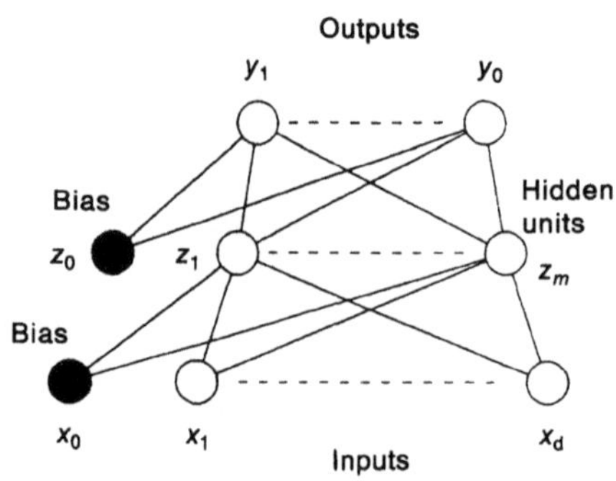


**Fig. 5.13** A multilayer perceptron neural network having two layers of weights. Such networks are capable of approximating any continuous nonlinear function to arbitrary accuracy provided the number $m$ of hidden units is sufficiently large.

system illustrated, the input, the hidden and the output layers are related where the internal mode thresholding of the weighted sums of the previous layer inputs, together with a bias term, is sufficient to replicate an overall mapping function on an essentially arbitrary nonlinearity. Any real function can be modeled with the use of the correct weights, after undergoing an appropriate 'training phase', where output node values are computed based on the input node value, and then compared with the 'true' value in the calibration process. The discrepancy is used to adjust the internal free parameter set, and with the use of many input data points and the correct repeated application of the process, convergence to values that accurately define the calibration mapping can occur. The NN can then be used to interpolate with the range, and thus the calibration is established. Such methods, in principle, can be applied to the calibration of fiber optic sensor systems where nonlinearities exist and a simple, linear calibration is inappropriate. Illustrations can include the mapping of a parameter such as displacement or temperature to an optical parameter such as optical path difference, intensity or luminescent decay. In particular, the incorporation of appropriate calibration is shown in the example where the output of an interferometer is measured (in the case mentioned earlier) when the refractive index of the air path in the system is environmentally sensitive. This is a good illustration of where nonlinearity can exist due to multivariate calibration factors if the relevant parameters are not held constant during the measurement process itself. Multiple linear regression techniques may be applied in such instances to provide an effective calibration of the sensor system.

### 5.10.2 Spectroscopic applications

An important use of NNs in optical sensing is in spectroscopic applications, which are well suited to fiber optics and the industrially important area of chemical and biomedical measurements. The use of broadband sources, coupling the light propagating in fiber optics to and from the sample where the species-sensitive interaction takes place, and detecting the dispersed light on a CCD array, for example, provides a large data set with each pixel potentially contributing a data point. This could be seen as a simplification of the problem typically experienced in video-image processing, where much larger data sets from 2-D CCDs have to be analyzed and image-matching data extracted. The use of a 1-D linear CCD is often sufficient for most applications in spectroscopy, thereby greatly simplifying the data set on which the neural network techniques are to be applied. Measurements are usually static in time, and the shape of the absorption or fluorescence spectrum may be used as the basis of the matching process.

An illustrative application of the technique is in the determination of pollutants in water using ultraviolet-visible (UV-vis) spectroscopy, seen for example in the work of Benjathapanum *et al.* [57]. Previously, optical spectral analysis had been accomplished with mathematical methods such as principal component analysis (PCA) using multiple linear regression, or the partial least-squares (PLS) algorithm [58]. Artificial neural networks were first used for spectroscopic analysis in 1990 by Ham *et al.* [59], and recently interest in their application in this area has been growing steadily, and there are examples of the successful use of NNs for the qualitative and quantitative resolution of multicomponents in analytical chemistry [60] and waste-water pollution research [61].

The use of data analysis methods can help reduce the complexity of monitoring system hardware. Early efforts in this direction have combined UV-vis spectroscopy with neural networks using stochastic back propagation neural networks (BPNN) [62] to examine data from water samples, and to determine from changes in the UV-vis spectra which information is characteristic of a pollution incident or 'breakthrough' of contamination, for example, of ammonia in the water. The approach developed by Benjathapanum *et al.* [57] centered around modeling the optical absorption expected from the process chemistry of a sample containing pollutant species at a given measurement site. The work resulted in a three-stage scheme for processing data. In the first stage, shape information in the spectra was derived by binary encoding segments of the second derivative [63] of the absorption spectra according to the shape, to help in the further classification of the spectra by a second stage using BPNN analysis. Finally, the classifications found in the second stage were used in a further BPNN analysis to predict the species concentrations. This combination of stages was shown to have a good predictive ability when used to examine spectra from a mixture of contaminants. Data for the reported study were obtained from laboratory-based measurements of UV-vis optical absorption of mixes composed of typical chemical pollutants in water at levels up to their legal limits, and from mathematical models deriving from these measurements, in order to characterize and evaluate the system. Principal component analysis (PCA) [64], second-derivative absorbance spectra [63] and second-derivative encoding were compared as preprocessing methods, with the latter being found to be the most effective in the work reported.

Specifically in that scheme, the absorption peak shape data for the various components of nitrate at 210 nm, hypochlorite at 290 nm and monochloramine at 245 nm were used to determine the input variables for the seven BPNNs used for estimating the concentration. Training and testing data for these BPNN were generated using extinction coefficient data for each of the expected components and varying the concentration of each species over the required ranges, and estimated absorption data for the expected species were used to train a second stage of BPNN. This overall scheme was shown in the study to have a good predictive ability when presented with spectra from a mixture of components. A further suggested use of this technique is in the important problem of trying to find a better way to determine the total organic carbon (TOC) content in water due to the absorption of UV radiation at a wavelength centered at 254 nm. This could be done using complex background spectra, which could include nitrates. The technique could also be used in colorimetric measurements, where it might help discern spectral features in the color change of dyes or fluorophores used in detecting important metal contaminants such as Hg, Cd and Cr(VI).

### 5.10.3 Sensor arrays

Sensor arrays are particularly useful for situations where interference is difficult to avoid, but where it may be possible to extract useful data through a multi-sensor approach. A simple example is the use of two sensors, one 'on-line' and the other also subject to extraneous perturbations with common-mode signal rejection techniques applied, now usually through software schemes. The electronic 'artificial nose' comprises an array of simple gas sensors where each element has a slightly different response to the gaseous sample present. In principle, any kind of sensor element may be used in this type of technology, including spectroscopic or thermal devices where fiber optics would be a valid technology, for example in chemical sensing. To a large extent the use of CCD arrays provides the essential type of output for the many optical sensors using such detectors. In such applications, 'perfect' selectivity is not essential, and the overlapping sensitivities of the different elements in the 'electronic nose' is a feature of what are now commercial devices, where the analytical techniques in the signal processing scheme are used to overcome any problems which may arise.

The advance of fiber optic smart structures has been greatly helped by the development of arrays of optical fiber sensors. In particular, Bragg grating devices [65] can be made responsive to many of the structural parameters required, such as strain, temperature and displacement, and as 'in-fiber' elements they can readily be incorporated in both civil structures and aerospace applications, for example. The requirements for effective monitoring of such structures will undoubtedly require the use of a large number of sensors, spread out within the system being interrogated. Techniques have been reported to multiplex small numbers (tens to hundreds) of these sensors, but for effective monitoring of real structures, it is likely that many thousands of sensors will be required. Further, given that the responses may be nonlinear, neural network methods may be one effective approach to the signal processing associated with these arrays. The systems could be calibrated during the fabrication phase of the structure when other aspects of the system are being tested, and data sets could be recorded for the training of the network, and for subsequent use in actual fiber optic sensing applications.

## 5.11 INTELLIGENT SENSORS

The status of 'intelligent' instrumentation has evolved from derision to acceptance in less than two decades [66], and with it the boundary between the sensor and the instrument has become more diffuse. The field has been driven by the processing power available from modern electronics incorporated in an integral way into what would otherwise be a passive sensor. Intelligent instrumentation can offer advantages in terms of accuracy and reliability by removing sensor imperfections such as cross-sensitivity or nonlinearity, and provide compensation for drift and degeneration. The associated costs need not be high if sufficient quantities are produced to justify the 'start up' funds required for both software and hardware development, but ultimately advantages in terms of superior performance must be seen in order to justify the concept.

Intelligent instrumentation can be defined in terms of the subsystems comprising it – the primary sensing element, and the use of mathematical techniques involved in excitation control, signal amplification, signal filtering, and data processing and compensation [67]. This is conventionally seen in terms of the integration of all components on a single chip. This is an elegant but not essential concept, and a hybrid approach may initially be more viable. Complex sensing tasks are potentially well suited to such instrumentation, e.g. the 'electronic nose', but the approach has considerable potential for fiber optic and integrated optical instrumentation. Many fiber optic sensors are relatively simple in concept and require signal processing of the output to extract the desired measurement from the output signal. The incorporation of an electronic subsystem, wholly or partially to undertake the range of signal processing tasks and data manipulation at the sensor head itself, is not incompatible with the usual view that optical sensor systems should be passive and nonelectrical. A number of optical sensor systems have been demonstrated where the electronic aspects of the measurement process have been powered by electro-optic conversion of a light signal from an LED or laser diode through a photodetector incorporated in the sensor head, for example in driven quartz resonant sensors for pressure measurement [68]. This retains the advantage of a nonelectrical transmission to and from the sensor head, and thereby the associated immunity to electromagnetic (e.m.) interference of the connecting optical fiber leads. In work by Pinnock [69] at Lucas, silicon resonant microsystems incorporating fiber optics 'on-chip' show the success of the integration method for resonant frequency-sensitive measurands. It is a small step to develop such a concept further, and in the silicon integration of the sensor system add the appropriate processing power and electro-optic conversion elements, thereby retaining the optical benefits but incorporating the potential for sensor intelligence. In this way the e.m. shielding of the sensor system can be located at the sensor head itself, and optical data and optical power provision can readily be multiplexed on the fiber optic cable. As a result fiber optic sensor systems can be configured with all the sophistication of electronic intelligent instruments, yet retain the benefits of optical data transmission.

This leads naturally to the application of the concepts of virtual instrumentation – a tool constructed as a development aid for testing and evaluating the performance of an intelligent instrument. This in many ways parallels the simulation of fiber optic sensors using conventional methods, but the software involved in the intelligent instrument forms part of the virtual instrument design [67]. In that way, with the aid of a graphical user interface (GUI) and associated mathematical software, the operation of the intelligent fiber optic instrument can be simulated in full, and development and design can be carried to a high level before more expensive fabrication is undertaken. Such concepts are under development for conventional electronic instrumentation, and in principle are also fully applicable in the fiber optic sphere.

## 5.12 SUMMARY

A wide range of mathematical and computer-based techniques can be applied to fiber optic instrumentation, from analysis and understanding of performance, to design and configuration through the use of advanced numerical methods, with signal processing and data manipulation via neural network methods for complex information processing. Finally, the concepts of intelligent instrumentation can be seen to be equally applicable to fiber optic instrumentation as to more conventional electronic sensor systems, and indeed virtual instrumentation methods may be considered as a sensor development tool for test and evaluation. With the ever increasing nature of computing power at lower cost, further developments in these areas in practical fiber optic sensor systems can undoubtedly be expected in the future.

## REFERENCES

1. Finkelstein, L. and Grattan, K. T. V. *Concise Encyclopedia of Measurement and Instrumentation*, Pergamon Press, Oxford, 1994.
2. James, J. R. and Gallett, I. N. L. Point-matched solutions for propagating modes on arbitrarily shaped dielectric rods, *Radio Electron. Eng.*, **42**, 103, 1972.
3. Chiang, K. S. Analysis of optical fibers by effective-index method, *Appl. Opt.*, **25**, 348–354, 1986.
4. Su, C. C. and Chen, C. H. Calculation of propagation constants and cutoff frequencies of radially inhomogeneous optical fibers, *IEEE Trans. Microwave Theory Technol.*, **MTT-34**, 328–332, 1986.
5. Rahman, B. M. A. and Davies, J. B. Finite-element solution of integrated optical waveguides, *J. Lightwave Technol.*, **LT-2**, 682–688, 1984.
6. Wang, D. N., Ning, Y. N., Grattan, K. T. V., Palmer, A. W. and Weir, K. Optimized multiwavelength combination sources for interferometric use, *Appl. Opt.*, **33**, 7326–7333, 1994.
7. Wang, D. N., Grattan, K. T. V. and Palmer, A. W. Dispersion effect analysis with multiwavelength combination sources in optical sensor applications, *Opt. Commun.*, **127**, 19–24, 1996.
8. Grattan, K. T. V., Selli, R. K. and Palmer, A. W. Ruby decay-time fluorescence thermometer in a fibre optic configuration, *Rev. Sci. Instrum.*, **59**, 1328–1335, 1988.
9. Zhang, Z., Grattan, K. T. V. and Palmer, A. W. A fibre optic high temperature sensor based on the fluorescence of alexandrite, *Rev. Sci. Instrum.*, **63**, 3869–3873, 1992.
10. Zhang, Z., Grattan, K. T. V. and Palmer, A. W. Temperature dependencies of fluorescence lifetimes in $Cr^{+3}$-doped insulating crystals, *Phys. Rev. B.*, **48**, 7772–7778, 1993.
11. Goell, J. E. A circular harmonic computer analysis of rectangular dielectric waveguides, *Bell Syst. Tech. J.*, **48**, 2133–2160, 1969.
12. Marcatili, E. A. J. Dielectric rectangular waveguide and directional coupler for integrated optics, *Bell Syst. Tech. J.*, **48**, 2071–2102, 1969.
13. Knox, R. M. and Toulios, P. P. Integrated circuits for the millimetre through optical frequency range, *Proc. MRI Symp. Submilimeter Waves*, Brooklyn Poly Inst., Brooklyn, pp. 497–516, 1970.

14. Sharma, E. K., Ghatak, A. K. and Goyal, I. C. Matrix method for determining propagation characteristics of optical waveguides *IEEE J. Quantum Electron*, **QE-19**, 1231–1233, 1983.

15. Stern, M. S., Kendall, P. C. and McIlroy, P. W. A. Analysis of the spectral index method for vector mode of rib waveguides, *Proc. IEE, J. Optoelectron.*, **137**, 21–26, 1990.

16. Rogge, U. and Pregla, R. Method of lines for the analysis of strip-loaded optical waveguides, *J. Opt. Soc. Am. B*, **8**, 459–463, 1991.

17. Bierwirth, K. N., Schulz, M. and Arndt, F. Finite difference analysis of rectangular dielectric waveguide structures, *IEEE Microwave Theory Technol.*, **MTT-34**, 1104–1114, 1986.

18. Dagli, N. and Fonstad, C. G. Analysis of rib dielectric waveguides, *IEEE J. Quantum Electron.*, **QE-21**, 315–321, 1985.

19. Huang, W. P. Methods for Modelling and Simulations of Guided-Wave Optoelectronic Devices. Part I. Modes and Couplings. Part II. Waves and Interactions. EMW Publishing, Cambridge, MA, 1995.

20. Rahman, B. M. A., Fernandes, F. A. and Davies, J. B. Review of finite element methods for microwave and optical waveguides, *Proc. IEEE*, **79**, 1442–1448, 1991.

21. Silvester, P. P. and Ferrari, R. L. *Finite Elements for Electrical Engineers*, 2nd edn, Cambridge University Press, Cambridge, 1990.

22. Yeh, C., Ha, K., Dong, S. B. and Brown, W. P. Single-mode optical waveguides, *Appl. Opt.*, **18**, 1490–1504, 1979.

23. Mabaya, M., Lagasse, P. E. and Vandenbulcke, P. Finite element analysis of optical waveguides, *IEEE*, **MTT-29**, 600–605, 1981.

24. Rahman, B. M. A. and Davies, J. B. Finite-element analysis of optical and microwave waveguide problems, *IEEE Trans. Microwave Theory Technol.*, **MTT-32**, 20–28, 1984.

25. Koshiba, M., Hayata, K. and Suzuki, M. Improved finite-element formulation in terms of magnetic field vector for dielectric waveguides, *IEEE Trans. Microwave Theory Technol.*, **MTT-33**, 227–233, 1985.

26. Koshiba, M., Hayata, K. and Suzuki, M. Approximate scalar finite-element analysis of anisotropic optical waveguides with off-diagonal elements in a permittivity tensor, *IEEE Trans. Microwave Theory Technol.*, **MTT-32**, 587–593, 1984.

27. Davies, J. B., Fernandez, F. A. and Philippou, G. Y. Finite element analysis of all modes in cavities with circular symmetry, *IEEE Trans. Microwave Theory Technol.*, **MTT-30**, 1975–1980, 1982.

28. Rahman, B. M. A. and Davies, J. B. Vector-H finite element solution of GaAs/GaAlAs rib waveguides, *Proc. IEE, J. Optoelectron.*, **132**, 349–353, 1988.

29. Katrisku, F., Rahman, B. M. A. and Grattan, K. T. V. Finite element analysis of diffused anisotropic channel waveguides, *J. Lightwave Technol.*, **14**, 780–786, 1996.

30. Austin, M. W. Theoretical and experimental investigation of GaAs/GaAlAs and $n/n^+$ GaAs rib waveguides, *J. Lightwave Technol.*, **LT-2**, 688–694, 1984.

31. Liu, Y., Rahman, B. M. A. and Grattan, K. T. V. Thermal stress induced birefringence in bow-tie fibers using finite element method, *Appl. Opt.*, **33**, 5611–5616, 1994.

32. Liu, Y., Rahman, B. M. A. and Grattan, K. T. V. Analysis of birefringence properties of optical fibers made by a preform deformation technique, *J. Lightwave Technol.*, **LT-13**, 142–147, 1995.

33. Rahman, B. M. A., Souza, J. R. and Davies, J. B. Numerical analysis of nonlinear bistable-optical waveguides, *IEEE Photonics Technol. Lett.*, **2**, 265–267, 1990.

34. Ettinger, R. D., Fernandez, F. A., Davies, J. B. and Rahman, B. M. A. Finite element analysis of slab-loaded nonlinear waveguide with saturable media, *IEEE Photonics Technol. Lett.*, **3**, 147–149, 1991.

35. Rahman, B. M. A., Liu, Y. and Grattan, K. T. V. Finite element modelling of one- and two-dimensional MQW semiconductor optical waveguides, *IEEE Photonics Technol. Lett.*, **5**, 928–931, 1993.

36. Fernandez, F. A. and Lu, Y. Variational finite element analysis of dielectric waveguides with no spurious modes, *Electron. Lett.*, **26**, 2125–2126, 1990.

37. Themistos, C., Rahman, B. M. A., Hadjicharalambous, A. and Grattan, K. T. V. Loss/gain analysis of optical waveguides, *IEEE J. Lightwave Technol.*, **13**, 1760–1765, 1995.

38. Feit, M. D. and Fleck, Jr. J. A. Light propagation in graded-index optical fibres, *Appl. Opt.*, **17**, 3990–3998, 1978.

39. Yevick, D. and Hermansson, B. Efficient beam propagation techniques, **QE-26**, 109–112, 1990.

40. Huang, W. P., Xu, C. L., Chu, S. T. and Chaudhuri, S. K. A finite difference vector beam propagation method: analysis and assessment, **JLT-10**, 295–305, 1992.

41. Buah, P. A., Rahman, B. M. A. and Grattan, K. T. V. Numerical simulation of pulse propagation in nonlinear tapered waveguide, *Proc. SPIE*, **2212**, 66–72, April 1994.

42. Hernadez-Figueroa, H. E. Nonlinear nonparaxial beam-propagation method, *Electronic Lett.*, **30**, 352–353, 1994.

43. Buah, P. A., Rahman, B. M. A. and Grattan, K. T. V. A split-step finite element scheme for spatio-temporal analysis of pulse propagation in nonlinear waveguides, *OSA Nonlinear Guided-Wave Phenomena*, **Vol. 15**, pp. 208–211, Cambridge, MA, 1993.

44. Rahman, B. M. A. and Davies, J. B. Analysis of optical waveguide discontinuities, *J. Lightwave Technol.*, **LT-6**, 52–57, 1988.

45. Rahman, B. M. A., Wongcharoen, T. and Grattan, K. T. V. Finite element analysis of nonsynchronous directional couplers, *Fiber Integrated Opt.*, **13**, 331, 1994.

46. Wongcharoen, T., Rahman, B. M. A. and Grattan, K. T. V. Power coupling for electro-optic directional coupler switch, in *Guided-Wave Optoelectronics* (Ed. T. Tamir), pp. 213–219, Plenum Press, New York, 1995.

47. Wongcharoen, T., Rahman, B. M. A. and Grattan, K. T. V. Accurate characterization of optical filters with two-dimensional confinement, *J. Lightwave Technol*, November 1996.

48. Mustacich, R. V. Scalar finite element analysis of electro-optic modulation in diffused channel waveguides and poled waveguides in polymer thin films, *Appl. Opt.*, **27**, 3732, 1988.

49. Young, T. P., Armstrong, G. A. and Smith, W. P. *SPIE*, **1141**, 103, 1989.

50. Rajarajan, M., Rahman, B. M. A. and Grattan, K. T. V. Accurate characterization of MMI devices with two-dimensional confinement, *J. Lightwave Technol.*, September 1996.

51. Hayata, K. and Koshiba, M. Numerical study of guided-wave sum-frequency generation through second-order nonlinear parametric process, *J. Opt. Soc. Am. B*, **8**, 449–458, 1991.

52. Bava, G. P., Montrosset, I., Sohler, W. and Suche, H. Numerical modelling of $Ti:LiNbO_3$ integrated optical parametric oscillators. **QE-23**, 42–51, 1987.

53. Wilt, D. P. and Yariv, A. A self-consistent static model of the double-heterostructure laser, *IEEE J. Quantum Electron.*, **QE-17**, 1941, 1981.

54. Tan, G., Bewtra, N., Lee, K. and Xu, J. M. A two-dimensional nonisothermal finite element simulation of laser diodes, *IEEE J. Quantum Electron.*, **QE-29**, 822–835, 1993.

55. Rahman, B. M. A., Lepkowski, P. and Grattan, K. T. V. Thermal modelling of vertical cavity surface emitting lasers, *Proc. IEE J. Optoelectron.*, 82–86, 1995.

56. Johnson, M. Neural network techniques in instrumentation, *Meas. Control*, **6**, 248–253, 1996.

57. Benjathapanum, N., Boyle, W. J. O. and Grattan, K. T. V. Binary encoded 2nd differential spectrometry using UV-vis spectral data and neural networks in the estimation of species type and concentration, *IEE Proc. Sci. Meas. Technol.*, **144**, 73–80, 1997.

58. Bhandare, P., Mendelson, Y., Peura, R. A., Janatsch, G., Kruse-Jares, J. D., Marbach, R. and Heise, H. M. Multivariate determination of glucose in whole blood using partial least squares and artificial neural networks using mid-infrared spectroscopy, *Appl. Spectrosc.*, **47**, 1214–1222, 1993.

59. Ham, F. M., Cohen, G. M. and Cho, B. Neural network based real-time detection of glucose using a non-chemical optical sensor approach. *Proc. 12th Annual Conf., IEEE Engineering in Medicine and Biology Society*, Vol. 2, LRW Associates, Arnold, MD, Philadelphia, PA, pp. 480–482, 1990.

60. Liu, Y., Upadhyaya, B. R. and Naghedolfeizi, M. Chemometric data analysis using artificial neural networks, *Appl. Spectrosc.*, **47**, 12–23, 1993.

61. Boger, Z. Application of neural networks to water and wastewater treatment plant operation, *Instrum. Soc. Am.*, **31**, 25–33, 1992.

62. Remelhart, D. E. and McClelland, J. L. in Forming internal representation by error propagation, *Parallel Distributed Processing, Vol. I. Foundation*, Chap. 8, MIT Press, Boston, MA, 1992.

63. Sommer, L. *Analytical Absorption Spectrophotometry in the Visible and Ultraviolet: The Principles*, Elsevier, Amsterdam, 1996.

64. Ares, R. E., Lidiard, D. P. and Spragg, R. A. Principle component analysis, *Chem. Br.*, 821–824, 1991.

65. Patrick, H., Williams, G. M., Kersey, A. D., Pedrazzani, J. R. and Vengsarkar, A. M. Strain/temperature discrimination using combined fiber Bragg and long-period grating sensors, *OFS 11th Int. Conf. Optical Sensors. Advanced Sensing Photonics*, pp. 96–99, May 1996.

66. Brignell, J. E. Measurement and control feature on intelligent instruments, *Meas. Control*, **29**, 164, 1996.

67. Tanner, A. H. and White, N. M. Virtual instrumentation: a solution to the problem of design complexity in intelligent instruments, *Meas. Control*, **29**, 165–171, 1996.

68. Grattan, K. T. V., Saini, D. P. S. and Palmer, A. W. Optical vibrating quartz crystal pressure sensor using frustrated-total-internal-reflection readout techniques, *J. Lightwave Technol.*, **LT5**, 972–979, 1987.

69. Pinnock, R. A. Micromachined silicon resonant sensors with robust optical interrogation, Paper presented at Sensors and their Applications VI, Manchester, 1993. Published in *Sensors VI: Technology Systems and Applications*, Institute of Physics Publishing, Bristol (Eds K. T. V. Grattan and A. T. Augousti), pp. 141–146, 1993.

# 6

# Polarimetric distributed optical fiber sensing

A. J. Rogers

## 6.1 INTRODUCTION

Distributed optical fiber sensing (DOFS) is a technique which utilizes the very special properties of the optical fiber to make simultaneous measurements of both the spatial and temporal behavior of a measurand field [1, 2]. As such, it provides an extra dimension in the measurement process, leading to finer monitoring and control, and to a new level of understanding, especially in regard to the behavior of large structures. Thus, we may expect to measure spatial distributions with a resolution 0.1–1 m over a distance of 100 m, to an accuracy of $\sim$1%.

The advantages of optical fibers for general measurement functions are well known: they rely on the insulating, dielectric passivity of the medium, allied to the sensitivity of its optical propagation properties to external influences, over a broad range. DOFS takes advantage of two additional properties of the fiber: its one-dimensional nature, and its mechanical flexibility.

With the aid of these it becomes possible, in principle, to determine the value of a wanted measurand continuously as a function of position along the length of a suitably configured fiber, with arbitrarily large spatial resolution; the normal temporal variation is determined simultaneously from the time-dependence of the signal.

Such a facility as this offers many attractive possibilities for industrial and research applications. The value of having access to the spatial/temporal behavior of strain and temperature in, for example, large, critical structures such as dams, aircraft, space-craft, bridges, multistorey buildings, electrical generators, boilers, chemical pressure vessels, etc. are clear from the points of view both of safety monitoring, and of improved understanding of behavior under anomalous conditions.

The flexibility of the fiber makes it relatively easy to install over the chosen measurement path and thus, unlike many other sensor systems, allows retrospective fitting. Furthermore, again in contradistinction to other sensor arrangements, DOFS offers a facility which is unique: there are no conventional techniques for acquiring the same information.

An important, recent application area is that of so-called 'smart' structures, where the design of large structures such as aircraft or buildings includes

*Optical Fiber Sensor Technology*, Vol. 3. Edited by K. T. V. Grattan and B. T. Meggitt.
Published in 1998 by Kluwer Academic Publishers, London. ISBN 978-1-4419-4736-9

sophisticated sensing systems capable of providing information which can be used, via actuators, to adjust the form of the structure in order to suit prevailing conditions. Thus, an aircraft wing might, for example, change shape to suit varying heights, speeds and turn radii. The passivity, freedom from electromagnetic interference and high spatial resolution of DOFS could well prove invaluable in this challenging application area.

A primary difficulty with DOFS systems (as with most measurement systems) is that of achieving high sensitivity together with high spatial resolution. The aim of 0.1 m spatial resolution (or better) with a measurement accuracy of 1% is difficult to achieve simply because the optical interaction length of 0.1 m allows relatively few photons to be modulated (per resolution interval) by the measurand field, unless optical powers are very high, and the power which can be propagated in an optical fiber is limited by the onset of nonlinear effects. There are two possible approaches to the solution of this problem: one is to use highly sensitive polarimetric/interferometric techniques; the other is actually to make use of those limiting, nonlinear effects. In addition, one can use the two simultaneously, i.e. nonlinear, polarimetric techniques! It is these methods with which this chapter is concerned.

## 6.2 BACKSCATTER DOFS

Most DOFS explorations to date have relied on backscatter arrangements [1–4]. In these, a pulse of light is launched into an optical fiber and is continuously Rayleigh-backscattered as it propagates (Fig. 6.1). Time resolution of the emerging backscattered light then reveals the spatial distribution of any external field which is capable of modulating, in a deterministic way, some characterizing property of the light, this latter being demodulated by the detector. Such a one-dimensional 'lidar' arrangement has the clear advantage of simplicity, but it has a number of disadvantages.

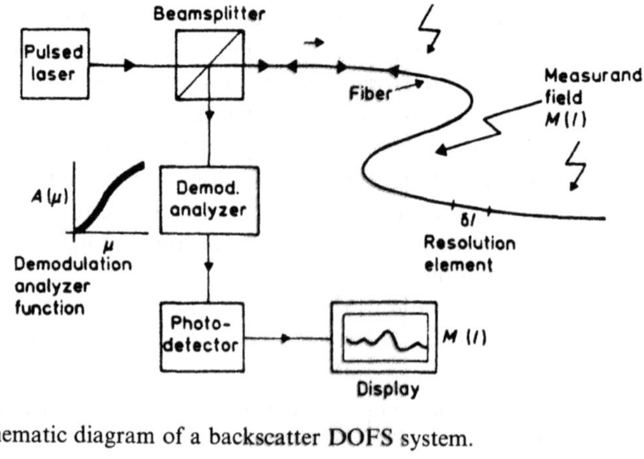


**Fig. 6.1** Schematic diagram of a backscatter DOFS system.

The primary disadvantage is that Rayleigh backscatter is at a low level, of the order of $10^{-5}$ of the forward traveling energy, per unit length of fiber. Moreover, since the fiber attenuation (in a propagation 'window') will depend almost entirely on Rayleigh scatter, there will be an optimum length, for a given scatter coefficient, with regard to both the sensitivity and the resolution of the DOFS system.

This can be given some quantification by examining the backscatter power equation [5]

$$p(t) = \frac{c}{2} E_0 \alpha s \exp(-\alpha c t), \tag{6.1}$$

where $p(t)$ is the backscatter power at time $t$, $c$ is the speed of light in the fiber, $\alpha$ is the attenuation coefficient, $s$ is the backscatter capture fraction and $E_0$ is the launch pulse energy.

In order to optimize the backscatter DOFS system, the requirement is for maximum backscattered power at maximum distance into the fiber, so as to achieve good sensitivity and resolution at maximum range. Thus the requirement is to maximize the power $\times$ distance product. The go and return passage to a distance $l$ takes a time $2l/c$, and thus from equation 6.1 we may write

$$p(l)l = l\frac{c}{2} E_0 \alpha s \exp(-2\alpha l).$$

Clearly this has a maximum value when

$$l = \frac{1}{2\alpha} = L \quad \text{(say)},$$

and hence the power received from the far end is given by

$$P_L = \frac{cE_0 s}{2Le},$$

where $e = \exp$ (equation 6.1). Now if we require a spatial resolution of $\Delta L$ at distance $L$ from the launch end, it is clear, firstly, that the launched pulse width must be less than $\Delta L$. Secondly, the achievable resolution is limited by the detector signal-to-noise ratio (SNR), where the energy received from $\Delta L$ at $L$ is given by

$$\delta E_L = P_L \frac{2\Delta L}{c} = \frac{E_0 s}{e} \times \frac{\Delta L}{L}. \tag{6.2}$$

Suppose that we need to measure the measurand to an accuracy (i.e. fractional error) of $\eta$ over $\Delta L$. If we assume that the detector is operating to the shot noise limit (Poisson statistics), we have

$$\frac{\delta E_L}{h\nu} > \frac{1}{\eta^2},$$

where $\nu$ is the optical frequency and $h$ is Planck's constant.

Hence, using equation 6.2,

$$\Delta L \geqslant \frac{eLh\nu}{E_0 s} \times \frac{1}{\eta^2} \tag{6.3}$$

$$\Delta L \times \eta^2 \geqslant \frac{eLh\nu}{E_0 s}. \tag{6.4}$$

We require both $\Delta L$ and $\eta$ to be as small as possible (equivalent to large spatial resolution and high measurement accuracy), and thus we may require the reciprocal of $\Delta L\eta^2$ as a performance figure of merit, $F$, where

$$F_{\mathrm{B/S}} = \frac{E_0 S}{eLh\nu}.$$

In order to provide some feel for the implications, let us take some 'typical' values for a silica fiber DOFS backscatter system: $L = 100\,\mathrm{m}$; $E_0 = 1\,\mathrm{nJ}$; $\nu = 5 \times 10^{14}\,\mathrm{Hz}$ ($\lambda = 600\,\mathrm{nm}$); $s = 5 \times 10^{-3}$. These give $F_{\mathrm{B/S}} = 5.6 \times 10^4\,\mathrm{m}^{-1}$.

This implies that, for an accuracy of 1% ($\eta = 0.01$), we have $\Delta L > 0.18\,\mathrm{m}$. (It should, however, be pointed out that the above calculation refers to 'single-pulse' performance, and significant improvement can result from integration over many pulses, if the consequent increase of measurement response time can be tolerated.)

Further to this, however, we must consider how the energy modulation (within the spatial resolution interval) is to be achieved by the measurand. There are two primary possibilities: the measurand can modulate either the amplitude (power) of the wave or its phase (polarization). In order to modulate the amplitude, the attenuation of the fiber must be influenced. To modulate the phase (polarization), it is the refractive index which must be affected.

If we choose to modulate the attenuation, we immediately see a further difficulty. Returning to the differentiation of equation 6.1, we find that

$$\frac{\mathrm{d}p}{p} = (1 - \alpha ct)\frac{\mathrm{d}\alpha}{\alpha},$$

and since for the optimum fiber length we have

$$ct = 2L = \frac{1}{\alpha},$$

it follows that $\mathrm{d}p/p$ is a decreasing fraction of the modulation as the pulse propagates along the fiber, and falls to zero at the optimum length.

This implies that only phase/polarimetric methods (in the linear domain) have any real promise for usable DOFS performance in backscatter. The first backscatter DOFS method to be explored was, indeed, a polarimetric arrangement.

### 6.2.1 Polarization-optical time-domain reflectometry (POTDR)

Where OTDR measures the distribution of the fiber's optical attenuation, polarization-optical time-domain reflectometry (POTDR) [5, 6] measures the

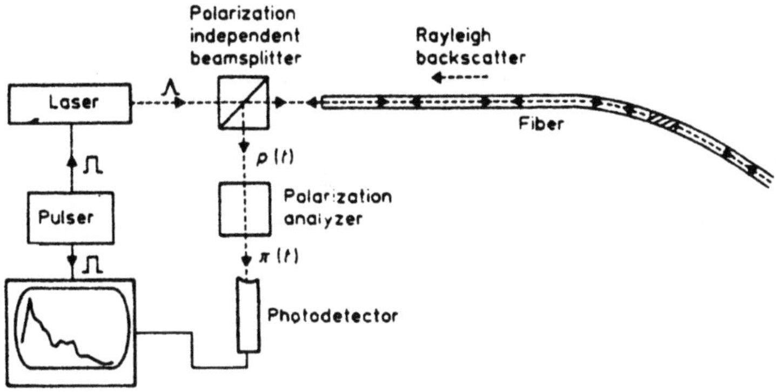


**Fig. 6.2** Arrangement for POTDR.

distribution of its optical polarization properties, again by time-resolving the Rayleigh backscatter from a propagating pulse. POTDR, of necessity, uses monomode fiber, since only with this type of fiber can one assign to the propagating light a single, identifiable polarization state at each point along the fiber.

The experimental arrangement is illustrated in Fig. 6.2. The polarization state of the backscattered light is determined continuously as a function of time (by measurement of the Stokes parameters [7]), allowing the required distribution of the polarization properties to be determined. It follows that the distribution of any external measurand field which can modify these properties (e.g. force/strain plus electric field, magnetic field, etc.) can now also be measured. Figure 6.3 shows the results obtained with POTDR when measuring the distribution of bend strain for a fiber wound on a drum of diameter 185 mm [8].

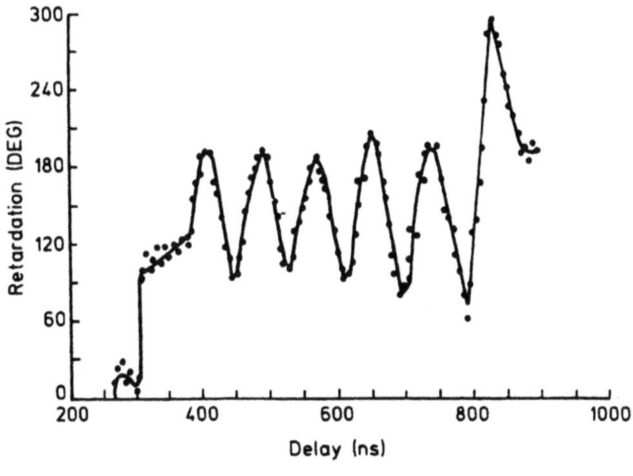


**Fig. 6.3** Measurement of distributed strain using POTDR.

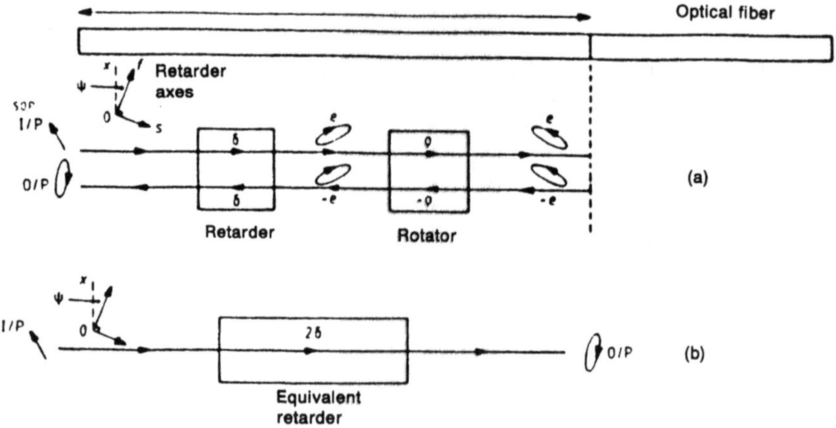


**Fig. 6.4** Loss of polarization information intrinsic to POTDR.

The primary disadvantages of POTDR are concerned with the interpretation of the returning polarization. The Rayleigh backscatter level is, of course, low, and the requirement is to measure and process the returning signal in the specified measurement response time, which is often less than 1 ms. The optical and electronic demands which this makes are considerable. Furthermore, not all the required information is available in backscatter, as is illustrated in Fig. 6.4. Any (loss-less) optical polarization element is equivalent to a retarder–rotator combination [9], and the effect of a go-and-return passage through the element, provided that the element is reciprocal (i.e. its polarization properties are independent of direction), is to cancel the equivalent rotation. Thus, of the three pieces of information needed to specify fully the polarization properties of the fiber (retardation of the retarder, orientation of the axes of the retarder, rotation), only two are available, and thus for full definition of the system *a priori* knowledge of the polarization properties is required. This requirement is best met by the development of fibers with special, controlled polarization properties.

## 6.3 FORWARD-SCATTER DOFS

Suppose that there is no necessity to rely on backscattered light. Consider the forward scatter schematic shown in Fig. 6.5. In this illustration a continuous wave (CW) source is launched at one end (F) of the fiber, into one of two possible modes, say T and R (Fig. 6.5(a)). These might be two propagation modes, two polarization modes, or even two separate cores within the same cladding. The important conditions which must obtain are that the two modes must be separately identifiable at the output from the fiber, and that light coupling between them must both be possible and dependent in some way on the external

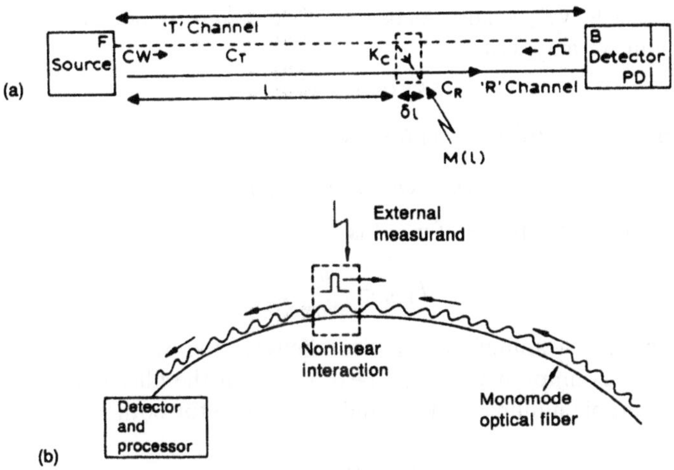


**Fig. 6.5** A forward-scatter DOFS. (a) Schematic diagram for an F/S DOFS. (b) One form of operational F/S DOFS using a nonlinear interaction between a pulse and a counter-propagating CW.

field to be measured (measurand). The mode into which the CW is launched will be labeled the 'filled' mode (in this case T), whilst the other is the 'free' mode (in this case R).

An optical pulse in now launched into the other end (B) of the fiber in such a way as to allow it, while propagating in the opposite direction to the CW, to couple the latter from the 'filled' to the 'free' mode. If, as was required above, the coupling mechanism is modulated by the measurand in a deterministic way, then by time-resolving the CW power level emerging from end B in the 'free' mode, one obtains the spatial distribution of the measurand field.

Figure 6.5(b) illustrates an operational arrangement where the pulse and the counterpropagating CW are coupled by means of a nonlinear interaction in the fiber at the point of overlap.

The important advantage which this method offers over the backscatter technique is that of increased received power level. There is no longer an optimum attenuation coefficient for a given fiber length, since one is not relying on back-scatter (although the mode-coupling requirement might impose other limitations on the minimum attenuation allowable). After a length $L$, the CW power will in this case be given by

$$p(L) = p_0 \exp(-\alpha L).$$

Suppose that a fraction $q$ of this light power is coupled from the filled to the free mode, by the passage of the counterpropagating pulse, in the absence of a measurand field. For a resolution length $\Delta L$ at distance $L$ from the CW launch end, the coupled energy will be

$$\delta E_L = q p_0 2\Delta L/c$$

(since $\alpha$ can now be very small). As before, if we require an accuracy of $\eta$, in the shot noise limit we may write

$$(\delta E_L/h\nu) \geqslant 1/\eta^2,$$

and thus, analogously to equation 6.3, we have

$$\Delta L \geqslant ch\nu/2qp_0 \times 1/\eta^2,$$

and the forward-scatter figure of merit is

$$F_{\mathrm{F/S}} = \frac{2qp_0}{ch\nu}.$$

In order to make a meaningful comparison between the two figures of merit we shall make two assumptions. (1) The peak power in the fiber is limited by the onset of undesirable nonlinear effects and is the same for both cases. Thus,

$$p_0' = E_0' \frac{c}{\Delta L},$$

where the primed quantities represent the maximum values. (Here we have also assumed that the pulse width is equal to the resolution length, which again represents the optimum condition.) (2) Over the full length, $L$, of the DOFS system, the CW is totally coupled to the free mode. In this case

$$q' = \frac{\Delta L}{L}.$$

Inserting these assumptions, the ratio of the optimum figures of merit is given by

$$\frac{F'_{\mathrm{F/S}}}{F'_{\mathrm{B/S}}} = \frac{2e}{s}.$$

For a typical value of the backscatter capture fraction, $s$, of $5 \times 10^{-3}$ we have

$$\frac{F'_{\mathrm{F/S}}}{F'_{\mathrm{B/S}}} \approx 10^3.$$

Thus, for a given accuracy of measurement (for example), the optimum resolution interval for the forward-scatter arrangement is reduced by three orders of magnitude, i.e. down to 0.18 mm. For the same resolution in the two cases, the accuracy of measurement in forward scatter is increased by a factor ~30.

Clearly, a study of the methods by which forward-scatter DOFS might be implemented must be worthwhile. Some of these methods will now be described.

## 6.4 STRATEGY FOR FORWARD-SCATTER DOFS

The first forward-scatter (F/S) method for DOFS to be studied relied on a Raman interaction between a counterpropagating pulse and a CW probe in a low birefringence fiber [10]. In this, the CW probe received Raman gain from the pulse, and this gain depended on the relative polarization states of the probe and the pulse, these, in turn, being dependent upon the strain distribution along the fiber (Fig. 6.6). As a demonstration of the basic principle of F/S DOFS this was satisfactory, but it was not developed further since the signal processing required to determine the strain distribution was too complex for practical implementation. However, polarization methods are to be preferred since they are both very sensitive and independent (to first order) of fiber attenuation. High birefringence (hi-bi) fibers are now quite readily available. Such fibers possess a natural linear birefringence which allows good control over the propagation state of the propagating light over long distances. Attention has recently concentrated on the use of these fibers for DOFS.

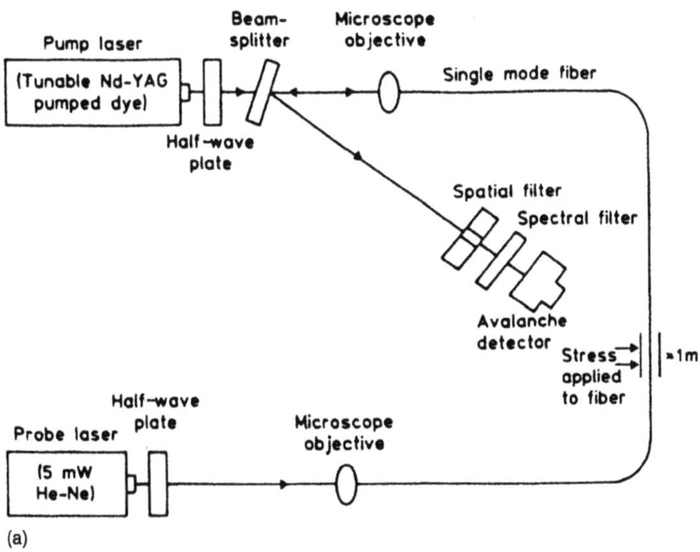


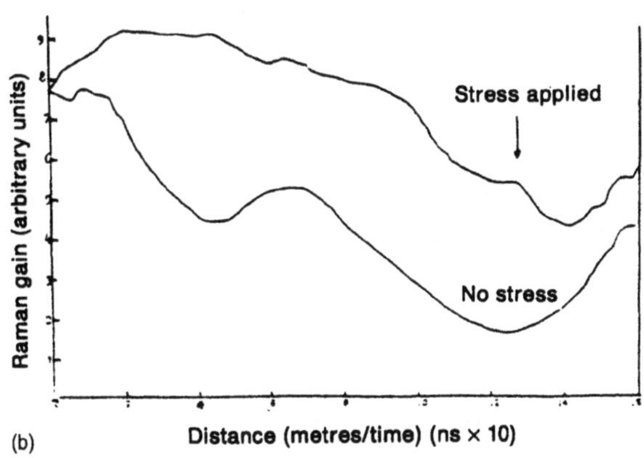


**Fig. 6.6** Raman-gain F/S DOFS. (a) Experimental arrangement. (b) Results for a stress field.

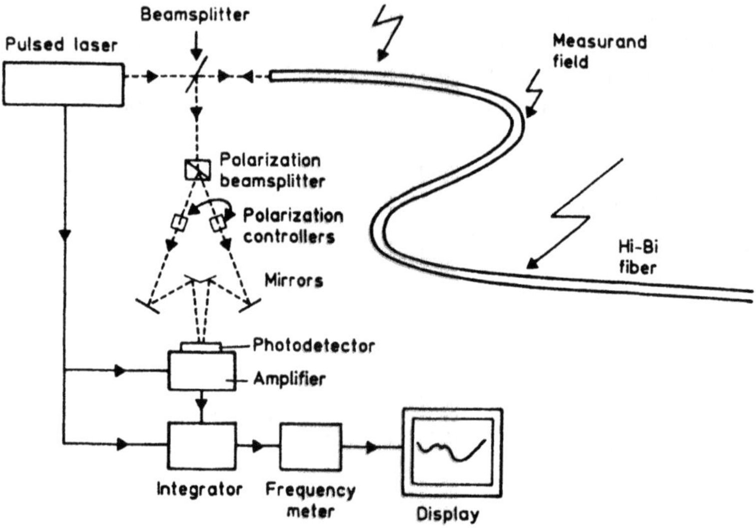


**Fig. 6.7** Schematic diagram for frequency-derived F/S DOFS.

The first attempt to use such fibers in DOFS involved a backscatter method [2, 11, 12] which was a variation on POTDR. In this method, the propagating pulse, equally disposed in each of the two hi-bi eigenmodes, accumulated a linearly increasing phase difference between its two modal components as it propagated, and this was doubled for backscatter from any given point. By mixing the two backscattered eigenmode signals on emergence, a frequency was generated, and the time-dependence of this frequency mapped the distribution of birefringence along the fiber, corresponding to the pulse's passage along it (Fig. 6.7). Any external field capable of modifying that birefringence was mapped correspondingly. Whilst this method worked encouragingly well, it still suffered from the fact that it was a backscatter method, and thoughts turned towards attempts to embody the same, or a similar, idea in a forward-scatter arrangement.

The first thoughts along these lines involved the use of 'resonant' coupling. If a pulse propagating in a hi-bi fiber could be arranged to impose a suitable periodic perturbation with a spatial period equal to the hi-bi beat length (distance over which the differing polarization mode group velocities impose a phase difference of $2\pi$ between mode components), then this would lead to resonant coupling, from one mode to the other, of a counterpropagating CW probe. The birefringence could then be mapped via the coupling efficiency (level of coupled light) as a function of time.

## 6.5 MODE COUPLING IN HI-BI FIBERS

The primary feature responsible for the polarization-holding property of hi-bi fiber is that of the difference in group velocities for the two eigenmodes. Any perturbation which acts to couple light between the modes therefore, in general, will couple components with phases which average the resulting amplitude to zero over one beat length. Since this latter is typically of the order of a few millimeters, only perturbations with spatial variations on this scale will cause significant coupling. Clearly, a periodic perturbation which has a spatial period equal to a beat length will couple optimally, since the points of maximum coupling will always lead to fully constructive interference. The most obvious coupling perturbation is a strain which, acting in a direction at a nonzero angle to the eigenaxes, will induce an extra birefringence and thus effectively rotate the eigenaxes. The result is a coupling of each of the pre-existing eigen-components into the new axes directions, a coupling which is not, in general, reversed when the perturbation ends.

These ideas may be summarily quantified as follows. Suppose that the complex amplitude components in the two orthogonal linearly polarized eigenmodes of a hi-bi fiber are

$$A_x = a_x \exp(i\beta_x z)$$
$$A_y = a_y \exp(i\beta_y z).$$

Suppose now that a perturbation acts so as to couple an amplitude fraction $k$ from mode $x$ to mode $y$ per unit length. It follows that a fraction $k^*$ (the complex conjugate of $k$) will be coupled from $y$ to $x$ (in order to conserve total power). Thus we may represent the coupling by means of a pair of coupled equations:

$$\frac{da_x}{dz} = ika_y(z)\exp(i\Delta\beta z)$$
$$\frac{da_y}{dz} = ika_x^*(z)\exp(-i\Delta\beta z),$$

where

$$\Delta\beta = \beta_y - \beta_x.$$

Clearly, in order to solve these equations it is necessary to prescribe the dependence of $k$ upon $z$. Suppose that $k$ is periodic, with period $L_p$, i.e.

$$k = k_0 \exp\left(\frac{i2\pi z}{L_p}\right)$$
$$k^* = k_0 \exp\left(\frac{-i2\pi z}{L_p}\right).$$

A solution of these equations now leads to the result that the fractional cross-coupled power over a distance $L$ of fiber is given by

$$\chi = k_0^2\left(\frac{\sin \alpha L}{\alpha}\right)^2, \tag{6.5}$$

where

$$\alpha = \left[k_0^2 + \left(\frac{\alpha\beta_0}{2}\right)^2\right]^{1/2}$$

with

$$\Delta\beta_0 = \beta_y - \beta_x - \frac{2\pi}{L_p} = 2\pi\left(\frac{1}{L_B} - \frac{1}{L_p}\right),$$

where $L_B$ is the intrinsic fiber beat length. From this we see that the coupling is maximized when $L_B = L_p$ ($\Delta\beta_0 = 0$), under which condition

$$\chi = \sin^2 k_0 L. \tag{6.6}$$

Further, it is easy to show that for an induced birefringence $B_I$ with axes at angle $\theta$ to the intrinsic axes,

$$K_0 = \frac{B_I}{\lambda_0}\sin 2\alpha,$$

and thus a measurement of coupled power as a function of position allows a measurement of the distribution of birefringence along the fiber, and thence also of any distributed field, such as stress, which modifies it.

## 6.6 KERR-EFFECT METHODS FOR IMPLEMENTING FORWARD-SCATTER DOFS

When an electric field is applied to an isotropic material (such as fused silica), linear birefringence is induced in the material, with the slow axis in the direction of the field and fast axis orthogonal to it. The birefringence is given by

$$B_k = \lambda_0 b E^2, \tag{6.7}$$

where $\lambda_0$ is the vacuum wavelength, $b$ the Kerr constant and $E$ the electric field.

In the optical Kerr effect, the electric field is that associated with an optical wave, so that the birefringence induced by an intense wave in a material may be probed via the effect it has on another wave, either co-propagating or counter-propagating.

Two F/S DOFS arrangements which use the optical Kerr effect have been explored.

### 6.6.1 Frequency-derived forward-scatter DOFS

Frequency-derived backscatter DOFS has already been described in Section 6.4 and has also been reported in the literature [e.g. 2, 11, 12]. The forward-scatter version of this idea uses the optical Kerr effect.

Suppose that an intense optical pulse is launched into a hi-bi fiber, with equal power in each of the two eigenmodes. As the pulse propagates, the two

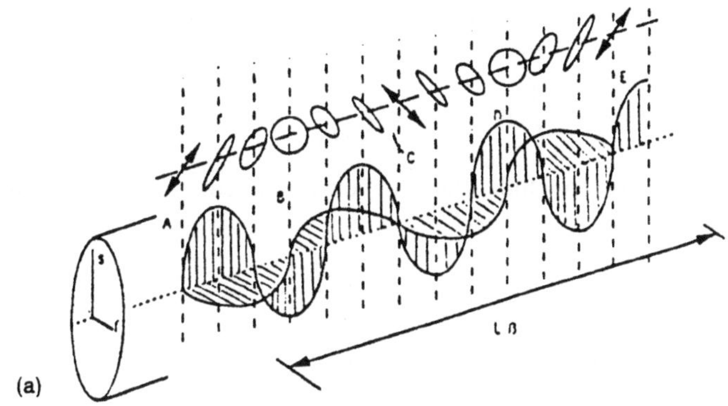


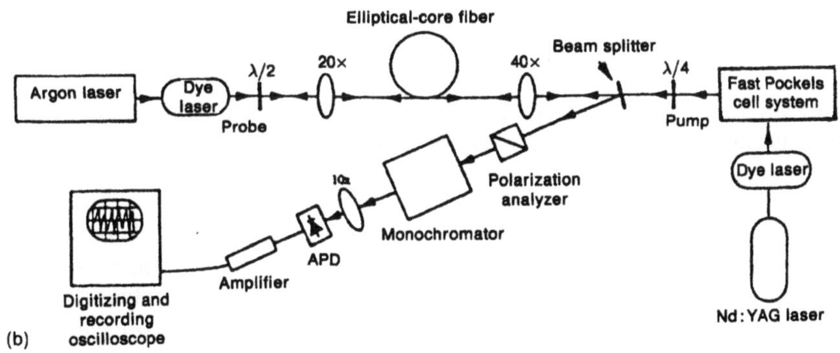


**Fig. 6.8** Optical Kerr-effect F/S DOFS. (a) Polarization-state progression down a hi-bi fiber. (b) Experimental arrangement.

components will come into phase once per beat length and into antiphase at the intervening half-beat lengths (Fig. 6.8(a)). The electric field of the resultant wave will thus maximize in a direction lying at ±45° to the intrinsic axes, every alternate half-beat length. The Kerr-effect-induced birefringence resulting from this field will consequently cause the resultant birefringence axes to rock around the intrinsic direction with a spatial period equal to the beat length.

Now consider a CW with the same wavelength as the pulse launched, again equally into the two eigenmodes, in the opposite direction to the pulse. When it encounters the pulse, the Kerr-induced rocking will produce optimal phase-matched coupling (since the rocking occurs naturally at the 'resonant' frequency), as a result of which probe power will be coupled from one mode to the other. When pump and probe are at the same wavelength there will be an exact correspondence in beat length, which will maintain the coupled power constant as the pulse propagates down a fiber with uniformly distributed birefringence. However, if pump and probe are at different wavelengths, the differing beat lengths will lead to the condition that optimum coupling will occur only when the rocking axes can couple in-phase probe components, i.e. with an effective beat-length given by $L_e$, where

$$\frac{1}{L_e} = \frac{1}{L_1} - \frac{1}{L_2}; \qquad L_e = \frac{\lambda_1 \lambda_2}{\Delta\lambda B_1} \qquad (\Delta\lambda = \lambda_1 - \lambda_2).$$

From this we may conclude that the optimum coupling occurs at a 'beat' frequency given by

$$f_D = cB_1 \times \frac{\Delta\lambda}{\lambda_1 \lambda_2}, \tag{6.8}$$

where $\lambda_1$ is the wavelength of the pulse light, $\lambda_2$ that of the probe light, $L_1$, $L_2$ the respective beat lengths, $c$ the velocity of light in the fiber and $B_I$ the intrinsic birefringence of the fiber. For small coupling, the power coupling efficiency, $k^2$, is governed by [13]

$$k^2 = \{\sigma^{2/}[\sigma^2 + (\theta/2)^2]\} \sin^2(L_{int}[\sigma^2 + (\theta/2)^2]^{1/2}), \tag{6.9}$$

where $\theta = (2\pi/L_e)$ is the dephasing parameter, and $L_{int}$ is the interaction length. Also, the coupling coefficient $\sigma$ is given by [14]

$$\sigma = \{\pi B_k / S\lambda_1\}, \tag{6.10}$$

where $B_k$ is the optical Kerr-induced birefringence. As the wavelength of the pump is shifted with respect to the probe, a phase mismatch will appear between the coupled wave elements along the fiber. As a result, the net coupling would be reduced compared with the case where both sources have the same wavelength, as expected from equation 6.9. The pump power cannot be raised indefinitely to compensate for this, because of a limitation on the peak pump power in the fiber due to the onset of spectral broadening effects (e.g. Raman, Brillouin), which would further reduce the coupling efficiency. A large wavelength shift between the pump and probe beams therefore cannot be achieved without paying a substantial penalty in terms of the coupling efficiency. Equation 6.8 shows that this, in turn, limits the tuning range of the derived probe frequency. To avoid this problem, the pump pulse should be short compared with the spatial beat length between the polarization states of the pump and probe beams, in order to avoid the 'smoothing' effect caused by the probe pulse embracing many beat lengths, i.e. to keep $L_{int}/L_e$ (in equation 6.9) small. The experimental setup is illustrated in Fig. 6.8(b). The fiber used was 24 m long with an elliptical core (supplied by Andrew Corporation) and was single mode at 633 nm, where it exhibited a beat-length of 4 mm. The pump pulses were generated by a dye laser pumped by a Q-switched, frequency-doubled Nd:YAG laser. Pulses of 8 ns full-wave half-maximum (FWHM) duration were generated with a repetition rate of 50 Hz. To allow useful differences between the pump and probe wavelengths, the duration of the pump pulse was required to be shorter than the pulse width actually available from the dye laser. Accordingly, a fast optical shutter was used to provide a sufficiently short pulse. The pump beam was chopped by a fast Pockels cell to provide pulses of 2 ns (FWHM), and was circularly polarized by the quarter-wave plate so as to excite equally the two eigenmodes of the fiber. The probe was provided by a CW Argon-pumped dye laser which was counter-propagating with respect to the pump, and linearly polarized at a small angle to the fiber's birefringence axes, with a launched power of ~300 mW. In principle, the strongest probe signal should be provided when the eigenmodes are equally excited by a launched polarization set at 45° to the fiber's principal axes. The signals would then be superimposed on a high mean received power. However, an avalanche photodiode detector was used, and this rendered high mean signal levels undesirable. After reflection from a beamsplitter at near-normal incidence, the probe beam was analyzed with a Glan Thompson analyzer. A monochromator was placed before the avalanche photodiode detector in order to block the large pulse predominantly produced by the front face reflection from the fiber. This pulse would otherwise have saturated the input amplifier. The pump and probe beams were set at various wavelength differences up to 2.4 nm apart, generating derived signal frequencies up to ~200 MHz. The signals were recorded and analyzed using a computer-interfaced digital storage oscilloscope, which was used to average the results from 256 pump pulses. The averaging process occupied approximately 10 s.

For the length of fiber used in the experiment, the spectrum began to broaden when the first stimulated Raman line was generated. This occurred at a pulse energy of about 36 nJ. The experimentally achieved coupling efficiency was about 1%. The variation of the derived frequency with wavelength offset was measured as follows. The pump and probe were initially set at the same wavelength of 646 nm. Then, by turning the pump wavelength with respect to the probe, the required wavelength shift was set. An example of a derived frequency signal averaged over 256 pump pulses is shown in Fig. 6.9. The wavelength shift between the pump and probe in this measurement is 1.5 nm, and the corresponding derived frequency is 71 MHz. The measured signal-to-noise ratio was 19.3 dB. The experimentally observed and theoretically calculated relationship between the frequency and wavelength shift are shown in Fig. 6.10. This figure shows

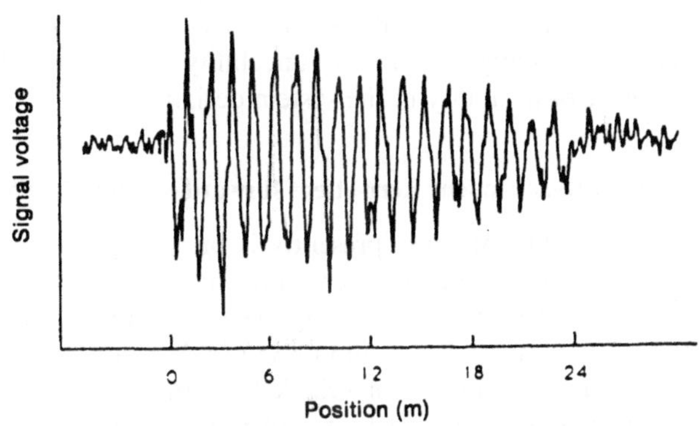


**Fig. 6.9** Derived-frequency signal from Kerr-effect F/S DOFS.

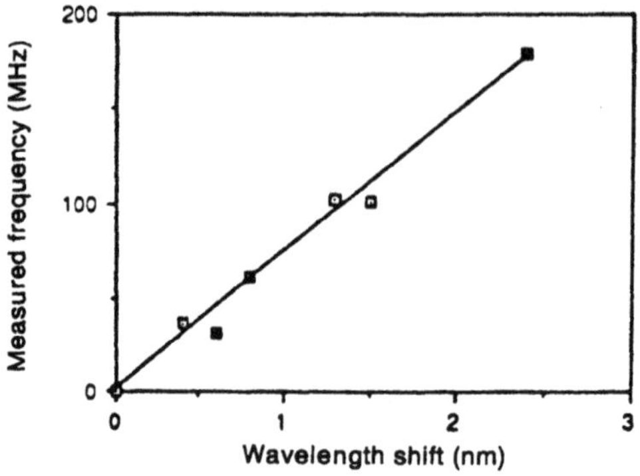


**Fig. 6.10** Derived frequency *vs.* wavelength difference between pump and probe.

the expected linear trend connecting the derived frequency with the wavelength shift. There was good agreement between the theoretical prediction represented by the line in the figure and the experimental values represented by the points when the wavelength of the probe was longer than that of the pump, but this relationship did not hold for the reverse case. Further tests made for the reverse case also showed a linear trend, but with a different slope. This may be due to systematic experimental error, but this remains to be confirmed. The tuneable frequency range offering strong signals was limited. This agrees closely with the theory given by equation 6.9, in that, for a 2-ns pump pulse, the first zero of the coupling response was expected to occur at a wavelength offset of $\sim$3.2 nm, and from equation 6.8 the equivalent signal frequency would then be $\sim$200 MHz.

Using the novel technique presented here, the spatial variation of the birefringence of a polarization maintaining fiber can be measured remotely in a short time, and since the signal is in the form of a frequency, it is immune from the common error sources present in intensity-coded systems. Any change in the applied stress or ambient temperature will tend to change the local birefringence and will therefore be indicated by a shift in the frequency due to the external perturbation. Thus, this technique may be applied to implement a distributed optical fiber sensor. In this application, the tuneability of the derived frequency could provide some extra flexibility in the system design.

### 6.6.2 Polarization-state-dependent Kerr-effect forward-scatter DOFS

In this second method, the emphasis presently is on the spatial location of a perturbation rather than on the measurement of its magnitude, although suitable processing is capable, in principle, of revealing the latter. In its present form it is capable of providing good spatial resolution and rapid response for application to, for example, intrusion monitoring or vehicle location.

The optical arrangement employs a length of polarization-maintaining fiber carrying two counterpropagating beams. A CW probe beam is launched from one end of the fiber so as to excite the two eigenmodes equally, and the polarization state of this beam is detected at the far end of the fiber by means of a beamsplitter and analyzer oriented at 45° to the birefringence axes. An intense, pulsed, pump beam is launched on one of the birefringence axes. This arrangement is similar to the well-known Kerr-effect shutter system [15].

As in the Kerr shutter, the pump pulse causes a phase shift between the eigenmodes of the probe beam, leading to a change in the output polarization state of the probe. This is detected as a sharp change in the probe intensity passed by the analyzer when the pump is initially launched into the fiber.

If, now, a force acts at an angle to the axes along a section of fiber, coupling of the pump light to the other axis will occur, and the Kerr effect on the probe will thus be modified. The probe light itself will also experience mode coupling, which will further modify the output polarization state. The actual change which occurs will depend, *inter alia*, on the states of polarization of the beams as they enter the perturbed region and thus, unless the birefringence perturbation is very small compared with the intrinsic birefringence, there will exist a mutual dependence of effects from different measurement locations, which only fairly complex signal processing would be able to resolve. However, it is clear that for any change in the direction of birefringence axes consequent upon the perturbation by a measurand, there will, in general, result a change in optical Kerr effect. A differentiated signal will thus at least indicate differential features of the measurand distribution, even though a fully quantified spatial distribution is more elusive.

The fiber used in the experiments [16] was a mono-mode high-birefringence fiber, manufactured by Andrew Corporation, with a diameter of 67 μm, attenuation of 35 dB km$^{-1}$ at 633 nm and core-to-cladding refractive index difference $\Delta n = 0.032$. The length of the fiber was about 100 m. The arrangement of the experiment is shown in Fig. 6.11. Pump pulses (617 nm) of 8 ns (FWHM)

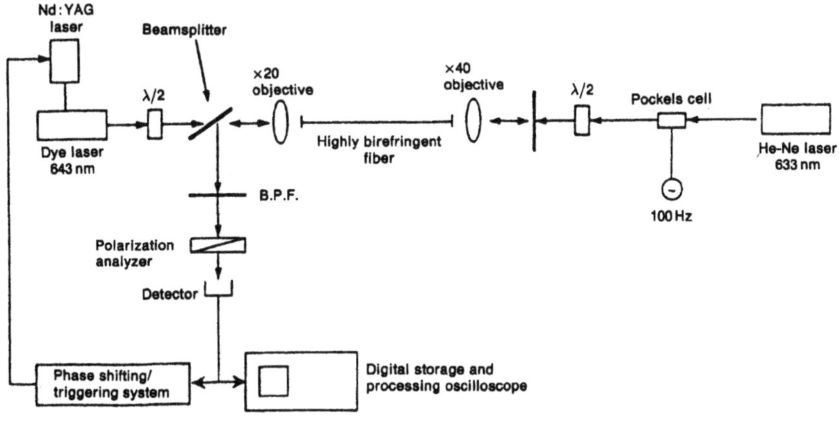


**Fig. 6.11** Experimental arrangement for polarization-state-dependent Kerr-effect F/S DOFS.

duration were generated in a dye laser with a repetition rate of 50 Hz. These pump pulses were launched on to one of the birefringence axes of the fiber with the help of a halfwave plate and with a peak power of 3 W measured at the output end of the fiber. The linearly polarized probe beam of wavelength 633 nm, from a He–Ne laser, was launched into the fiber at 45° to the birefringence axes. On emergence, the probe beam was directed by a beam splitter to the detector via a polarization analyser; its average power at the detector was about 25 μW. The He–Ne laser and the detector were protected from the pump light by use of band-pass filters at 633 nm. Force was exerted by pressing metal rods on the fiber. The received signals were recorded, averaged and differentiated using the functions of a digital storage oscilloscope.

In the absence of any measurand-induced perturbation, the Kerr effect of the pulse is to modify the local value of birefringence as it propagates, a modification which is sensed by the probe beam as a phase shift between the eigenaxes, and which is, in principle, constant for the duration of the pulse's passage through the fiber. The effect of this phase shift on the optical signal passing through the analyzer is shown in Fig. 6.12. In practice, the slow fall in the value of the phase shift is due to the attenuation of the pump pulse with distance along the fiber. Figure 6.12 also shows the effect of differentiating this signal with respect to distance.

Figure 6.13 shows the fluctuating analyzer signal when the fiber was perturbed at two points, and its differential with respect to distance. The points at which the weights were applied are clearly evident. Such a system, even as it stands, could be used as an intruder alarm or as an indication of anomalous disturbance of almost any kind.

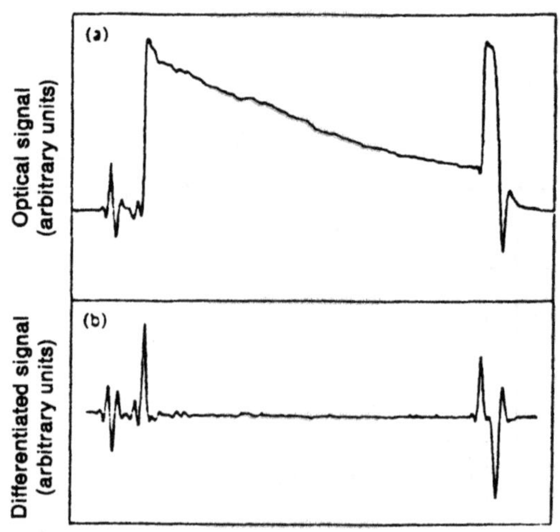


**Fig. 6.12** Kerr-effect phase shift for counterpropagating pulse and probe.

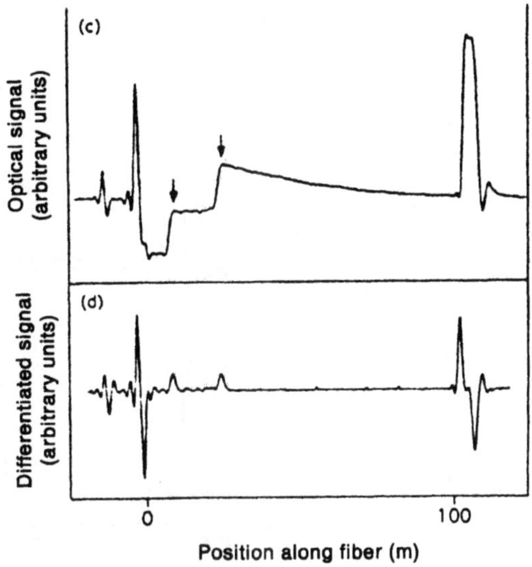


**Fig. 6.13** Effect of stress perturbations on Kerr-effect phase shift.

A further development is shown in Fig. 6.14. In order to guard against the possibility of the detector system possessing low sensitivity to the polarization change brought about by the perturbation (by virtue of the polarization bias from the emerging state), diversity polarization detection is employed: three different detectors are employed, each with a different inserted bias, corresponding to an effective measurement of the Stokes parameters [7] for the emerging polarization state. The processing then takes the modulus of each detected signal and adds them to produce a final output, as shown in Fig. 6.15.

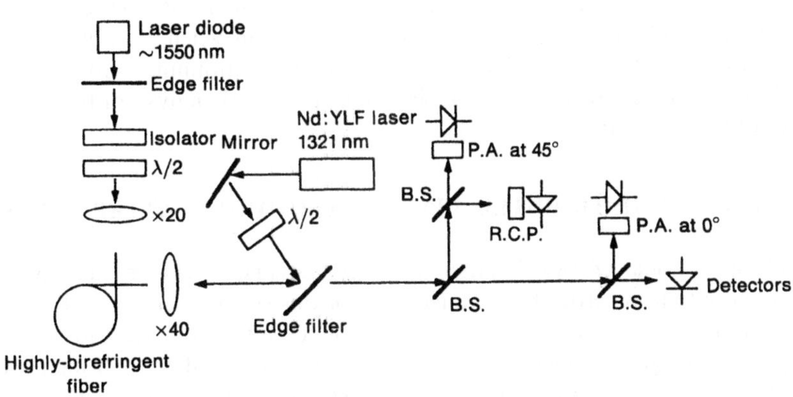


**Fig. 6.14** Experimental arrangement for polarization-diversity detection.

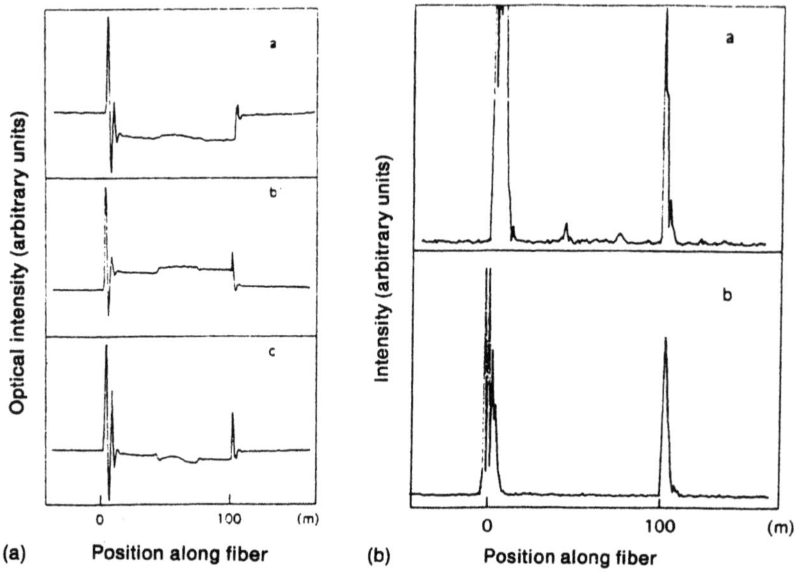


**Fig. 6.15** Perturbation signals for polarization diversity detection. (a) Individual detector outputs. The output of detectors with: a, right-circular polarizers; b, 0° linear polarizers; c, 45° linear polarizers. (b) Summed moduli signals.

Of course, it is true that if the perturbing force acts along one of the eigenaxes, no rotation of the axes occurs and there is no resultant polarization perturbation. Hence, either the direction of the perturbing force must be known (e.g. a vertical weight), or two fibers may be used; these should run closely parallel and be oriented with their birefringence axes at 45° to each other.

This section has demonstrated the use of the optical Kerr effect to determine the locations of discrete mode coupling points spaced along a polarization-maintaining fiber. Differentiation of the received signal with respect to time provides a simple way to reduce confusing interactions when multiple coupling points are present. Further work on this method is continuing, with a view to developing practical real-time processing and theoretical analysis of the potentialities of the technique.

## 6.7 A QUASI-DISTRIBUTED LINEAR FORWARD-SCATTER SYSTEM

Finally, we consider a forward-scatter system of a rather different type: it is optically linear and interferometric. The arrangement for this system is shown in Fig. 6.16. Each of the discrete sensors is a Mach–Zehnder interferometer with its own unique path-length difference, and with each path-length difference being greater than the coherence length of the source [17]. The returning signal is demodulated by means of a set of corresponding Mach–Zehnder interferometers, each of

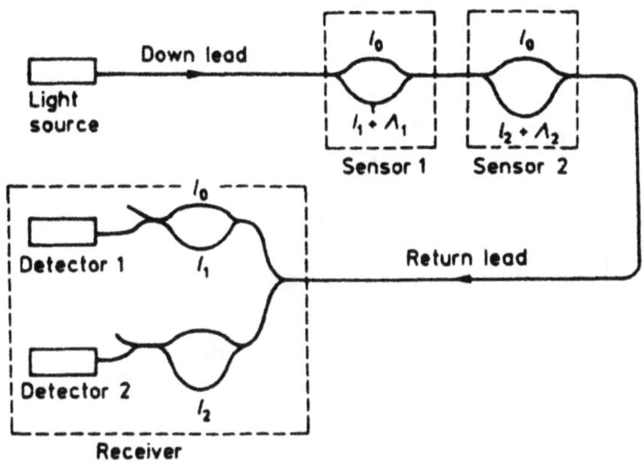


**Fig. 6.16** Forward-scatter quasi-distributed Mach–Zehnder interferometers.

which has the effect of cancelling the path delay for one particular measuring transducer. The cancellation of the delay allows the optical signals from the two arms to interfere, and the interference pattern is sensed with the aid of a photodiode. Any small variation in the path-length difference due to the action of the measurand will be registered as a change of photodiode output. Each detector thus registers an output only from its corresponding transducer. An alternative detection scheme is to use one scanning, demodulating Mach–Zehnder interferometer.

This is an attractive scheme in that its performance is independent of the lead fiber characteristics and their variations. However, the light is subject to multiple splitting functions, and the coherence length of the source has to be very short in order to avoid ambiguity resulting from one or more of the many possible paths when more than about three sensors are used. In turn, a very short coherence length requires a high stability of path difference in the demodulating interferometer, and thus the system becomes somewhat temperamental. The low coherence interferometry concept is a valuable one, however, and several other arrangements which use it are under investigation.

One of these is illustrated in Fig. 6.17. This is referred to as electronically scanned white light interferometry [18–20]. The pressure points cause coupling

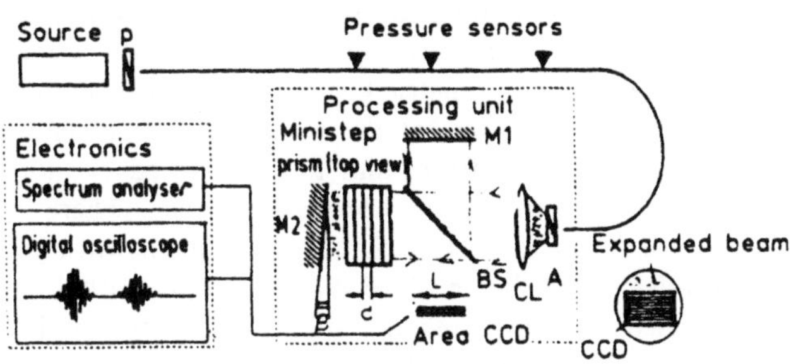


**Fig. 6.17** Electronically scanned white-light interferometer.

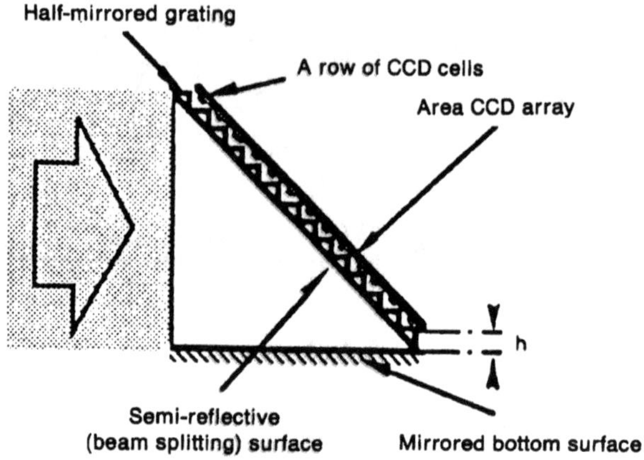


**Fig. 6.18** Stepped-reflector CCD interferometer.

of the low coherence light from one eigenmode of a hi-bi fiber to the other. A phase delay is thus inserted between these modes, of an amount which depends on the distance of the coupling point to the exit end of the fiber. The emergent light is passed into a Michelson interferometer which consists of a stepped reflector (Fig. 6.18) in the reference path and which feeds the interference pattern on to a CCD array. The effect is to spread the correlogram across the CCD, which then gives an output corresponding to the distribution of coupling strain.

This approach to low-coherence interferometry could also be useful in a fully distributed (as opposed to quasi-distributed) arrangement, with suitable electronic processing.

## 6.8 CONCLUSIONS

Distributed optical-fiber measurement systems offer an extra level of information gathering for large structures. The potential for applications in civil engineering, aerospace engineering, mining, petrochemical engineering, etc., and in research diagnostics, is immense.

In order to achieve high spatial resolution with good sensitivity and relatively simple signal processing, it is very fruitful to look towards systems which use forward-scatter techniques, polarimetric techniques, or both simultaneously.

This chapter has indicated some of the possible arrangements of this type which are under active investigation, and has illustrated the kind of system which might prove viable in certain application areas. There is continuing development of these ideas. It is clear that there is much work yet to be done before such methods can be shown to be viable for operational measurements. In particular, it is clear that great reliance will, of necessity, be placed upon the fabrication of special fibers for enhancement and control of nonlinear optical interactions, for control of the polarization state and for optimized interfacing with the measurand field. The potential pay-offs in terms of a generic range of fully distributed optical-fiber sensors capable of providing for the majority of requirements in industrial measurement (in addition to allowing access to an extra level of information) are deemed sufficient to make the research investment conspicuously worthwhile.

## ACKNOWLEDGMENTS

The author is very grateful for the contributions to this work made by V. A. Handerek, S. U. Ahmed, F. A. Parvaneh, I. Cokgor and L. C. G. Valente. Funding for various parts of the research has been provided by the Royal Society of London, the Engineering and Physical Sciences Research Council (EPSRC) and the UK Optical Sensors Collaborative Association (OSCA), and this is gratefully acknowledged.

## REFERENCES

1. Rogers, A. J. (1986) Distributed optical fibre sensors, *J. Phys. D., Appl. Phys.*, **19**, 2237–2255.
2. Rogers, A. J. (1988) Distributed optical-fibre sensors for measurement of pressure, strain and temperature, *Phys. Rep. (Phys. Lett.)*, **169**, 99–143.
3. Dakin, J. P. (1987) Multiplexed and distributed optical-fibre sensor systems, *J. Phys. E.*, **20**, 954–967.
4. Culshaw, B. (1986) Distributed and multiplexed fibre-optic sensor systems, *Proc. NATO Adv. Study. Inst.*, Erice (Italy), Nijhoff, Erice.
5. Rogers, A. J. (1981) Polarization-optical time domain reflectometry, *Appl. Opt.*, **20**, 1060–1074.
6. Rogers, A. J. (1980) Polarization-optical time domain reflectometry, *Electron. Lett.*, **16**, 489–490.
7. Born, M. and Wolf, E. (1959) *Principles of Optics*, Pergamon Press, New York, Sections 1–4.
8. Ross, J. N. (1981) Birefringence measurements in optical fibres by POTDR, *Appl. Opt.*, **21**, 3489–3495.
9. Clark-Jones, R. (1948) A new calculus for the treatment of optical systems, *J.O.S.A.*, **38**, 671–685.
10. Farries, M. C. and Rogers, A. J. (1984) Distributed sensing using stimulated Raman action in a monomode optical fibre, *Proc. 2nd Int. Conf. Optical-Fibre Sensors*, Stuttgart, pp. 121–132.
11. Handerek, V. A., Parvaneh, F. and Rogers A. J. (1991) Frequency-derived distributed optical-fibre sensing: Single frequency downshifting, *Electron. Lett.*, **27**, 394–396.
12. Rogers, A. J. and Handerek, V. A. (1992) Frequency-derived distributed optical-fibre sensing: Rayleigh backscatter analysis, *Appl. Opt.*, **31**, 4091–4095.
13. Ahmed, S. U., Handerek, V. A. and Rogers, A. J. (1992) Phase-matched polarization coupling in high-birefringence fibres through the optical Kerr effect, *Opt. Lett.*, **17**, 643–645.

14. Parvaneh, F., Valente, L. C. G., Handerek, V. A. and Rogers, A. J. (1992) Forward-scatter frequency-derived distributed optical-fibre sensing using the optical Kerr effect, *Electron. Lett.*, **28**, 1080–1082.
15. Dziedzic, S. M., Stolen, R. and Askhin, A. (1981) Optical Kerr effect in long fibres, *Appl. Opt.*, **20**, 403–406.
16. Handerek, V. A., Rogers, A. J. and Cokgor, I., Detection of localized polarization mode coupling using the optical Kerr effect, *Proc. 8th Int. Conf. Optical Fibre Sensors* (OFS 8), Monterey, CA, January 1992, pp. 250–253.
17. Brooks, J. L. (1983) Coherence multiplexing of fibre-optic interferometric sensors, *JLT*, LT3, 1062–1072.
18. Chen, S., Meggitt, B. T. and Rogers, A. J. (1990) Novel electronic scanner for coherence multiplexing in a quasi-distributed pressure sensor, *Electron. Lett.*, **26**, 1367–1369.
19. Gerges, A. S., Farahi, F., Newson, T. P., Jones, J. D. C. and Jackson, D. A. (1987) Interferometric fibre-optic sensor using a short-coherence-length source, *Electron. Lett.*, **23**, 1110–1111.
20. Gerges, A. S., Farahi, F., Newson, T. P., Jones, J. D. C. and Jackson, D. A. (1988) Fibre-optic interferometric sensor using low-coherence-length source-resolution enhancement, *Electron. Lett.*, **24**, 472–474.

# 7

# Optical current sensor technology

K. T. V. Grattan and Y. N. Ning

## 7.1 INTRODUCTION

The principles of optical and optical fiber current sensor technology have been known for some considerable time, and some of the earliest papers on optical fibre measurement techniques have considered this topic. The general advantages of the use of optical technology were discussed by Rogers [1] in an earlier volume, in which the essential principles of the methods available and a description of some of the essential technologies were described. This builds upon that introduction, and discusses in some detail the optical current sensor devices and technology advances which have been developed in recent years. Optical current sensors (OCSs) show several important features when compared with conventional current transformers (CTs), such as their having highly effective isolation from high line potentials offered by the dielectric nature of the optical fibers, freedom from the saturation effect which may be observed in conventional transformers, the potential to make measurements in high voltage and/or high magnetic induction noise fields, a high linear response over a wide frequency bandwidth, a remote, high-speed measurement capability for monitoring or metering purposes, and the fact that they are compact and light-weight measuring devices, available at potentially low cost.

Considerable effort has been devoted to the exploitation of different types of optical current sensing mechanisms (such as the Faraday effect, and magnetostriction), and various materials (such as optical fibers, bulk glass or yttrium iron garnet (YIG) materials), in order to investigate the implementation of high-performance OCSs which can be used to replace the CT, and thus may be deployed in the presence of electrical noise or in the very high-potential fields experienced in the power distribution industries [2]. In general, the device measures the integral of the magnetic field along a closed (or nearly closed) optical loop around the current to be measured, and of an optical fiber link which connects the sensing element with an opto-electronic control and processing unit, which is used to launch a light beam into the optical fiber and detect and decode the modulated optical signal collected from the fiber. As a result of its intrinsic features of high electrical insulation and immunity to electro-magnetic

*Optical Fiber Sensor Technology*, Vol. 3. Edited by K. T. V. Grattan and B. T. Meggitt.
Published in 1998 by Kluwer Academic Publishers, London. ISBN 978-1-4419-4736-9

interference (EMI), optical fiber has been a strong candidate to form the basis of current sensing elements, and the development of optical fiber current sensors has been the subject of many research programmes. Although a current measurement scheme based upon the use of an optical fiber as the sensing element was proposed as early as in 1973 [3], and a number of new methods which employ optical fiber sensing elements have been explored and reported, a truly viable or widely accepted commercial product has not yet emerged, despite the considerable level of research that has been devoted to the concept. It is well known that the major difficulty in using optical fiber sensing elements is caused by the presence of various sources of birefringence *inside* the sensing fiber. When the value of the Faraday rotation or the current-induced azimuth variation, which is governed by the Verdet constant of the optical fiber, is comparable to that of the azimuth caused by the birefringence within the fiber, the former tends to be quenched by the latter. As the Verdet constant of an ordinary silica fiber is very low ($4.68 \times 10^{-6}$ rad $A^{-1}$ at 633 nm), in order to obtain high sensitivity, a fiber optic sensing element requires a large number of fiber turns, which will further increase both the intrinsic birefringence and the bend-induced linear birefringence, making the sensitivity considerably less than that predicted by the theory. The output of the sensor is also subject to error due to the position of the current-carrying conductor inside the sensing element itself and the presence of the fields induced by other currents, which are not part of the measurement process [4]. The environmental sensitivity of the device, caused by the fiber birefringence increase, is a further complication. As a result, in order to improve the performance of this type of sensor, it is important to minimize the effects of all sources of birefringence in order to design a practical optical fiber current sensor.

This chapter mirrors and expands somewhat upon a recent review by Ning *et al.* [5] on progress in optical current sensing techniques, illustrating the diversity and wide applicability of such techniques for current monitoring. Taking into account the sensing mechanisms employed and the sensing materials used, optical current measurement devices may be categorized into four main groups: (i) OCSs employing optical fiber as their sensing element; (ii) OCSs using bulk glass to sense the current; (iii) OCSs using electro-optic hybrid devices; (iv) OCSs using magnetic field sensing devices. In general, the operational principle of the first two groups of sensors is based upon the so-called 'Faraday magneto-optic effect' [6], whilst the third group of sensors is often constructed from a conventional CT, interrogated using different electro-optic or fiber optic sensing schemes. Optical-fiber current sensors forming this group have been intensively discussed and reviewed by a number of authors [2, 5, 7, 8]. Hence, in this chapter, the possible solutions for dealing with the stability problems associated with bend-induced linear birefringence in optical fiber current sensors will be summarized (Section 7.3), and recent developments in the area of current measurement using bulk optic glasses (Section 7.4) and electro-optic techniques (Section 7.5) will also be reviewed and discussed. Summary data on the performance of these sensors are given in Table 7.1.

**Table 7.1** Summary of some results, with literature reference

| Sensor type | Size (mm) | Bandwidth (KHz) | Dynamic range (KA)[a] | Sensitivity (rad $A^{-1}$) | Temperature range (°C) | % error in T (A per °C) | Resolution (A per $\sqrt{Hz}$) | Ref No. |
|---|---|---|---|---|---|---|---|---|
| Fiber sensing element | | | | | | | | |
| SEP fiber | $\phi 25$ | 1 | ±1.5 | 69.25±0.3% of $V_{ideal}$ | 20–70 | ~0.05 | 0.001 | 19 |
| Annealing coil | $\phi 5$–$\phi 10$ | | | | −20–120 | $1.7 \times 10^{-4}$ | | 22 |
| Separation of $\phi_F$ and $\delta$ | $\phi 220$ | 4 | 8 | $5.76 \times 10^{-6}$ | | | | 23 |
| Square topology | 50 × 50 | 33 | 2.5 | 83.3% of $V_{ideal}$ | 5–10 | | | 24 |
| Using an OCR | $\phi 250$ | | | $4.1 \times 10^{-6}$ | | | | 27 |
| Bulk-glass sensing element | | | | | | | | |
| Square-shaped | 98 × 98 × 33 | | 0.2–2 | | −20–90 | ±0.003 | | 34 |
| Triangular-shaped | 300 × 250 | 10 | $3 \times 10^{-5}$–3 | $2.35 \times 10^{-5}$ | | | 0.02 | 11 |
| Ring-shaped | $\phi 80$ | 10 | 1–600 | $3.75 \times 10^{-4}$ | | | 1 | 10 |
| Openable | 89 × 75 | 11 | $4 \times 10^{-4}$–1.5 | $2.21 \times 10^{-5}$ | | | 0.007 | 38 |
| Hybrid sensing element | | | | | | | | |
| Phase modulation | | 10 | 1–140 | | | | 0.086 | 43 |
| Intensity modulation | | | $6 \times 10^{-4}$–1.1 | | 20–44 | ~0.5 | 0.45 | 44 |
| Frequency modulation | | | $1 \times 10^{-3}$–1.5 | | | | 0.32 | 44 |
| Polarization modulation | | | 3 | $5.3 \times 10^{-6}$[b] | | | | 42 |
| Chromatic modulation | | 10 | 0.2 | 0.45 nm $A^{-1}$ | | ~0.5 | | 49 |

[a]The maximum currents reported for several of the sensors described here were limited by the local current supply rather than by the sensor itself.
[b]Interferometric sensitivity.

By measuring the magnetic field at a location near the current-carrying conductor, it is also possible to monitor or determine the current if the system has previously been calibrated. Hence, a number of magnetic sensing schemes may be used directly or indirectly for the purpose of current measurement. The details of these schemes will be discussed and summarized.

## 7.2 FARADAY EFFECT AND ASSOCIATED DETECTION SCHEMES

### 7.2.1 The Faraday effect

In the Faraday magneto-optic effect, the polarization azimuth, $\Phi_F$, of a linearly polarized light beam propagating inside an optical material is rotated under the influence of a magnetic field, $H$, generated by the electrical current to be measured, and this may be given by

$$\Phi_F = \int_l V\bar{H}\,\mathrm{d}\bar{l} \tag{7.1}$$

where $V$ is the Verdet constant of the optical material and $l$ is the interaction length. When the optical path forms a closed loop, equation 7.1 then becomes

$$\Phi_F = V\oint \bar{H}\,\mathrm{d}\bar{l}. \tag{7.2}$$

If the medium is subject to the magnetic field within a long solenoid, this expression then becomes

$$\Phi_F = VN_l N_c I \tag{7.3}$$

where $N_l$ and $N_c$ are the number of passes of the light through the solenoid and the number of turns of wire in the solenoid, respectively, and $I$ is the electrical current to be measured. Although equation 7.2 has the same form as Ampere's law, i.e. the integral of the magnetic field along a closed path around a conductor is dependent only on the current in the conductor, it only is valid under the condition that the state of polarization (SOP) of a linearly polarized light beam along the optical path is preserved: that is, when the linear polarized light propagates inside the sensing element, its azimuth angle has to be maintained unchanged unless a current flows through the center of the sensor itself. Otherwise, different sections of the optical path inside the sensing element will each have different sensitivities. As a result, the overall sensitivity of the device may be modulated by the position of the current-carrying conductor with respect to the sensing optical device and the presence of the fields induced by other currents [5].

The types of OCS which use bulk glass as the current-sensitive device do not suffer the problems associated with the presence of intrinsic birefringence and the bend-induced linear birefringence inside an optical fiber sensing element; however, a problem does arise in the internal reflections which are required to allow the light beam to encircle the current-carrying conductor. This is the so-called 'reflection-induced phase difference' problem. According to electromagnetic theory,

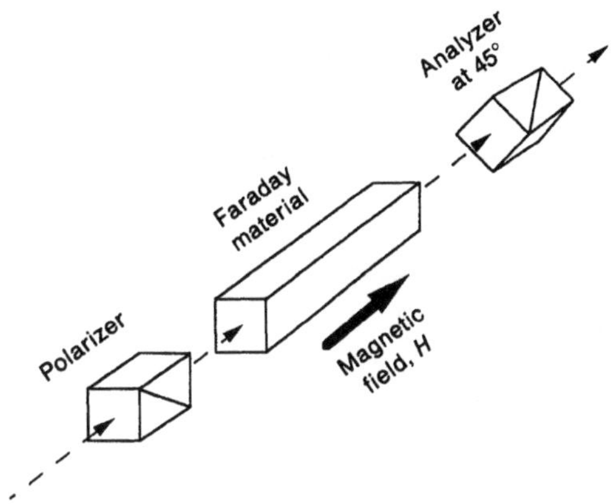


**Fig. 7.1** A basic polarimetric detection scheme.

when linearly polarized light is reflected at an angle of incidence which is larger than the critical angle, there will be a phase difference between the two orthogonal components of the linearly polarized light after the reflection, and the SOP of the light will vary. This variation will affect the sensitivity of the device in a similar way as does the birefringence in an optical fiber sensing element. Hence, finding a method to reduce this reflection-induced phase difference becomes a key issue in designing this type of sensor effectively. In the past few years, a number of sensor schemes have been proposed and reported to try to overcome this problem [9–11], a detailed discussion of which is in Section 7.4.

### 7.2.2 Basic polarimetric detection scheme

In order to measure the rotation of the azimuth, $\Phi_F$, of the output light from an optical Faraday current sensor, as shown in equation 7.1, a considerable number of different detection schemes have been proposed and reported. The arrangement of the basic detection scheme is illustrated in Fig. 7.1, where the difference between the transmission axes of the polarizer and analyzer is 45°, and hence the intensity of the light which is detected after propagating through the system is given by

$$I = (I_0/2)(1 + \sin 2\Phi_F), \tag{7.4}$$

where $I_0$ is the input intensity of the light.

In order to eliminate the above dependence on the light intensity, the final output can be obtained by dividing the a.c. component of equation 7.4 by the d.c. component, and $S$, the ratio, can be given by

$$S = \sin 2\Phi_F. \tag{7.5}$$

Since the above result is obtained without considering the effect of the birefringence within the sensing material, some errors due to this effect may be introduced in practice.

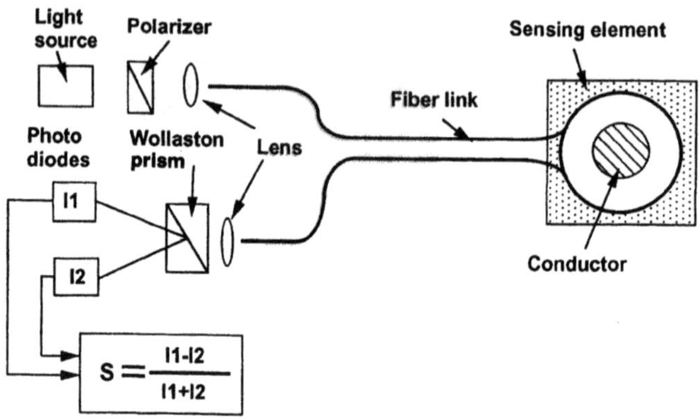


**Fig. 7.2** Schematic diagram of dual quadrature polarimetric detection.

### 7.2.3 Dual quadrature polarimetric detection scheme

One of the most commonly used detection schemes is the so-called dual quadrature polarimetric configuration [11], as shown in Fig. 7.2, in which the output light from the current sensing element is divided into two orthogonal linearly polarized beams, through a Wollaston prism. The intensities of the two beams are detected separately, and the two output signals may be fed into an analogue electronic circuit which computes the output function

$$S = (I_1 - I_2)/(I_1 + I_2), \tag{7.6}$$

where $I_1$ and $I_2$ are the intensities of the two beams. After the Wollaston prism is aligned so as to make the output, $S$, fall to zero when there is no magnetic field present, the system can be used to measure the current-induced Faraday rotation, $\Phi_F$, and equation 7.6 becomes

$$S = \sin 2\Phi_F. \tag{7.7}$$

When the Faraday rotation is relatively small, the output signal obtained is directly proportional to the value of the rotation of the azimuth of the linear polarized output light.

If there is a birefringence effect, $\delta$, due to the sensing element itself, the output signal from the detection system discussed above becomes [7]

$$S = 2\Phi_F \frac{\sin\delta}{\delta} \qquad \text{if } \delta \gg 2\Phi_F \tag{7.8}$$

and

$$S = \sin 2\Phi_F \qquad \text{if } \delta \ll 2\Phi_F, \tag{7.9}$$

where $\delta$ is the linear birefringence inside the fiber coil. This birefringence not only reduces the sensitivity of the device, but also reduces the stability of the output signal.

### 7.2.4 Detection schemes for eliminating the birefringence effect

In order to overcome the difficulties induced by the presence of birefringence, Ulmer [12] has introduced two novel detection schemes. In the first, the polarizer transmission axis is set at an angle of 22.5° with respect to the characteristic direction (equivalent to a principal axis in crystals) in the Faraday material, and a polarization beam splitter is used to give two output signals in the polarizer-transmission directions of 67.5° and 22.5°, respectively. With the use of these two outputs, the value of the birefringence can be determined, and thus the Faraday rotation can be obtained by modifying the output with an appropriate correction factor. A second scheme was developed, based on the fact that the relative magnitude of the output signal is a periodic function of the orientation of the polarizer with respect to this characteristic direction. Hence, there are a number of orientations (eight such angles for a polarization azimuth rotation of 360°) at which the relative magnitudes of the output signal are independent of the birefringence. By setting the polarizer transmission at those angles (for instance, 10.3° or 33°), and the analyzer at 45° to that of the polarizer, and then using the ratio of the d.c. component of the signal to the a.c. component (to remove the light intensity effect) for a measurement of current of up to 4000 A, the errors introduced due to birefringence (in a range from 0.24 to 0.34 radians) are within the narrow range of +0.03% to −0.07% of the ideal case [13].

### 7.2.5 Heterodyne detection scheme

The schemes described above are essentially homodyne-based. By using an optical fiber interferometric system, it is possible to synthesize an output with a 'phase carrier', the relative phase variation of which is directly modulated by the magnetically induced circular polarization mode-phase delay in the fiber, and which may be detected with a heterodyne signal recovery technique [14–16]. This phase carrier may be generated by using an unbalanced Michelson interferometer [14] or a polarization interferometer [15], illuminated with a laser diode which is frequency-modulated when driven with a sawtooth-modulated current, or by employing a Sagnac interferometer with part of the fiber coil being wrapped onto a sinusoidally driven cylindrical piezo-electric phase modulator [16]. The basic mode of operation of the first two schemes is based upon the 'pseudo-heterodyne' signal recovery technique [17], in which two circularly polarized beams in the Michelson interferometer or two polarization modes in a highly birefringent (hi-bi) fiber are used to introduce a phase difference at the output of the system. Thus if the optical frequency of the laser diode is swept, the azimuth of the linearly polarized output light will be rotated, resulting in a so-called 'optical frequency-to-polarization' modulation. When this light beam is analyzed by a polarizer, the spinning polarization state is transduced to a heterodyne carrier, which can be phase-modulated by the Faraday rotation. The principles of the operation of these schemes are discussed in more detail in the references cited [12–17].

## 7.3 OCS SCHEMES USING OPTICAL FIBER AS THE SENSING ELEMENTS

For those OCS devices using optical fiber as the sensing element, the condition exists under which the SOP of a linearly polarized light beam along the optical path is often not preserved due to the presence of intrinsic birefringence induced by core ellipticity and asymmetric stress, or the extrinsic birefringence caused in deploying the fiber sensing element, and hence the sensitivity of this type of device is less than would be predicted from equation 7.2. Furthermore, since the birefringence is temperature-dependent, the device sensitivity is also subject to environmental temperature perturbations. In order to improve the performance of this type of sensor, it is important to reduce the effects of all extraneous sources of birefringence in designing optical fiber current sensors. On the basis of the results, a number of sensing schemes which have been reported. Some of the possible solutions for dealing with the problem of birefringence are summarized below.

### 7.3.1 Suppressing the linear birefringence by using fiber with a large degree of circular birefringence

It is well known that if a large amount of circular birefringence is introduced into the fiber, the bend-induced linear birefringence can be suppressed whilst the Faraday rotation, which is a magnetically induced circular birefringence effect, is superposed onto this circular birefringence, and hence the measuring sensitivity may be retained. A special type of fiber with a large degree of circular birefringence can be obtained by twisting the fiber [18], or by using spun elliptically birefringent (SEB) fiber [19]. The upper range of the twist is limited by the mechanical fatigue limit of the fiber. The main problem of sensors constructed from this type of fiber is that they are temperature-dependent, which is largely due to the fact that the circular birefringence is subject to such a wide temperature variation. However, by employing an active temperature compensation scheme, Laming and Payne [19] have shown that over a 20–70°C temperature range, a fiber current sensor made from SEB fiber exhibited only a 0.05% per °C change in its sensitivity.

### 7.3.2 Removing the linear birefringence by annealing the fiber coil

One of the attractive approaches to remove the bend-induced linear birefringence by means of annealing was introduced by Day and Etzel [20]. In their approach, a fiber coil is heated to 800–900°C for about 24 h and then cooled slowly. By using this technique, an all-fiber current sensor with a coil of one or only a few centimetres diameter and more than 100 turns of fiber has been demonstrated [21]. A disadvantage of this approach is that high-temperature annealing removes the protective coating of the fiber and tends to weaken the fiber; hence the annealed fiber coil must be packaged without inducing additional linear birefringence. By

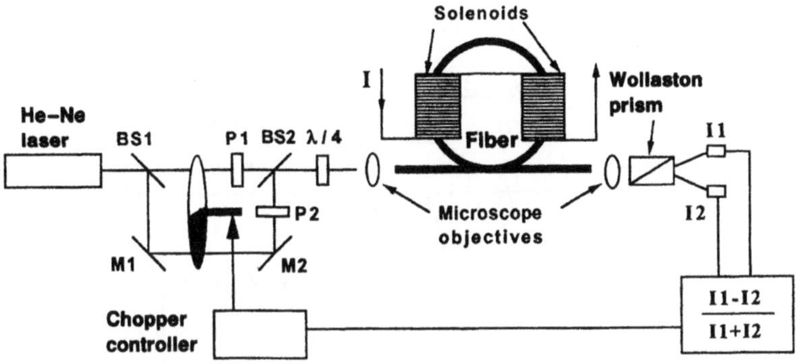


**Fig. 7.3** Schematic diagram of current measurement using time polarization multiplexing. BS, beamsplitter; P, polarizer; M, mirror; $\lambda/4$, quarter waveplate [23].

embedding an annealed fiber coil in a high-viscosity lubricant containing Teflon, a temperature coefficient of its current sensitivity of about $1.7 \times 10^{-2}\%$ per °C over a temperature range from −10°C to 120°C was reported by Tang *et al.* [22].

### 7.3.3 Separating the Faraday rotation from linear birefringence by using time multiplexing of two different states of polarization of the input light

Instead of removing the bend-induced linear birefringence, Ren and Robert [23] introduced a new detection scheme in which the Faraday rotation is separated from the bend-induced linear birefringence by using a time multiplexing of two different states of polarization at the input light. In their system, both linearly and circularly polarized light are launched into the fiber sensor alternately, as shown in Fig. 7.3. For linearly polarized input light, the sensor output becomes

$$S_1 = 2\Phi_F \frac{\sin(\sqrt{\delta^2 + (2\Phi_F)^2})}{\sqrt{\delta^2 + (2\Phi_F)^2}}, \tag{7.10}$$

where $\Phi_F$ and $\delta$ are as defined above. For circularly polarized input light, the output obtained is

$$S_2 = \delta \frac{\sin(\sqrt{\delta^2 + (2\Phi_F)^2})}{\sqrt{\delta^2 + (2\Phi_F)^2}}. \tag{7.11}$$

Therefore, the Faraday rotation can be directly obtained from the two outputs determined by equations 7.10 and 7.11 and this is given by

$$2\Phi_F = \frac{\arcsin\left(\sqrt{S_1^2 + S_2^2}\right)}{\sqrt{1 + (S_2/S_1)^2}}. \tag{7.12}$$

Hence, for the ideal case where the input states are perfectly linearly and circularly polarized, respectively, the effect of the linear birefringence can be eliminated

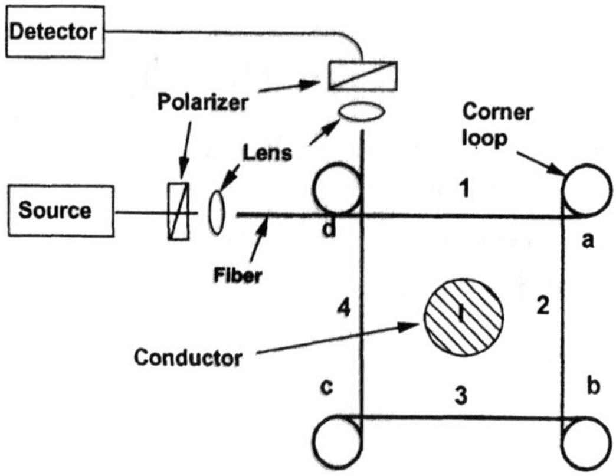


**Fig. 7.4** Layout architecture of a sensing head [24].

by using such a scheme. If the input states are not perfect, there will be an apparent sensitivity which will not be quite the same as is predicted by equation 7.12.

### 7.3.4 Separating the Faraday rotation from linear birefringence geometrically, by winding a special topology of sensing element

In this design [24], the bend-induced linear birefringence is employed as a phase retarder. By carefully designing the layout architecture of the sensing element, the bend-induced linear birefringence can be 'tied together' at the four corners of this square-shaped sensing element with no bend-induced linear birefringence present in the four sensing arms, as shown in Fig. 7.4. At each corner of this sensor, there are three fiber loops with a 10-mm bending radius yielding a $2\pi$ retardation between the two orthogonal components of the linearly polarized light. Since the four sensing arms of the sensing element consist of straight fiber, forming closed loops around the current conductor, the output signal gives only the Faraday rotation defined by equation 7.2. The advantages of this scheme are that the size of the sensor can be very small, and its sensitivity can be increased linearly with an increase in the number of fiber loops. The main disadvantage of the sensor is that its sensitivity is subject to temperature variations due to the presence of a large degree of bend-induced linear birefringence inside the sensing element.

### 7.3.5 Completely analyzing the polarization state of the output light by using a polarimeter

If the temperature variation occurs over a small range, typically 5–10°C, it is possible to distinguish the current-induced variation in the SOP of the output light from the influence of temperature by analyzing the complete SOP [25], i.e. by measuring its azimuth and ellipticity simultaneously, and then evaluating the

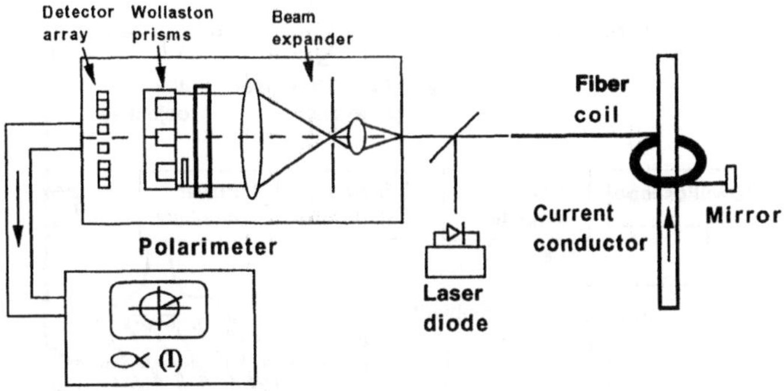


**Fig. 7.5** Complete polarization analysis of a magneto-optical current transformer with a polarimeter [25].

instant current and temperature by using a published table. In this detection scheme, three Wollaston prisms and a six-detector array with a polarization filter array have to be used in order to generate six output signals, which are needed in calculating the azimuth angle and the ellipticity of the output signal. Since the magnetically induced azimuth variation is superposed on the temperature-induced variation caused by the twist-induced circular birefringence, it is not possible to separate these two variations by using a polarimeter. In order to eliminate the temperature perturbation on the twist-induced circular birefringence, a back-reflection configuration was used by Ahlers and Bosselmann [25], as shown in Fig. 7.5. By using this system to analyze the complete polarization state, a resolution in azimuth and ellipticity of 0.1° was obtained.

### 7.3.6 Mode coupling detection by using a highly birefringent (hi-bi) fiber

As discussed above, any polarization variation which is affected by environmental perturbations via different sources of birefringence limits the performance of low-birefringence fiber current sensors. However, in contrast, this effect (including the Faraday effect) will be suppressed in highly birefringent (hi-bi) fibers. If a hi-bi fiber is subject to a spatially periodic perturbation which has a period of the fiber beat length, linearly polarized light inside one of the two orthogonally polarized eigenmodes of the hi-bi fiber can be coupled into another mode by the effect of the Faraday rotation. By configuring an elliptically cored hi-bi fiber into a periodic conductor array, a new type of optical fiber current sensor was demonstrated by Chu *et al.* [26]. In their system, shown in Fig. 7.6, linearly polarized light was launched into one of the two orthogonally polarized eigenmodes with an extinction ratio of 30 dB. The Faraday rotation-induced coupling light was detected at another mode through a Glan–Thompson analyzer. It was shown that the output signal from this system was proportional to the current applied. Since the mode-coupling coefficient of the sensor is strongly dependent on the period of the applied field, which is defined by the configuration of the sensor, the

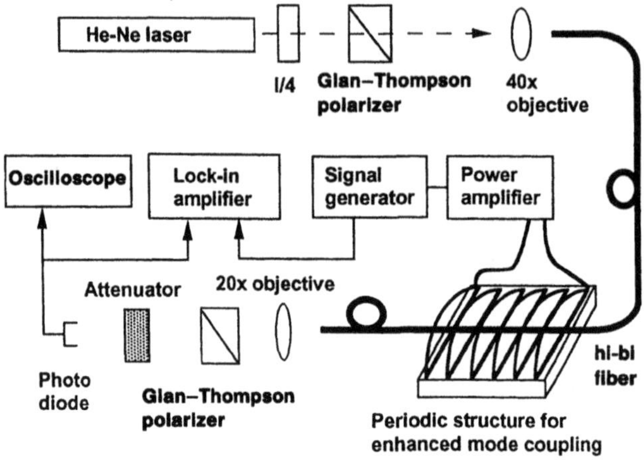


**Fig. 7.6** Experimental arrangement using a hi-bi fiber for current measurement [26].

performance of this device is such that it will be less sensitive to environmental perturbations, because they do not appear in a periodic form along the sensing fiber.

### 7.3.7 Interferometric detection schemes

The Faraday effect can also be represented in terms of the circular birefringence, i.e. the current-induced azimuth variations can be effectively described as a current-induced circular birefringence change or a differential refractive index modulation and therefore a phase change. It is possible to detect such a phase change using interferometric detection schemes. Numerous optical fiber interferometric current sensors have been investigated and reported [15, 16] in the past years. Among them, the Sagnac interferometer, which permits the measurement of nonreciprocal effects independently of the influence of any reciprocal perturbations, has several fundamental advantages in current detection due to the nonreciprocal nature of the Faraday effect. In a Sagnac interferometer, apart from the temperature-induced variation in the circular birefringence of the fiber which is irrecoverable, the effect of the perturbation on the light intensity and input polarization state variation can be alleviated by using a common mode rejection scheme. Further, since all sources of perturbation show a low frequency variation, they can be avoided by using a heterodyne approach in detecting the Faraday rotation. By using a sinusoidally driven cylindrical piezo-electric phase modulator in a Sagnac interferometer, Leilabady *et al.* [16] have shown a pseudo-reciprocal current sensor which is able to detect a minimum a.c. current of 10 mA.

### 7.3.8 Reducing the effect of birefringence in fiber leads by using an orthoconjugate reflector (OCR)

For almost all types of OCS, optical fiber leads are required to connect the sensing element to the light source and the detection unit within the measurement

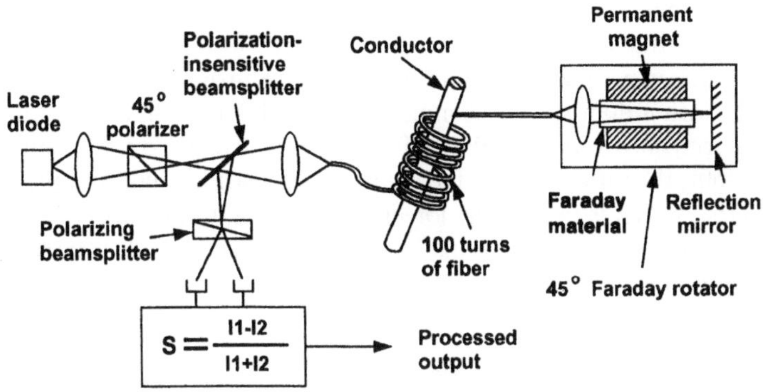


**Fig. 7.7** Current measuring system with an orthoconjugate reflector [27].

apparatus. In any practical optical fiber current measurement system, the environmental perturbations such as temperature fluctuations and mechanical vibration on the fiber leads will reduce the signal-to-noise ratio (S/N) of the output and affect the performance of the sensing system. In order to overcome this problem, a novel passive birefringence compensation scheme was introduced, that uses an orthoconjugate reflector (OCR) [27] or a mirrored Faraday rotator (MFR) [28]. In the system discussed, as shown in Fig. 7.7, an OCT or an MFR, which consists of a Faraday rotator which rotates the azimuth of the light by 45° and a reflection mirror, is used in conjunction with an optical fiber current sensing circuit which is made of a highly twisted low-birefringence optical fiber (40 turns $m^{-1}$, 0.17 rad $m^{-1}$ of linear retardance). Since the OCR is able to return the orthogonal polarization state back along the fiber so that all reciprocal linear and circular birefringence in the fiber can be 'unwrapped', with the returned SOP being linear and oriented at 90° to the launch state, the effect of any reciprocal birefringence in the fiber leads may be suppressed provided that the ratio of the circular to linear birefringence is maintained at a high level everywhere along the fiber circuit. By using this arrangement, the estimated rejection of the vibration noise was of the order of 30 dB.

### 7.3.9 Fiber sensing element using low stress-optic coefficient fibers

In order to overcome the bend-induced birefringence in conventional single-mode fibers, a considerable amount of effort has been expended to develop flint glass fiber for optical current sensing applications [29, 30]. There are two types of flint fiber, classified according to the transmitting wavelength: 1550 nm or 850 nm. The former was developed for sensing applications where the current is over 180 kV, whilst the latter can be used for conventional current measurement. The photo-elastic constant of the core is $0.45 \times 10^{-9}\,cm^2\,kg^{-1}$, which is 1/780 that of fused silica fiber, and the Verdet constant is $0.86°\,A^{-1}$ which is six times higher than that of fused silica fiber [29]. Since the flint fiber is made of glass which is composited with a high percentage of lead, the fiber shows a very low level of bend-induced birefringence because of its low stress-optic coefficient. According to the experimental results reported by Kurosawa *et al.* [30], for one loop of fiber with a diameter of 80 mm, the azimuth variation due to the bend-induced birefringence was only 0.5°. If the fiber is twisted over a range from −360° to +450°, the ratio between the rotated polarization angle and the twisted angle is 0.11%, which is two orders lower than that of a silica fiber. As a result, when a sensor made of flint glass is installed in the field, it exhibits little sensitivity to vibration or shock [31].

### 7.3.10 Reciprocal compensation techniques

Since the Faraday rotation is a nonreciprocal effect whilst both linear and circular birefringence are reciprocal, it is possible in theory to compensate the effect due to birefringence variation by using counterpropagating beams. Based on this technique, Rogers *et al.* [32] reported an optic fiber current sensor with a vibrational noise rejection of 54 dB. In their system, two counterpropagating light beams were launched into a toroidal fiber loop at the two ends of the fiber, and the polarization state of the modulated light from the two fiber ends were then separately detected and analyzed. With the use of this system, a current of up to 600 A at 50 Hz was measured, with the imposed vibration being suppressed (the vibrational frequency region was from 0.1 Hz to 5 kHz, and the maximum vibrational amplitude was about 3 mm).

Again, by making use of the nonreciprocal nature of the Faraday effect, a novel Brillouin fiber current has been developed and tested by Kung *et al.* [33]. Brillouin techniques have been widely applied in various optical schemes, especially narrow-linewidth lasers, microwave optical phase modulators and Brillouin fiber optic gyroscopes (BFOGs), for example. In the latter devices, two counterpropagating Brillouin lasers share the same fiber ring resonator and experience different apparent cavity lengths under rotation. Thus they oscillate at different frequencies, and the beat frequency produced is proportional to the resolution rate. The technique is sensitive not only to the Sagnac effect in the BFOG, but also to the Faraday effect, another nonreciprocal optical effect. In the Brillouin fiber optic current sensor (BFOCS) thus produced, the beat frequency due to the nonreciprocal phase shift induced by the Faraday effect is monitored, and has maximum sensitivity as soon as the polarization states of the circular waves are kept circular all along the fiber resonance, when using, for example, highly twisted or spun highly birefringent (hi-bi) fiber.

The system is shown schematically in Fig. 7.8, using light from a Nd:YAG laser at a wavelength of 1320 nm. Polarization controllers $PC_1$ and $PC_2$ are used to adjust the input polarization states to the same eigenmode as the resonator. The couplers $C_1$ and $C_2$ allow measurement of the input pump power and the frequency locking of the pump waves to the resonant frequency of the fiber ring cavity by using a FM sideband technique, thus allowing the generation of the two counterpropagating Brillouin waves, $B_1$ and $B_2$. The frequency shift

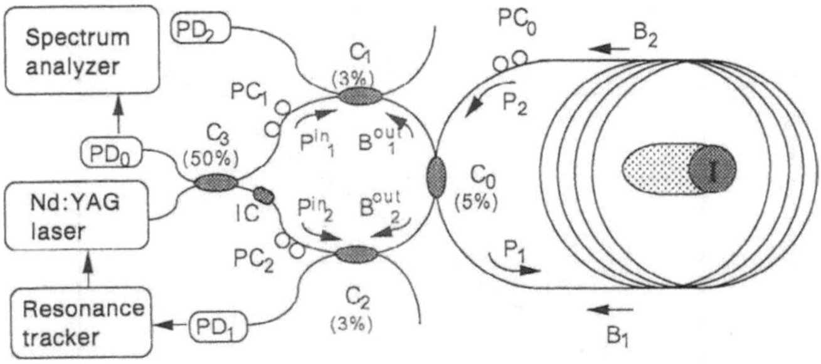


**Fig. 7.8** Sketch of a Brillouin fiber optic current sensor (BFOCS). PC, polarization controller; C, coupler; IC, intensity controller; P, pump waves; B, Brillouin waves [33].

between the two outputs $B_1^{\text{out}}$ and $B_2^{\text{out}}$ is measured on the photodiode, $\text{PD}_0$. A 14-cm coil is used, of 40 turns of highly twisted fiber. A polarization controller, $\text{PC}_0$ is used to compensate the birefringence in the coupler, $\text{C}_0$.

The phase shift, $\phi$, induced by the Faraday effect in the situation where the polarization state is kept circular all along the fiber coil is given by

$$\phi = 2NVI_{\text{el}}, \tag{7.13}$$

where $N$ is the number of turns of the fiber coil, $V$ is the Verdet constant and $I_{\text{el}}$ is the electric current. It is important to note here the factor of two between the nonreciprocal phase-shift and the usual Faraday rotation angle, $\theta = NVI_{\text{el}}$, of linearly polarized light. A phase-shift of $2\pi$ gives rise to a frequency shift, $\Delta f$, equal to one value of the free spectral range (FSR) between the two counter-propagating Brillouin lasers, leading to the transfer function given by

$$\Delta f = \frac{FSR}{2\pi}\phi = \frac{c}{2\pi n_g L} 2NVI_{\text{el}}, \tag{7.14}$$

when $n_g$ is the fiber group index, $c$ is the vacuum velocity of light and $L$ is the resonator length. The beat frequency gives no information on the current direction.

The scale factor $S = \Delta f / \Delta I_{\text{el}}$ of the sensor is estimated to be $128\,\text{Hz}\,\text{A}^{-1}$ taking into account the experimental values of $\text{FSR} = 9.62\,\text{MHz}$, $V = 1.04 \times 10^{-6}\,\text{rad}\,\text{A}^{-1}$ and $N = 40$. This sensitivity can be improved by increasing the number of turns of the fiber coil whilst keeping the FSR constant. However, this means that the coil radius, $R$, will decrease accordingly, and consequently that the linear birefringence will increase according to the $1/R^2$ law. In practice a tradeoff has to be found between the bent induced birefringence and the fiber twist amount in order to guarantee a low residual linear birefringence.

The characteristic of the sensor is shown in Fig. 7.9 over the current range 0–200 A. The nonlinear behavior for currents $<30$ A is due to the frequency locking of the two laser beams through Rayleigh backscattering and reflection sites in the system. The authors report that this could be overcome using an optical

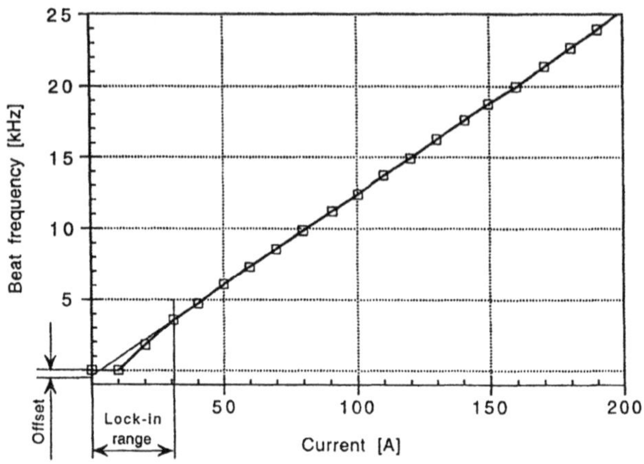


**Fig. 7.9** Transfer function of a Brillouin fiber optic current sensor and its linear fit [33].

dithering technique. An offset at zero current of 390 Hz is seen, and work is continuing to optimize the system.

## 7.4 OCS TYPES USING BULK GLASS AS THE SENSING ELEMENTS

Whilst optical fiber current sensors continue to remain under development, a range of OCS devices based on bulk materials have been intensively investigated and developed in recent years. These sensors, fabricated from a single glass block with a high Verdet constant, are light-weight, small in size, extremely robust, and they do not suffer from problems associated with the presence of intrinsic birefringence and bend-inducer linear birefringence inside optical fiber sensing elements. However, another problem arises in the internal reflections that are required to allow the light beam to encircle the current-carrying conductor. This is the so-called 'reflection-induced phase difference' problem. According to electromagnetic theory [4], when a linear polarized light beam is reflected at an incident angle which is larger than the critical angle, there will be a phase difference between the two orthogonal components of the linearly polarized light after the reflection, and the SOP of the light will be varied. This variation affects the sensitivity of the device in a similar way to the birefringence in an optical fiber sensing element. Hence finding a way to reduce this reflection-induced phase difference becomes a key issue in design of this type of sensor. In order to overcome this problem, two methods have been investigated: the use of dual-quadrature reflection [9], and critical angle reflection [10, 11]. Based upon these approaches, a number of novel OCS types have been developed. The details of the configurations and the performance of these sensors are discussed in the following sections.

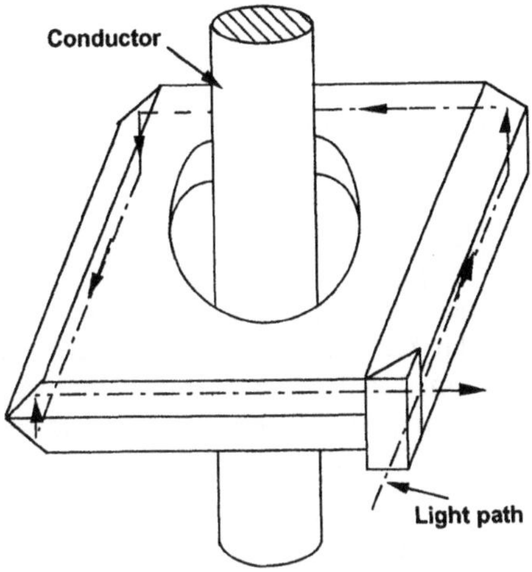


**Fig. 7.10** 3-D square-shaped current sensing element [9].

### 7.4.1 OCSs using a square-shaped sensing element

This type of sensor topology was first introduced in a patent by Sato *et al.* in 1983 [9]. In their design, the reflection-induced phase differences were eliminated by using dual-quadrature reflections. Figure 7.10 shows this square-shaped sensing element, in which four sensing arms form a closed optical path via three reflection corners at which a linear polarized light beam is totally reflected twice. Under certain conditions, the reflection-induced phase differences can cancel each other out through the double quadrature reflections.

Writing the incident field as $\mathbf{E_{in}}$, Jones calculus can be used to predict the ellipticity and azimuth of the output electric field, $\mathbf{E_n}$

$$\mathbf{E_n} = \mathbf{P}(\phi)\mathbf{F_4R_sF_3'R_pF_3R_sF_2'R_pF_2R_sF_1'R_pF_1E_{in}}, \tag{7.15}$$

where $\mathbf{P}(\phi)$ is the polarizer matrix, $\mathbf{R_{p,s}}$ are the two quadrature reflection matrices at the sensor corners, and $\mathbf{F_{1,2,3,4}}$ and $\mathbf{F'_{1,2,3}}$ are the Faraday rotation matrices corresponding to the Faraday effect in the four sensing arms and in the optical path between two reflection points at each corner. These are given by

$$\mathbf{P}(\phi) = \begin{pmatrix} \cos^2\phi & \sin\phi\cos\phi \\ \sin\phi\cos\phi & \sin^2\phi \end{pmatrix} \tag{7.16}$$

and

$$\mathbf{R}_\mathrm{p} = \begin{pmatrix} \rho_x & 0 \\ 0 & \rho_y \end{pmatrix}, \qquad \mathbf{R}_\mathrm{s} = \begin{pmatrix} \rho_y & 0 \\ 0 & \rho_x \end{pmatrix} \tag{7.17}$$

$$\mathbf{F}_\mathrm{i} = \begin{pmatrix} A_i & -B_i \\ B_i & A_i^* \end{pmatrix}, \qquad (i = 1 \ldots 4), \tag{7.18}$$

where

$$A_i = \cos(\psi/2) + i\sin(\psi/2)\cos(\chi)$$
$$B_i = \sin(\psi/2)\sin(\chi)$$
$$\tan(\chi) = 2\Phi_i/\gamma_i, \qquad (\psi/2)^2 = (\gamma_i/2)^2 + \Phi_i^2,$$

and $\rho_x$ and $\rho_y$ are the reflectances for the orthogonal components of the field. For total reflection, $|\rho_x| = 1$ and $|\rho_y| = 1$, $\Phi_i$ and $\gamma_i$ are the Faraday rotation angle and the birefringence respectively, generated in the $i$th sensing arm, and $\phi$ is the angle of the output polarizer. If the birefringence inside the sensing element can be ignored and it is assumed that $\chi = \pi/2$, equivalent to a pure Faraday rotation, then for the $i$th sensing arm, the Faraday rotation matrix, $\mathbf{F}_i$, may be represented by a pure rotation matrix, and given by

$$\mathbf{F}_i = \begin{pmatrix} \cos(\Phi_i) & -\sin(\Phi_i) \\ \sin(\Phi_i) & \cos(\Phi_i) \end{pmatrix}. \tag{7.19}$$

When the sensor is working under the conditions that (i) the current to be measured is passing through the center of the sensor in a direction which is perpendicular to the sensor plane, and (ii) there is no external field presence near the sensor, or the effect of these external fields is small enough to be ignored, the Faraday effect on the optical path between the two reflection points at each corner is zero, and hence

$$\mathbf{F}'_i = \begin{pmatrix} 1 & 0 \\ 0 & 1 \end{pmatrix}, \qquad (i = 3). \tag{7.20}$$

Therefore, when a linearly polarized light beam is reflected twice at each corner, the electrical field of the reflected light becomes

$$\mathbf{E}_{\text{reflected}} = \begin{pmatrix} \rho_x & 0 \\ 0 & \rho_y \end{pmatrix} \begin{pmatrix} 1 & 0 \\ 0 & 1 \end{pmatrix} \begin{pmatrix} \rho_y & 0 \\ 0 & \rho_x \end{pmatrix} \mathbf{E}_{\text{in}}$$
$$= \rho_x\rho_y \begin{pmatrix} 1 & 0 \\ 0 & 1 \end{pmatrix} \mathbf{E}_{\text{in}} \tag{7.21}$$

for the total reflection, $|\rho_x\rho_y|^2 = 1$, and this implies that the SOP of a linear polarized light beam can be preserved through the use of these double quadrature reflections. In practice, the optical path between the two reflection points should be as small as possible in order to meet the second condition mentioned above.

The sensing elements of the OCSs manufactured in this topology have been extensively field-tested at several locations [34–36]. The results of these tests show that the performance of such an optical current measurement system compares favorably with that of a conventional CT [35].

### 7.4.2 OCS using a triangular-shaped sensing element

According to electromagnetic theory, the SOP of a light beam can be preserved when the beam is reflected at its critical angle [4]. Based on this principle, a new

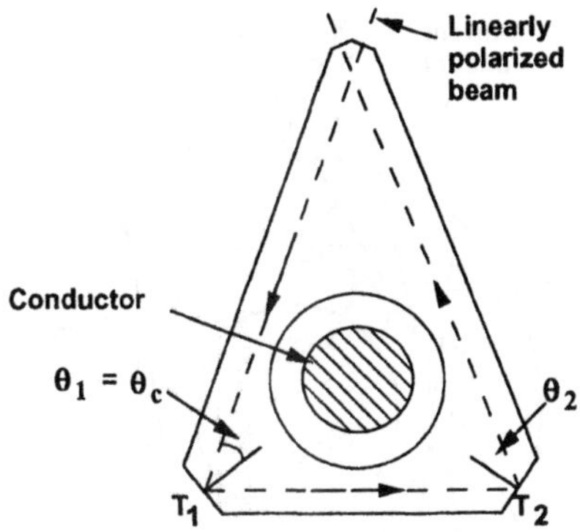


**Fig. 7.11** Topology of a triangular sensing element [11].

topology for a bulk optical Faraday current sensor with a triangular-shaped sensing element has been introduced by Chu and co-workers [10]. The sensing element is designed in such a configuration that the light beam propagating inside it is internally reflected twice at the critical angle, forming a triangular optical path which encircles the current to be measured, as shown in Fig. 7.11.

The propagation of the incident electric field, $\mathbf{E}_{\text{in}}$, along the sensing element can be described by using Jones calculus. The output electric field, $\mathbf{E}_{\text{out}}$, is given by

$$\mathbf{E}_{\text{out}} = \mathbf{P}(\phi)\mathbf{F}(\Phi_3)\mathbf{R}_2\mathbf{F}(\Phi_2)\mathbf{R}_1\mathbf{F}(\Phi_1)\mathbf{E}_{\text{in}}, \qquad (7.22)$$

where $\mathbf{F}(\Phi_1)$, $\mathbf{F}(\Phi_2)$ and $\mathbf{F}(\Phi_3)$ are the Faraday rotation matrices corresponding to the Faraday effect in the first, second and third sensing arms of the sensing element, respectively, given by equation 7.19, and $\mathbf{P}(\phi)$ is the polarizer matrix given by equation 7.16. $\mathbf{R}_1$ and $\mathbf{R}_2$ are the reflection matrices for the first and second corners, respectively, and are defined by equation 7.17. If the incident angle, $\theta_1$ or $\theta_2$, at the two corners $T_1$ and $T_2$ are exactly at the critical angle, $\theta_c$, then $\rho_s$ and $\rho_p$ are equal to unity, and the scale factor for the Faraday rotation can be predicted from the known Verdet constant of the glass. By using this sensing element, combined with a dual quadrature polarimeter detection arrangement [11], a d.c. current measurement of up to 3000 Amp-turns was achieved, with a sensitivity of $2.3 \times 10^{-5}\,\text{rad A}^{-1}$ and a resolution of 20 mA per $\sqrt{\text{Hz}}$ over a frequency region from d.c. to 10 kHz [10].

By comparison with a square-shaped sensing element, the configuration of this triangular-shaped sensor is much simpler, and hence it is easier to manufacture at low cost. However, care must be taken in polishing the two reflecting surfaces, since their angular tolerance is relatively small (the angles of internal reflection are within $\pm 0.01°$), in order to ensure that the light is reflected at the critical angle $\theta_c$. If the incident angles $\theta_1$ and $\theta_2$ are not equal to $\theta_c$, there will be errors in the magnitude of the rotation of the output beam. The sensitivity of the device is subject to any variations of the input azimuth [37]. Hence the sensing system requires to be calibrated before being used for practical current measurements.

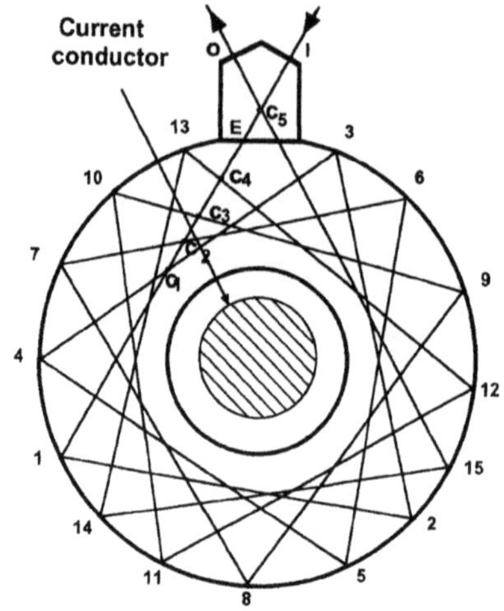


**Fig. 7.12** Topology of a ring-shaped sensing element [10].

### 7.4.3 OCS using a ring-shaped sensing element

According to equation 7.3, if the number of turns of the optical loop, $N_1$, is increased, the sensitivity of the sensing element will also increase by the same factor. Based upon this principle, a new topology of a ring-shaped sensing element was introduced by Ning *et al.* [10]. This novel sensor is made of a bulk-glass ring with an optically polished edge of high circularity and a small 'roof-top' prism, as shown in Fig. 7.12, where a computer simulation of the optical path taken by the light is plotted. The current-carrying wire passes through the central aperture. A linearly polarized light beam is coupled into the sensing element via the 'roof-top' prism and undergoes sequential internal reflections at positions 1, 2, 3, 4 etc., finally emerging through the prism after the 15th reflection. In passing through the ring, the light will have effectively encircled the current five times, thereby increasing the current sensitivity by the same factor.

Writing the incident electric field as $\mathbf{E}_{\text{in}}$, the output electric field, $\mathbf{E}_{\text{out}}$, is given by

$$\mathbf{E}_{\text{out}} = \mathbf{P}(f)\mathbf{F}_n\mathbf{R}\mathbf{F}_{n-1}\mathbf{R}\ldots\mathbf{F}_2\mathbf{R}\mathbf{F}_1\mathbf{E}_{\text{in}} \qquad (n = 15), \tag{7.23}$$

where $\mathbf{P}(\phi)$, $\mathbf{F}_i$ and $\mathbf{R}$ are as defined above.

By using a simple polarization detection scheme, i.e. inserting an analyzer with its polarization direction at 45° with respect to that of the output linearly polarized light between the sensing element and the detector, a sensitivity of 1 A per $\sqrt{\text{Hz}}$ for a.c. current measurement and a flat frequency response over the range of 10–$14^4$ Hz were obtained [9].

### 7.4.4 OCS using an openable sensing element

At the present time there are many applications for high-voltage current sensors which can be either permanently installed for metering purposes or demountable

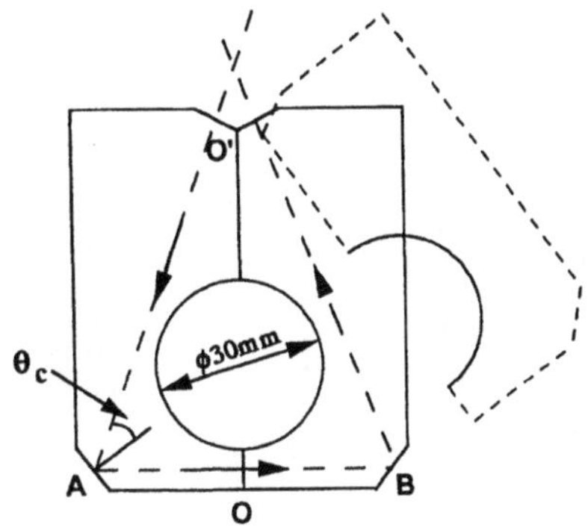


**Fig. 7.13** Topology of an openable current sensing element [38].

for diagnostic investigations where it is desirable to make the measurement without interrupting the current or disconnecting power lines. Clearly an openable optical current sensor is necessary if optical systems are to be competitive with present commercial systems, such as electrical current clamps. For this purpose, a demountable (i.e. openable) optical current clamp was introduced by Ning and Jackson [38]. The sensing element is constructed from two identical pieces of high Verdet constant bulk-glass, as shown in Fig. 7.13. A linearly polarized beam is launched into the first half of the sensing element through the input surface and directed towards surface A, where it is reflected at the critical angle, and after this reflection the beam passes at normal incidence through the interface between the two pieces of glass. After a second critical angle reflection at surface B, it emerges at the output face of the second half of the sensor. Hence the light encircles the conductor by means of two reflections at the critical angle, undertaking a closed triangular route. Since the sensor can be opened when it is deployed on the conductor, the measurement can be achieved without the need to interrupt the current.

Compared with the triangular-shaped Faraday current sensor described in Section 7.4.2, this openable optical current clamp exhibits a number of additional features. As the sensing element is composed of two identical pieces of glass, they can be used with independent beam launch conditions to ensure that the beam reflected at A and B are at exactly the critical angle, and hence the polishing tolerances of the angles at A and B can be relaxed. Furthermore, if the wavelength of the light source is changed, for example, to a slightly longer wavelength, the value of the refractive index at this wavelength will decrease, and hence the value of the critical angle will increase. Thus by adjusting the optical beam conditions in the two sensor halves, it is possible to make the sensor operate over a small wavelength range. Hence, in a commercial implementation, the glass blocks may all be identical in manufacture so that the sensors can all operate at the same nominal wavelength and over a small wavelength range, ensuring a minimum cost of production. The design of the openable mechanical clamp would ensure

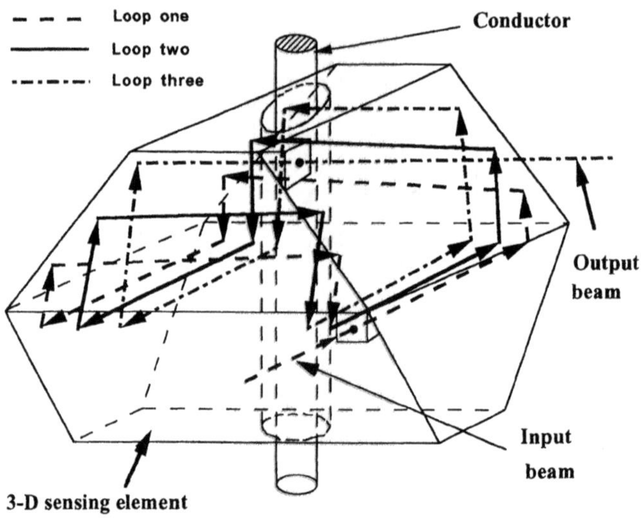


**Fig. 7.14** Topology of a multi-optical-loop sensing element [39].

that the mechanical orientations of the two halves of the sensor could be optimized for any specific wavelength over the range. The results from an initial test show a demonstrated resolution of 7.2 mA per $\sqrt{\text{Hz}}$ over a measurement range from 1 to 500 A. The sensitivity of the system was $2.21 \times 10^{-5}$ rad $\text{A}^{-1}$, which is comparable to that with the triangular sensor [10] and represents a major advance in the implementation of detachable devices in the field of optical current sensing.

### 7.4.5 OCS using a multi-optical-loop sensing element

As indicated in equation 7.3, the actual sensitivity of the current sensing element is proportional to the number of turns of the optical loop, $N_1$. Based on this principle, a new 3-D topology (using three optical loops) in the sensing element has been introduced by Ning *et al.* [39]. This optical element, shown in Fig. 7.14, is made of SF6 Schott glass (of refractive index 1.798), of size 100 mm × 100 mm ×60 mm, and has a 35-mm-diameter central aperture through which the current-carrying conductor passes. A linearly polarized beam from the light source is coupled into the sensing element through a small input prism. When the beam is injected at specific angles, the light in the sensor can undergo as many as 22 internal reflections, at angles which are slightly offset from 45° (by about 0.90° with ±0.09° of tolerance), forming three optical loops around the current-carrying conductor, and finally emerging at the other output prism, as shown in Fig. 7.14. In this new topology, each optical loop needs eight interreflections (both the first and the last loops require seven reflections each), where the number of loops is determined by the offset angle and the size of the sensing element. Since each interreflection is not at 45°, the reflection-induced optical phase difference will not be canceled to zero by each pair of 'complementary reflections'. However, according to theoretical results obtained with the use of Fresnel's equations and Jones calculus, the maximum phase shift induced by a pair of reflections is 0.019 rad, giving an ellipticity as small as $8.08 \times 10^{-3}$. As a result, the overall phase shift is 0.022 rad, yielding an ellipticity of $1.07 \times 10^{-2}$. Hence, the phase shift between the two orthogonal states of electric field after 22 reflections may be ignored. Although a maximum number of three transits of the beam can be achieved using this sensing element, one and two loops may also be obtained by altering the incident angle appropriately, and hence the sensitivities of the device with one and two loops could be measured.

The experimental results show that the measured sensitivities for one, two and three loops were $2.19 \times 10^{-5}$, $4.35 \times 10^{-5}$ and $6.57 \times 10^{-5}$ rad $A^{-1}$ ($\pm 1\%$ in each case), respectively. Their ratio is 1:1.99:3, which agreed well with the designed specifications. The measured Verdet constant in each case was also in good agreement with the manufacturer's data of $2.23 \times 10^{-5}$ rad $A^{-1}$, at the wavelength of operation, 633 nm. With the use of a spectrum analyzer, an a.c. current of 11.1 mA (rms) at 360 Hz can be seen, giving a minimum detectable current of 11.3 mA (rms) per $\sqrt{Hz}$.

### 7.4.6 Suppression of vibration-induced noise in a fiber lead

In an optical current measurement system, the essential element of which is a bulk-glass sensing element, the light beam is often delivered by an optical fiber lead from a source to the sensing element and linearly polarized by a polarizer before it enters the sensing element itself. The azimuth of this linearly polarized beam may then be altered within the sensing element as a result of the Faraday effect. Hence, the measurement of the current not only depends on the variation of the output azimuthal angle from the sensing element, which is proportional to the magnetic field produced by the current to be measured, but also on the SOP of light at the input of the sensing element, i.e. any change in the SOP of the input light will induce an error signal at the output of the system.

Hence, optical fiber links between the sensing element and the opto-electronic signal processing unit which connects the light source and the photodetectors can make the overall output of the bulk-glass-based sensing system subject to the presence of vibration on the leading fibers, especially when a single-mode fiber is used as an input fiber. When the fiber is perturbed by a vibration, such as is experienced when the device is used in any practical sensing system, birefringence variations within the fiber will be induced due to the photo-elastic effect. Thus the SOP of the output light from the fiber will be changed accordingly, resulting in polarization noise in the light signal at the output of the fiber or in the optical input to the sensing element. Therefore, in order to obtain a high accuracy of measurement, it is important to maintain this input SOP unchanged when the system is in operation. This means that any SOP variation induced due to the inevitable vibration of the leading input fiber should be suppressed to a level which is as low as possible.

Recently, the results of an experimental investigation of the possibility of the suppression of such polarization noise induced from the leading input fiber were reported by Zhang *et al.* [40]. It was shown in the work reported that the input polarization noise can be considerably reduced when a multimode fiber of a long length, and also a light source with a short coherence length, are employed. This noise suppression effect may be due to two main factors, i.e. the depolarization effect and the decoherence effect.

In the first of these effects, when all the modes of a multimode fiber are excited by the injected beam, due to the presence of various forms of dispersion such as mode dispersion, the optical power is split among the various modes, each propagating at slightly different velocities, and traveling along the fiber for a certain length. Since each mode experiences a different optical path and multi-total reflections at different incident angles within the fiber core, phase differences between the two orthogonal states of each mode will be randomly distributed from 0 to $2\pi$, and the polarization state of each mode will no longer be that of linear polarization. Since the output of a multimode fiber is the sum of all such beams, no dominant polarization direction may be formed. As a result, a uniform level of depolarization of the light will be generated after the beam travels in a multimode fiber of such a relatively long length, and this result is widely applicable in this type of polarization-based sensor.

The second effect may be considered in terms of the ray-optical method, i.e. the guided modes supported by a multimode fiber can be represented geometrically by a group of allowable rays that will propagate along the different paths at different velocities depending on the injected angles. The different velocities resulting will lead to different times of arrival of the modal energy components at the far end of the fiber. As a consequence, an optical path difference (OPD) will be introduced between the two adjacent modes at the far end of the fiber. If this OPD is larger than the coherence length of the source used, these modes become incoherent with each other. If there is any energy coupling between them (such as would result from a perturbation of the fiber), the value of the intensity components between them will be altered incoherently. As a result, the output beams will add incoherently at the detector, and the interference noise due to the intermodal coupling can be suppressed. Hence, by increasing the fiber length or reducing the coherence length of the source used, the total level of vibration-induced noise may be considerably reduced [41].

In any practical system, the input fiber may be considered as being equivalent to a lead fiber and a delay fiber. Since the delay fiber may perform the functions of decoherence and depolarization, a reduction of the overall noise induced in the input fiber can be achieved, even if the entire lead fiber is subject to a vibrating environment. It should be pointed out that with the use of a long-delay fiber as part of the input fiber, the optical intensity may be slightly reduced due to the attenuation of the fiber. However, this effect can easily be overcome as multimode laser diodes with high output optical power are now commercially available.

With the use of a low-coherence source and a long multimode fiber of 2200 m [40], the experimental results obtained have shown an improvement of more than 40 dB of the output S/N ratio between conditions of nonvibration and vibration of the fiber, and thus a significant reduction of the vibration-induced noise has been observed. The results were demonstrated with a known and reproducible perturbation applied over a small region of the fiber, but given the nature of the optical interaction caused, it is reasonable to extrapolate the benefits seen to cases where the perturbation is caused over a wider region, or indeed the whole lead part of the input fiber. This would cause a similar change in the nature of the propagation of the optical signal through the fiber, and hence the results can be directly cross-compared. Such results have already indicated that it is possible to eliminate the effect of a vibrating fiber lead on the output, which is a very important result for the use of optical current sensors in practical situations.

## 7.5 OCS DEVICES USING ELECTRO-OPTIC HYBRID SENSING DEVICES

Although the conventional CT, with its comprehensive measurement range, is well developed and widely accepted in electrical distribution industries, its applications are limited to use at a relatively low potential. For applications in a high potential field, the requirement of a high level of electric insulation makes the CT both heavy and expensive. Targeting those applications, it has been felt worthwhile to develop several 'optical demodulation' techniques incorporating a fiber optic link, and to introduce several hybrid sensing schemes to minimize the requirements on the insulation level by interrogating the excited sensing systems with an optical fiber sensing/communication link. The main advantage is that as the insulation is provided by the fiber optic itself, the cost of the insulation needed in a CT would be greatly reduced.

OCS systems using electro-optic hybrid sensing devices have also been developed and introduced in recent years. This type of sensor consists of a conventional electrical or electronic current sensor, such as a magnetic-field concentrator [12, 42], a current transformer (CT) [43] or a photo-electronic device [44], but one which is interrogated using an optical fiber system. In such a typical hybrid sensing system, the current-modulated output signal from a conventional current sensor or a CT is converted into a modulated optical signal, which is then transmitted from a high potential area to the ground via an optical fiber link for further processing. Hence, this system not only takes advantage of the high level of electrical insulation offered by optical fibers, and is comparable to the conventional CT which is well accepted by industry, but also avoids the difficulties associated with birefringence in the sensing fiber itself.

By using different optical modulation schemes, such as optical phase modulation [43, 45], intensity modulation [44], frequency modulation [46–48], polarization modulation [12, 42] and chromatic modulation [49], different types of electro-optic hybrid current sensing schemes have been introduced. Examples of those schemes are now discussed.

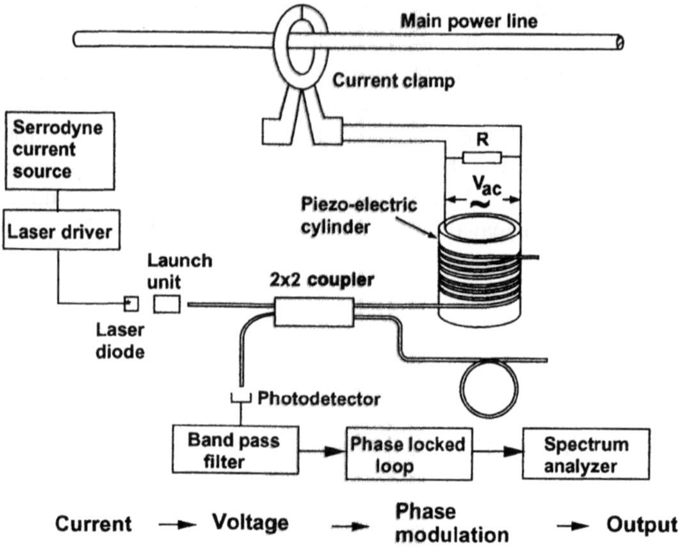


**Fig. 7.15** Hybrid current sensing system based on optical phase modulation [43].

### 7.5.1 Hybrid OCS using a phase-modulation scheme

For the measurement of large currents at high potential, a new approach using an optical phase-modulation scheme to form an electro-optic hybrid current sensor has been developed and reported [43]. In this scheme, a conventional CT is used as the primary transducer, and a fiber optic Michelson interferometer acting as a secondary sensor is employed, both to measure the voltage signal appearing at the output of the secondary of the CT and to transfer the information to ground potential for subsequent processing, thereby combining the virtues of a conventional CT with the high resolution and electrical insulation achievable with a fiber optic system.

The experimental arrangement is shown in Fig. 7.15, where the CT is a standard 'clip-on type' model (supplied by Northern Design Ltd., Model No HP30); this unit has a specified working range of up to 3000 A, with a current conversion ratio of 3000:1. The quoted accuracy of this unit for current measurement is ±1% and the frequency response exceeds 10 kHz. A high-wattage resistor is connected to the output of the secondary of the CT to convert the fluctuating current into a corresponding voltage signal, which is then applied directly to the piezo-electric cylinder (PZT).

A fiber optic Michelson interferometer with a path imbalance of a few centimetres is illuminated with light from a low-power single-mode laser diode. The injection current of the laser diode may be modulated by a serrodyne waveform at a frequency of 1 kHz with an amplitude sufficient to sweep the output of the interferometer over 1 cycle of the transfer function. The PZT (supplied by Vernitron), with 12 turns of optical fiber in tension wrapped around it, is employed to convert the voltage into a phase-modulated signal in the interferometer. Hence,

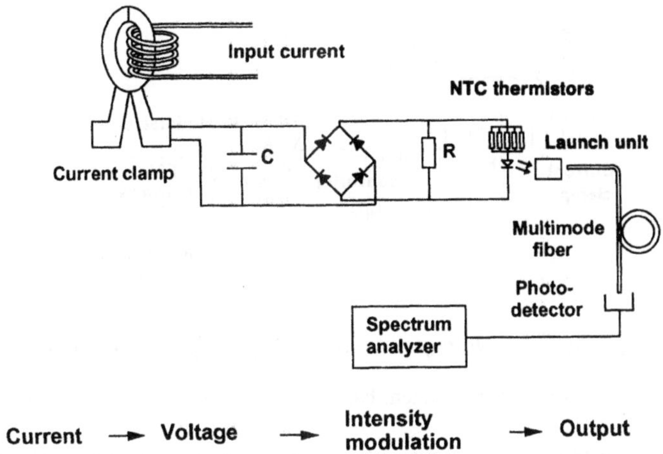


**Fig. 7.16** Hybrid current sensing system based on optical intensity modulation [44].

the variations in the length of the wrapped fiber and the concomitant phase changes induced in the interferometer are linearly related to the current changes induced in the secondary of the CT. The output of the interferometer is detected by a photodiode and then filtered with a low-Q bandpass filter, centred at 1 kHz, to generate a pseudo-heterodyne carrier. Thus the current-induced phase changes at 50 Hz in the interferometer are transposed to a phase modulation on the 1 kHz electronic carrier signal generated at the photodetector output. By using this pseudo-heterodyne signal recovery technique, a current range of 1–140 A with a resolution of $>0.1$ A per $\sqrt{\text{Hz}}$, over a frequency range $30\text{–}10^4$, has been demonstrated.

### 7.5.2 Hybrid OCS using an intensity modulation scheme

One of these sensors, shown schematically in Fig. 7.16, has been constructed with a conventional CT and a high-power light-emitting diode (LED) which is connected with the secondary of the CT and is used to convert the electrical signal into an optical intensity signal. If the LED is working within its linear region (for example, from 1 to 500 mA), its output intensity is proportional to the value of the current to be measured. By using this system, the current at high potential may be measured directly with the CT, whilst the output of the CT is then converted into an optical intensity signal which may be detected by the photodiode in a low potential region, via the multimode fiber link. Since the LED device is temperature-dependent, as its temperature increases, the output intensity of the LED will be reduced by typically 0.5% per °C, resulting in a sensitivity variation of around 11% (of the value at 25°C) over a temperature range from 0 to 50°C [44]. Hence, it is necessary to monitor the temperature of the LED and then modify the corresponding results in the signal processing scheme if an accurate measurement is to be obtained.

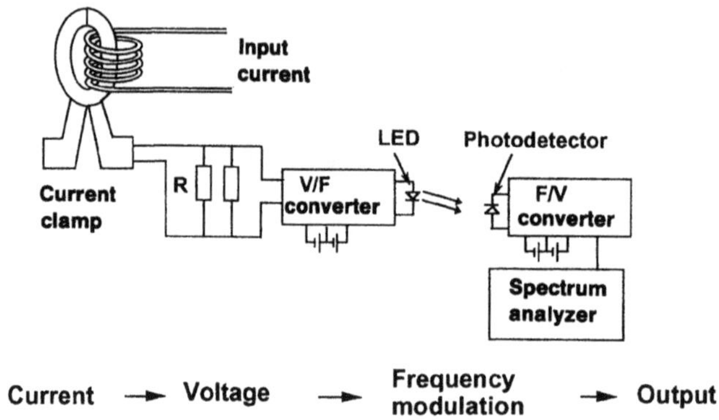


**Fig. 7.17** Hybrid current sensing system based on frequency modulation [44].

It can be seen that a hybrid current sensor using an optical intensity modulation offers the advantages of simplicity and low cost, in a robust and easy-to-implement configuration. This system would be particularly useful for indicating whether the conductor is carrying current or not, or for making lower accuracy measurements but at extremely high potentials. However, care must be taken in shielding the sensing section in order to prevent possible damage caused by a large fault current.

### 7.5.3 Hybrid OCS using a frequency-modulation scheme

One of the methods reported to overcome the temperature effect on the LED is the use of a frequency-modulation scheme [44], where the output of the CT is employed to generate a voltage signal by the use of two high-wattage resistors (4.7 ohm, 21 W) in parallel across the secondary of the CT, as illustrated in Fig. 7.17. This voltage is then converted into a frequency signal by using a voltage/frequency converter (supplied by Burr–Brown Corporation, VFC 100AG). The frequency signal, with its frequency being proportional to the current to be measured, is then converted into an optical signal by using a low-power LED, and it is transmitted via a fiber to the control room in the form of an optical pulse. The frequency-modulated light signal is detected by a photodiode and converted into a voltage output with another similar converter operating in a reciprocal mode. By using this system, a minimum detectable signal of 1 A at a frequency of 50 Hz with a bandwidth of 9.55 Hz (which is equivalent to 0.32 A per $\sqrt{\text{Hz}}$) has been achieved [44]. Because the VFC circuit requires a power supply, the sensing head has to be constructed as an active device, and some form of power source such as rechargeable batteries (which may be recharged by the use of a solar cell panel), or a second small conventional CT which may be employed to obtain power from the bus bar is required. An attractive and relatively simple approach would be to convert optical power from an inexpensive solid-state laser, transferred from the low potential area by a multimode fiber, into a voltage to drive the circuit directly [46, 47].

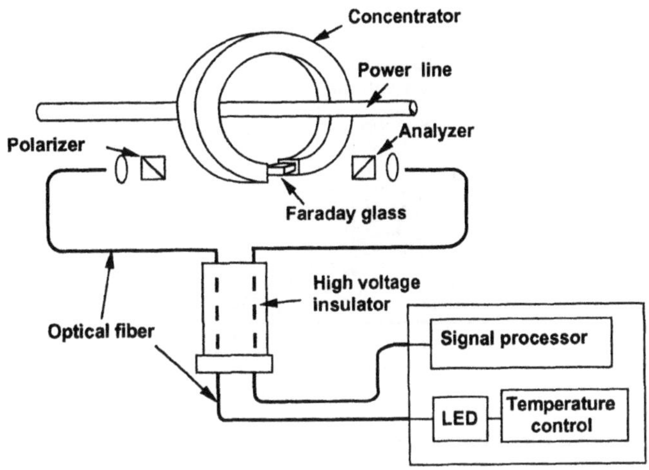


**Fig. 7.18** Current measurement system using a magnetic-field concentrator with a piece of Faraday glass [12].

### 7.5.4 Hybrid OCS using a polarization modulation scheme

By using a magnetic field concentrator made of high-permeability magnetic material to generate a relatively uniform magnetic field throughout the Faraday sensing element which is placed in the gap of the concentrator, as shown in Fig. 7.18, a nearly closed integration of the current-induced magnetic field, as required by Ampere's law, can be formed. The rotation of output azimuth is then proportional to the magnetic field and the length of the sensing element, and hence the corresponding current can be determined through measuring the rotation of the output azimuth [12]. To improve the sensitivity of this type of configuration, an interferometric detection scheme has been developed [42], where the sensing glass with a high Verdet constant is manufactured, such that a laser-sensing beam undergoes a multireflection inside the sensing element, as shown in Fig. 7.19. In the presence of a magnetic field, a circular birefringence proportional to the axial value of the field can be detected by using a multiphase detection scheme, and a dynamic range of more than 300 : 1 can be achieved with an interferometric sensitivity of only 0.1 cycle of phase for a maximum operating current of 3000 A, i.e. $5.3 \times 10^{-6}$ rad $A^{-1}$.

### 7.5.5 Hybrid OCS using a chromatic modulation scheme

By using liquid crystals (LC) as modulators to transduce a voltage related to the current to be measured into an optical signal, an electro-optic hybrid current sensor employing a chromatic modulation scheme has been introduced [49]. In this system, as shown in Fig. 7.20, a CT is employed both to provide power for an LC-drive electronic circuit and to generate a voltage proportional to the line

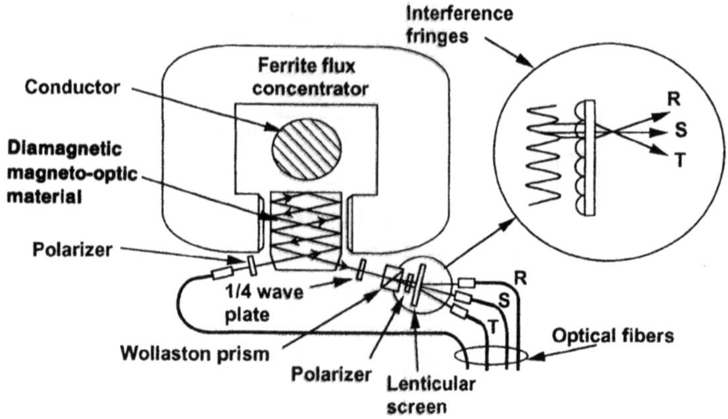


**Fig. 7.19** Layout of a current sensor with fiber links [42].

current. A broadband light source is used to illuminate the LC via an optical fiber, the transmitted light then being collected by another fiber and detected by a double-layer photodiode of which the layers have different but overlapping spectral sensitivities. The ratio of the short-circuit currents of the photodiode is defined as the chromaticity, and varies according to the dominant wavelength of the detected optical signal. Since the LC modulator attenuates a certain portion of the source spectrum by an amount which depends on the voltage applied to it, the ratio of the photodiode short-circuit currents may be used to represent the current to be measured. With this measurement system, within the dominant wavelength spread of approximately 90 nm, an average sensitivity of 0.45 nm $A^{-1}$ (i.e. the change in the dominant wavelength per A) over a dynamic range of 200 A rms was achieved. The main advantages of the system are its optical intensity-independent characteristics, the structural simplicity and the relatively low cost, although the temperature dependence of the system needs to be further reduced in order to achieve a better accuracy than has been reported to date.

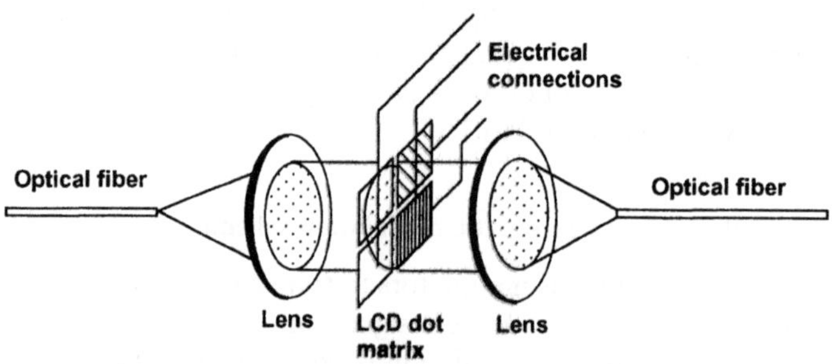


**Fig. 7.20** Four-element LC modulator used in a hybrid current sensing scheme based on chromatic modulation [49].

## 7.6 OCS SCHEMES BASED UPON MAGNETIC FIELD SENSING TECHNIQUES

As discussed earlier, with the use of a magnetic field sensor at a location near the current-carrying conductor, it is also possible to monitor or measure the variations of the current-induced magnetic field. As a result of intensive research, a number of magnetic sensing schemes have been proposed and reported for various applications in condition monitoring and fault-location systems, as well as in current measurement, if the systems are calibrated. Among these sensing schemes, fiber optic current sensors using the magnetostrictive effect and the Faraday effect are areas of active investigation. Several detailed reviews on developments in some of those fields have been published by a number of researchers [1, 50, 51]. In this paper, however, only a brief summary is given, together with some illustrative examples, in order to reflect the salient features of this important research area in the field of optical current measurement.

### 7.6.1 OCS devices based upon the magnetostriction effect

It is well known that when a magnetostrictive material is exposed to a magnetic field, its dimensions (and thus its shape) will vary as a function of the change of the magnetic field applied. If a fiber is bonded to a sample, or jacketed with a magnetostrictive material, the strain inside the fiber will be changed accordingly in response to the change in the shape of the sample or jacket, and this can be measured with the use of an optical fiber interferometer, usually a fiber Mach–Zehnder interferometer. The most commonly used magnetostrictive sensing elements have been constructed in the forms of either (a) a metallic glass cylinder wrapped around with a section of the fiber in the sensing arm of the interferometer [52], (b) a metallic glass ribbon [53] or an amorphous metal wire [54] bonded with a section of the sensing fiber, or (c) a section of the sensing fiber coated with magnetostrictive film [55, 56]. Since longer lengths of fiber (a few tens of metres) may be used to wrap around a cylindrical transducer in the first sensing configuration, the sensitivity of the device is thus substantially higher than that obtained with the second form of sensing element, although the latter has the advantages of smaller size, being easy to manufacture and inexpensive. As both sensor configurations require the use of epoxy adhesive to bond the fiber onto the magnetic material, the so-called 'loading effect' [52], which will reduce the magnetostrictive effect of the sensing material, is one of the important factors that has to be considered in the sensor design. The coated or jacketed fiber sensing elements, on the other hand, eliminate the loading effect and make the magnetostrictive strain couple more efficiently onto the fiber itself.

The major problems of the use of magnetostrictive sensing elements are that the presence of the sensor can modify the magnetic field to be measured, and the output of the sensor can be varied by the field generated by other currents. In addition, the sensing element itself becomes electrically conductive, which negates the major benefit of the nonconductive nature of optical fibers in problem current monitoring and measuring applications. Since the sensing elements have higher relative sensitivities to the magnetic field and a limited linear dynamic range, they are only suitable for the measurement of relatively small currents. Furthermore, the current measurements require the use of an a.c. dither or a d.c. bias, which makes the system more complicated.

### 7.6.2 OCS devices employing magneto-optic materials as sensing elements

The prime operating principle of the OCS devices using sensing elements which are made of magneto-optic materials is via the Faraday effect, as discussed in Section 7.2. The available optical materials which can produce a Faraday rotation angle large enough to allow a small current-induced magnetic field to be measured may be classified into three groups.

In the first group [57] are: (1) optical glasses, where the Verdet constant correlates well with the reciprocal dispersion; (2) diamagnetic doped glasses, where the glasses, which contain high concentrations of ions, give rise to large Verdet constants; (3) rare-earth glasses, i.e. paramagnetic ion-doped glasses which have the largest Verdet constants. In general, the Verdet constants are in the region of $10^{-5}$ to $10^{-4}$ rad $G^{-1}$ cm. The most commonly used materials are Schott glass SF6 ($2.6 \times 10^{-5}$ rad $G^{-1}$ cm at a wavelength of 633 nm), dysprosium alumina silicate glass ($-6.4 \times 10^{-5}$ rad $G^{-1}$ cm at a wavelength of 633 nm), and Hoya FR-5 glass ($-7.3 \times 10^{-5}$ rad $G^{-1}$ cm at a wavelength of 633 nm). By using an FR-5 glass rod as the sensing element, and a common-path optical fiber heterodyne interferometer to detect the current-induced Faraday rotation, Bartlett *et al.* [58] have demonstrated a remote current sensor with a sensitivity of 21.4° $A^{-1}$ over a region of 10–150 mA.

In the second group are the rare earth iron garnets such as yttrium iron garnet, $Y_3FE_5O_{12}$ (YIG), which is a widely available commercial crystal, showing transparency in the near infrared region. When a magnetic field is applied to this ferrimagnetic material, the measured azimuth rotation will depend upon the contribution from the actual alignment of the magnetization of the domains along the optical path. When a light beam with a large diameter is used to illuminate a crystal with a thickness of approximately 300 μm, it is possible to measure an average rotation of many domains [59], making the effect of the movement due to the individual domain walls less significant, and the results obtained will have a higher degree of reproducibility. In general, the Verdet constant of a YIG material is two to three orders of magnitude larger than that for typical diamagnetic glasses. By substituting some elements into a YIG crystal, the Verdet constant and the temperature coefficient of the material can be greatly improved. For instance (according to the results reported in [60]), for an unsubstituted YIG, the Faraday rotation angle and the temperature coefficients are 240° $cm^{-1}$ at a wavelength of 1.15 μm and ±7.0% over a temperature range of −20 to +60°C respectively; however, for Bi-substituted YIG, the Faraday rotation angle and the temperature coefficients are 2200° $cm^{-1}$ at the same wavelength and ±2.5%

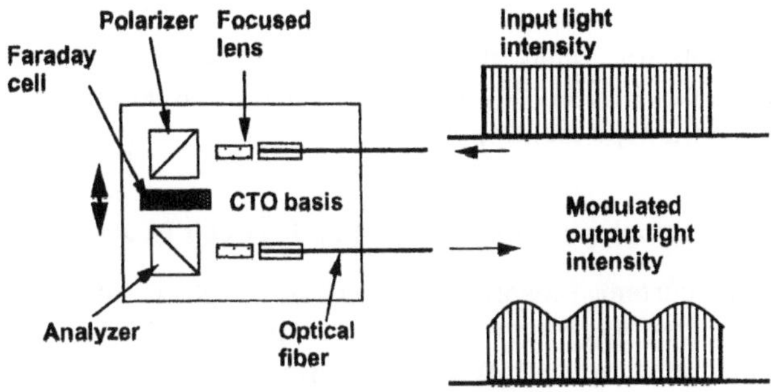


**Fig. 7.21** Principle and structure of an optical magnetic field sensor [60].

over the same temperature range, respectively, and when using Gd substitution (with a Verdet constant of $3.07 \times 10^{-2}$ rad $A^{-1}$ at a wavelength of 1.15 μm), the temperature coefficient can be further improved to ±0.5%. A schematic of one of the optical magnetic field sensors used for fault current detection is shown in Fig. 7.21. With the use of this sensor, the measurement of a maximum load line current of 600 A and a minimum zero-phase-sequence current of 0.1 A by ground fault have been demonstrated [60].

The third group of sensing materials comprises garnet films, which can be in the form of epitaxial thin films (a few micrometers in thickness) with a planar antisotropy [61], or thick films (one to a few hundred micrometers in thickness) with uniaxial anisotropy perpendicular to the surface [62]. The operational principles and the performance of those films are discussed in detail in the literature

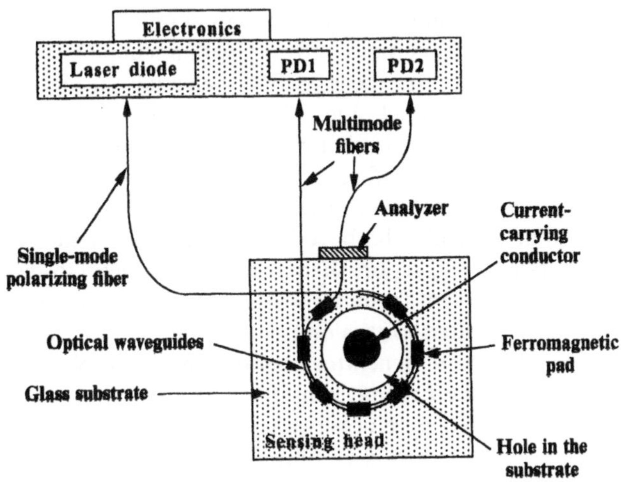


**Fig. 7.22** Schematic diagram of an integrated optical current sensor [63].

[61, 62]. The advantages in using garnet films, such as high Faraday rotation and high frequency response, make them attractive for applications such as fault current detection, or as a Faraday rotation element in an electro-optical hybrid current sensor, of the type mentioned earlier.

### 7.6.3 OCS devices employing an integrated optic circuit as a sensing element

A Faraday-effect current sensor, fabricated on a glass substrate with ion-exchanged waveguides, has been reported by Minier *et al.* [63] as a development of an earlier integrated optic (IO) current sensor [64]. The new configuration, shown in Fig. 7.22, shows periodic magnetic screening of the optical waveguides. In laboratory tests with a 0–1500-A, 50-Hz current generator, an accuracy of ±0.5% at 600 A has been demonstrated, but at lower currents, the accuracy is limited by the noise of the laser diode used to ±1% at 300 A and ±5% at 60 A.

## 7.7 COMMERCIALIZATION OF OPTICAL CURRENT SENSORS

As discussed previously, from the earliest work on optical methods of current sensing, meeting the durability and accuracy requirements of the electrical power industry have been prime concerns of the developers of optical devices. The route to commercial success has been difficult, and Day *et al.* [65] have discussed a number of aspects of the commercialization of research work. Several companies, in various parts of the world, are now producing optical current sensors based on bulk optical materials, mainly designed for the high voltage (>100 kV) region, where the costs are competitive with those of conventional devices, and typical weights are about 100 kg compared with about 7000 kg for a 500-kV conventional device. Market data are limited, but Day *et al.* [65] suggest that these sensors are being shipped in the hundreds of units per year, world-wide.

In Japan, the market appears more buoyant, with a shipping of some several thousand units per year for inexpensive fault-detecting current sensors. An important technology here is thick films of high magneto-optic constant, such as substituted iron garnets. If relatively low accuracy and linearity are tolerated, devices can be produced at prices of a few hundred dollars per unit, with the cost of electronics extra, thereby making these devices attractive for more widespread use. Sato *et al.* [66] of Hitachi, who have been working for some time on bulk optic current sensors [67, 68], report that there is still a need to develop these sensors to a more rugged form for more widespread use, although such a bulk optic sensor has been operated for the first time to measure the rotor excitation current of an adjustable speed generator operating at 400 MW [69]. Field tests with a flint glass fiber as the Faraday sensor element have been carried out by Tokyo Electric and reported by Kurosawa [29] using an rms current of 800 A and two sensing units. The report is of 'normal operation' over a test period of some 6 months. Ishizuka *et al.* [70] of Matsushita report on the performance of iron garnet crystals for power distribution fields which offer high sensitivity, linearity and stable temperature characteristics in measurements on a 6.6-kV power line, although quantitative data are not given. Katsukawa and Yokoi [75] have given details of an OCT manufactured by NGK Insulators in Japan using bismuth silicon oxide (BSO) single crystals meeting the metering requirements of 0.3% accuracy, using BSO with right optical rotating power combined with BSO with left optical rotatory power, and improved temperature characteristics. It was reported that over 1000 OCTs were in service in substations in Japan, from 1996. Sone [72] has discussed the installation of a ring-type Faraday current sensor being tested at a 77-kV substation, operating in a satisfactory way over 20 months, with ratio errors in general below 1%. This has led to the development of three-phase OCT devices for future substations [73]. A sensor of dimensions 250 mm × 250 mm × 40 mm and inner diameter 165 mm has been designed for low frequency (<5 Hz) currents, to accommodate a 100 mm × 100 mm conductor at rms currents of 8000 A, and has been in apparently successful operation since 1993 [74]. These recent results, coupled with continuing research on the important aspects of vibration compensation [75] and temperature drift compensation [76], are continuing to underpin the proven potential of these devices for current monitoring in real measurement situations.

## 7.8 SUMMARY

The operational modes of a range of optical current sensors have been systematically grouped, dismissed and reviewed, with specific representative examples given. The performance of some of these sensors is summarized in Table 7.1, enabling a cross-comparison and evaluation to be made more easily. This clearly shows that the research activities on current measurement using optical techniques has now been extended from the area where only optical fibers are employed as the sensing elements, to a wider region such as using bulk-glasses or hybrid devices to sense currents. It can be seen that as a result of the intensive degree of research on a number of different approaches, substantial progress in the optical current sensing field has been achieved in the past few years. The main problems in using optical fiber current sensors have been identified and now are better understood. Although considerable research effort has been expended, and a large number of the schemes have been introduced to solve the problems associated with the presence of birefringence in the case of OCS types using an all-optical fiber sensing element, these devices are still not widely suitable for use unless the difficulties caused by the various sources of birefringence within the sensing fiber are more effectively overcome. More recently, a Faraday sensor element made from flint glass fiber has been reported by Kurosawa *et al.* [77], and according to their report, the flint glass fiber (the Verdet constant is around $4.0 \times 10^{-2}$ min Oe cm at 850 nm) shows very low intrinsic birefringence and hence excellent polarization-maintaining properties. With the use of reciprocated light reflection by a mirror at the end of the fiber, the variation of the polarization plane caused by fiber bend-induced birefringence can be cancelled. This result shows a good example of a recent advance in the field of optical fiber current sensors, to overcome a major design problem.

With careful design, the reflection-induced phase shift of the sensing beam in those OCSs using bulk glass sensing elements can be cancelled or avoided by the use of the dual-quadrature reflection and critical angle reflections. Thus, these devices may operate according to Ampere's circular law, such that the induced Faraday rotation is not corrupted by the position of the current inside the sensor or the magnetic fields induced by neighboring conductors. These designs, being based upon various topologies using high Verdet constant bulk glasses, show superiority for the measurement of large currents at high potential. In addition to the advantages of being miniature, robust and low-cost systems, both high sensitivity and the ability to be demounted from a conductor can be achieved with the use of different topologies for the sensing elements. These sensors can be either permanently installed for metering purposes or demountable for diagnostic investigations in the electrical distribution industry.

Those OCS devices that are electro-optic hybrid sensing schemes represent another practical approach for improved current measurement systems. These systems benefit from the advantages of the well-characterized conventional current sensing devices which have already been accepted by industry. By taking advantage of various optical signal modulation techniques, together with all the advantages offered by the use of an optical fiber system, such as immunity to electromagnetic interference combined with a high level of electrical insulation, these systems allow a much less expensive CT to be used at very high potential, and offer the same security from EMI of any optical transmission system. From the point of view of the development of OCS techniques as a whole, the continuing research on OCS devices using electro-optic hybrid sensing elements may represent a necessary preliminary step towards all-optical current sensors being fully developed for, and accepted by, the power industry.

By using magnetic field sensors for line monitoring and current measurements, it is possible to achieve low-cost, simple and robust sensor configurations which may be very useful for indicating whether the conductor is carrying current, for providing diagnostic information and for warning of incipient failures.

Although the technology of the optical current sensor has been gradually developed, in order to ensure that the operational characteristics of the OCS devices are well matched with the standards accepted for current transformers for metering and protection applications, problems such as how to interface to and implement the OCSs in the power network, and the calibration of this kind of instrument are important aspects which still need to be more fully addressed. A more detailed discussion of those aspects may be found in a recent publication [78] and the references therein.

It would appear from the discussion that practical optical current measurement devices may be expected to be produced in different forms for different applications. However, a fully developed all-optical current sensor may be used for applications to both fault monitoring and power metering. The differences between devices in the laboratory and in practice are gradually disappearing.

## REFERENCES


1. Rogers, A. J. (1995) Optical fiber current measurement, *Optical Fiber Sensor Technology* (Eds K. T. V. Grattan and B. T. Meggitt). Chapman & Hall, London, pp. 421–439.
2. Day, G. W. (1989) Recent advances in Faraday effect sensors, *Springer Proceedings in Physics*, **44**, *Optical Fiber Sensors*. Springer, Berlin, p. 250.
3. Rogers, A. J. (1973) Optical technique for measurement of current at high voltage, *Proc. IEE*, **120**(3), 261–267.
4. Bush, S. P. and Jackson D. A. (1992) Numerical investigation of the effects of birefringence and total internal reflection on Faraday effect current sensors, *Appl. Opt.*, **31**(25), 5366.
5. Ning, Y. N., Wang, Z. P., Palmer, A. W., Grattan, K. T. V. and Jackson, D. A. (1995) Recent progress in optical current sensing techniques, *Rev. Sci. Instrum.*, **66**, 3097–3111.
6. Born, M. and Wolf, E. (1986) *Principles of Optics*, 6th edn., Pergamon Press, Oxford.
7. Rogers, A. J. (1977) Optical methods for measurement of voltage and current on power systems, *Opt. Lasers Technol.*, 273.
8. Donati, S., Annovzzi-Lodi, V. and Tambosso, T. (1988) Magneto-optical optical current sensors for electrical industry: analysis of performances, *IEE Proc.*, J, **135**(5), 372.
9. Sato, T., Takahashi, G. and Inui, Y. (1983) Method and apparatus for optically measuring a current, *Europe Patent*, Publication No. 0088419 A1.
10. Ning, Y. N., Chu, B. C. B. and Jackson, D. A. (1991) Miniature Faraday current sensor based on multiple critical angle reflections in a bulk-optic ring, *Opt. Lett*, **16**, 1996.
11. Chu, B. C. B., Ning, Y. N. and Jackson, D. A. (1992) Faraday current sensor that uses a triangular-shaped bulk-optic sensing element, *Opt. Lett.*, **17**, 1167.
12. Ulmer, E. A. Jr. (1988) High accuracy Faraday rotation measurement, *OSA/IEEE 1988 Technical Digest of Optical Fiber Sensors Topical Meeting*, 27–29 January, New Orleans, LA, p. 288.
13. Ulmer, E. A. Jr. (1990) A high-accuracy optical current transducer for electric power systems, *IEEE Trans. Power Delivery*, **5**(2), 892.
14. Kersey, A. D. and Jackson, D. A. (1986) Current sensing utilizing heterodyne detection of Faraday effect in single-mode optical fiber, *IEEE J. Lightwave Technol.*, **4**(6), 2084.
15. Kersey, A. D. and Davis, M. A. (1989) All-fiber Faraday-rotation current sensor with remote laser-FM based heterodyne detection, *Springer Proceedings in Physics*, **44**, *Optical Fiber Sensors*, Springer, Berlin, p. 285.
16. Leilabady, P. A., Wayte, A. P., Berwick, M., Jones, J. D. C. and Jackson, D. A. (1986) A pseudo-reciprocal fiber-optic Faraday rotation sensor: current measurement and data communication applications, *Opt. Commun.*, **59**(3), 173.
17. Jackson, D. A., Kersey, A. D., Corke, M. and Jones, J. D. C. (1982) Pseudo-heterodyne detection scheme for optical interferometers, *Electron. Lett.*, **18**, 1081.
18. Ulrich, R. and Simon, A. (1979) Polarisation optics of twisted single-mode fibers, *Appl. Opt.*, **18**(13), 2241.

19. Laming, R. I. and Payne, D. N. (1989) Electric current sensors employing spun highly birefringent optical fibers, *IEEE J. Lightwave Technol.*, **7**(12), 2084.

20. Day, G. W. and Etzel, S. M. (1985) Annealing of bend-induced birefringence in fiber current sensors, *Tech. Digest Int. Conf. Integrated Optics and Optical Fiber Communication. European Conf. Optical Communication*, Venice, p. 871.

21. Tang, D. and Day, G. W. (1988) Progress in the development of miniature optical fiber current sensors, *IEEE Lasers and Electro-Optics Society LEOS'88 Annual Meeting*, Conference Proc., pp. 306.

22. Tang, D., Rose, A. H. and Day, G. W. (1990) Practical considerations in the design of optical fiber current sensors SPIE, **1267**, *Fiber Optic Sensors IV*, p. 29.

23. Ren, Z. B. and Robert, Ph. (1989) Input polarization coding in fiber current sensors, *Springer Proceedings in Physics*, **44**, *Optical Fiber Sensors*, Springer, Berlin, p. 261.

24. Ben-Kish, A., Tur, M. and Shafir, E. (1991) Geometrical separation between the birefringence components in Faraday-rotation fiber-optic current sensors, *Opt. Lett.*, **16**(9), 687.

25. Ahlers, H. and Bosselmann, Th. (1990) Complete polarization analysis of a magneto-optic current transformer with a new polarimeter, *Conf. Proc. 7th Optical Fiber Sensors Conference*, p. 81.

26. Chu, W., McStay, D. and Rogers, A. J. (1991) Current sensing by mode coupling in fiber via the Faraday effect, *Electron. Lett.*, **27**(3), 207.

27. Laming, R. I. and Payne, D. N. (1989) Electric current sensors employing spun high birefringence optical fibers, *IEEE J. Lightwave Technol.*, **7**(12), 2084.

28. Pistoni, N. C. and Marttinelli, M. (1993) Vibration-insensitive fiber-optic current sensor, *Opt. Lett.*, **18**(4), 314.

29. Kurosawa, K. (1996) Optical current transformers using flint glass fiber as a Faraday sensor element. *Conf. Proc. 11th Optical Fiber Sensors Conference*, Sapporo, Japan, pp. 134–139.

30. Kurosawa, K., Yoshida, S. and Sakamoto, K. (1994) Polarization maintaining properties of the flint fiber for the Faraday sensor element, *Conf. Proc. 10th Optical Fiber Sensors Conference*, Glasgow, SPIE, **2360**, 28–31.

31. Yoshida, S., Kurosawa, K. and Sano, O. (1996) Development of an optical current transformer using a flint glass fiber for a gas circuit break, *Conf. Proc. 11th Optical Fiber Sensors Conference*, Sapporo, Japan, pp. 172–175.

32. Rogers, A. J., Xu, J. and Yao, J. (1994) Vibration immunity for optical-fiber current measurement, *Conf. Proc. 10th Optical Fiber Sensors Conference*, Glasgow, SPIE, **2360**, 40–44.

33. Kung, A., Nicati, P. A. and Robert, P. A. (1996) Brillouin fiber optic current sensor, *Conf. Proc. 11th Optical Fiber Sensors Conference*, Sapporo, Japan, pp. 156–159.

34. Kanoi, M., Takahashi, G., Sato, T., Higaki, M., Mori, E. and Okmura, K. (1986) Optical voltage and current measuring system for electric power systems, *IEEE Trans. Power Delivery*, **1**, 91.

35. Caese, T. W. and Johnston, P. (1990) A magneto-optic current transducer, *IEEE Trans. Power Delivery*, **5**, 548.

36. Cease, T. W., Driggans, J. G. and Weikel, S. J. (1991) Optical voltage and current sensors used in a revenue metering system, *IEEE Trans. Power Delivery*, **6**, 1374.

37. Chu, B. C. B., Ning, Y. N. and Jackson, D. A. (1992) Polarization analysis of bulk optic Faraday current sensor with a triangular configuration, *San Diego '92 – 37th Annual International Symposium on Optical and Optoelectronic Applied Science and Engineering*, SPIE, **1746**, 21–23.

38. Ning, Y. N. and Jackson, D. A. (1993) Faraday effect optical current sensor using a bulk glass sensing element, *Opt. Lett.*, **18**(10), 835.

39. Ning, Y. N., Wang, Z. P., Grattan, K. T. V. and Palmer, A. W. (1995) Faraday current sensor using a novel multioptical-loop sensing element, *Meas. Sci. Technol.*, **6**(9), 1339–1342.

40. Zhang, W., Ning, Y. N., Chu, B. C. B., Grattan, K. T. V. and Palmer, A. W. (1996) Vibration-induced noise in a fiber lead of an optical current measurement system, *Rev. Sci. Instrum.*, **67**(2), 553–557.

41. Ning, Y. N., Liu, Y., Grattan, K. T. V., Palmer, A. W. and Weir, K. (1994) The relation between the coherence length and modal noise in a graded index multimode fiber for white light interferometric systems, *Opt. Lett.*, **19**(6), 372–374.

42. Hercher, M. (1991) Ultra-high resolution interferometric sensors, *Optical Photonics News*, **11**, 24.

43. Ning, Y. N., Chu, B. C. B. and Jackson, D. A. (1991) Interrogation of a conventional current transformer via a fiber optic interferometer, *Opt. Lett.*, **16**(18), 1448.

44. Ning, Y. N., Liu, T. Y. and Jackson, D. A. (1992) Two low-cost robust electro-optic hybrid current sensors capable of operation at extremely high potential, *Rev. Sci. Instrum.*, **63**(12), 5771.

45. Jackson, D. A., Ning, Y. N., McGarrity, C. and Santos, J. L. (1992) Three phase current measurement using a hybrid current sensing technique, *Conf. Proc. 8th Optical Fiber Sensors Conference*, Monterey, CA, p. 426.

46. Kirkham, H. and Johnston, A. R. (1989) Optically powered data link for power system applications, *IEEE Trans. Power Delivery*, **4**(4), 1997.

47. Adolfson, M., Einnvall, C. H., Lindberg, P., Samuelson, J., Ahlgren, L. and Ediund, H. (1989) EHV series capacitor banks. A new approach to platform to ground signalling, relay protection and supervision, *IEEE Trans. Power Delivery*, **4**(2), 1369.

48. Tonnesen, O., Beatty, N. and Skilbreid, O. (1989) Electrooptic methods for measurement of small DC currents at high voltage level, *IEEE Trans Power Delivery*, **4**(3), 1568.

49. Pilling, N. A., Holmes, R. and Jones, G. R. (1993) Optical fiber current measurement system using liquid crystals and chromatic modulation, *IEE Proc.*, C, **140**(5), 351.

50. Bucholtz, F., Koo, K. P., Kersey, A. D. and Dandridge, A. D. (1986) Fiber optic magnetic sensor development, *Fiber Optic and Laser Sensors IV, SPIE*, **718**, 56, 1986.

51. Bucholtz, F., Dagenais, D. M., Koo, K. P. and Vohra, S. (1990) Recent developments in fiber optic magnetostrictive sensors, *Fiber Optic and Laser Sensors VIII, SPIE*, **1367**, 226.

52. Bucholtz, F., Dagenais, D. M. and Koo, K. P. (1989) High-frequency fiber-optic magnetometer with 70fT/$\sqrt{\text{Hz}}$ resolution, *Electron. Lett.*, **25**(25), 1719.

53. Bucholtz, F., Koo, K. P., Sigel, G. H. Jr. and Dandridge, A. D. (1985) Optimization of the fiber/metallic glass bond in fiber-optic magnetic sensors, *IEEE J. Lightwave Technol.*, **3**(4), 814.

54. Koo, K. P., Bucholtz, F., Dagenais, D. M. and Dandridge, A. D. (1989) A compact fiber-optic magnetometer employing an amorphous metal wire transducer, *IEEE Photonics Technol. Lett.*, **1**(12).

55. Jarzynski, J., Cole, J. H., Bucaro, J. A. and Davis, C. M. Jr. (1980) Magnetic field sensitivity of an optical fiber with magnetostrictive jacket, *Appl. Opt.*, **19**(22), 3746.

56. Bibby, G. W., Larnson, D. C., Tyagi, Y. and Bobb, L. C. (1992) Fiber optic magnetic field sensors using metallic-glass-coated optical fibers, *Conf. Proc. 8th Optical Fiber Sensors Conference*, p. 161.

57. Borrelli, N. F. (1964) Faraday rotation in glasses, *J. Chem. Phys.*, **41**(11), 3289.

58. Bartlett, S. C., Farahi, F. and Jackson, D. A. (1990) A common path optical fiber heterodyne interferometric current sensor, *Conf. Proc. 7th Optical Fiber Sensors Conference*, p. 85.

59. Svantesson, K., Sohlstrom, H. and Holm, U. (1990) Magneto-optical garnet materials in fiber optic sensor systems for magnetic field sensing, *Electro-Optic and Magneto-Optic Materials II, SPIE*, **1274**, 260.

60. Imaeda, M. and Kozuka, Y. (1992) Optical magnetic field sensors using iron garnet crystals, *Conf. Proc. 8th Optical Fiber Sensors Conference*, p. 386.

61. Wolfe, R. and Lieberman, R. A. (1991) Fiber optic magnetic field sensor based on domain wall motion in garnet film waveguides, *Appl. Phys. Lett.*, **58**(16), 1733.

62. Wolfe, R., Gyorgy, E. M., Lieberman, R. A., Fratello, V. J., Licht, S. J., Deeter, M. N. and Day, G. W. (1992) High frequency magnetic field sensors based on the Faraday effect in garnet thick films, *Conf. Proc. 8th Optical Fiber Sensors Conference*, p. 390.

63. Minier, V., Persegol, D., Lovato, J. L. and Kevorkian, A. (1996) Integrated optical current sensor for a high-power system, *Conf. Proc. 11th Optical Fiber Sensors Conference*, Sapporo, Japan, pp. 164–167.

64. Minier, V., Danel, A., Persegol, D. and Kevorkian, A. (1995) *Proc. ECIO'95*, pp. 379, Delft, The Netherlands.

65. Day, G. W., Rochford, K. B. and Rose, A. H. (1996) Fundamentals and problems of fibre current sensors, *Conf. Proc. 11th Optical Fiber Sensors Conference*, Sapporo, Japan, pp. 124–129.

66. Sato, T., Sone, I., Hayashida, H. and Nakagama, Y. (1996) Development and applications of bulk-optic current sensors, *Conf. Proc. 11th Optical Fiber Sensors Conference*, Sapporo, Japan, pp. 130–133.

67. Kanoi, M. *et al.* (1986) Optical voltage and current measuring system for electric power system, *IEEE Power Delivery*, **PWRD-1**(1), 91–97.

68. Kobayashi, S. *et al.* (1991) Development of optical current transformers with the portions of free gas space propagation of light, *T. IEE Japan*, **111-B**(9), 999–1006.

69. Fujimoto, T. *et al.* (1994) Commercial operation of optical current transformers for overcurrent detection, Power and Energy Division Convention Records of IEE Japan, No. 641, pp. 862–863.

70. Ishizuka, S., Itoh, N. and Minemoto, H. (1996) Optical fibre current sensors using garnet crystal for power distribution fields. *Conf. Proc. 11th Optical Fiber Sensors Conference*, Sapporo, Japan, pp. 140–141.

71. Katsukawa, H. and Yokoi, S. (1996) Optical current transducer with a bulk type BSO Faraday sensor for power systems, *Conf. Proc. 11th Optical Fiber Sensors Conference*, Sapporo, Japan, pp. 142–143.

72. Sone, I. (1996) Ring glass type Faraday effect current sensor, *Conf. Proc. 11th Optical Fiber Sensors Conference*, Sapporo, Japan, pp. 144–145.

73. Takagi, H. *et al.* (1994) Future-oriented substation using computer technologies, *International Conference on Large High Voltage Electric System* (CIGRE), 23/13-04.

74. Kuwabara, T. *et al.* Design and dynamic response characteristics of 400 MW adjustable speed pumped storage unit for Ohkawachi power station, IEEE PES 1995 Summer Meeting, No. 95SM615-5EC.

75. Willsch, M. and Bosselmann, T. (1996) Vibration compensation for a glass ring type magneto optic current sensor, *Conf. Proc. 11th Optical Fiber Sensors Conference*, Sapporo, Japan, pp. 148–151.

76. Niewisch, J., Menke, P., Krammer, P. and Bosselmann, T. (1996) Temperature drift compensation of a potential transformer using a BSO Pockels cell, *Conf. Proc. 11th Optical Fiber Sensors Conference*, Sapporo, Japan, pp. 152–155.

77. Kurosawa, K., Yoshida, S. and Sakamoto, K. (1994) A method for improvement of immunity from environment in Faraday effect current using the flint glass fiber, *Conf. Proc. 10th Optical Fiber Sensors Conference, SPIE*, **2360**, 430–433.

78. Emerging Technologies Working Group, Power Systems Instrumentation and Measurements Committee, The Fiber Optic Sensors Working Group, Fiber Optics Subcommittee, Power Systems Communications Committee, Optical current transducers for power systems: a review, *IEEE Trans. Power Delivery*, **9**(4), 1778–1788, 1994.

# 8

# Microbend fiber optic sensors

John W. Berthold III

## 8.1 INTRODUCTION

The microbend sensor was one of the earliest fiber optic sensors. Microbend losses have always been a curse to the fiber optic cable designer, but it is this very same microbend loss effect in optical fibers which was exploited by the microbend sensor designer who adapted the microbend effect to the measurement of many physical parameters and physical variables such as temperature and pressure.

Microbend sensors are very interesting sensors with some outstanding performance characteristics that have made them successful in several commercial applications. These characteristics and applications, and some problems inherent to microbend sensors, are discussed in this paper.

Sensors based on microbend loss in optical fibers were first proposed and demonstrated in 1980 [1, 2], although careful experiments done earlier had demonstrated vibration-induced intensity modulation of light in bent fibers [3]. Since this early work, much has been done to understand microbend sensors and to investigate how to increase dynamic range and improve sensitivity to the measurement parameter of interest, while reducing sensitivity to unwanted variables.

The early interest in microbend sensors was for hydrophone applications [4–6], and this work was driven by the Navy's FOSS (fiber optic sensor systems) program. Since that time, over 100 different studies on microbend sensors have appeared in the literature, and the sensors have been adapted to many different measurement applications.

In addition to the general advantages of fiber optic sensors so often touted (EMI resistance, electrical isolation and freedom from ground loops, small size and weight), microbend sensors have been demonstrated to have their own unique set of advantages. These advantages include:

1. mechanical and optical efficiency that leads to low parts count and low cost;
2. easy mechanical assembly that does not require fiber bonding to other components and thus avoids differential thermal expansion problems;



*Optical Fiber Sensor Technology*, Vol. 3. Edited by K. T. V. Grattan and B. T. Meggitt.
Published in 1998 by Kluwer Academic Publishers, London. ISBN 978-1-4419-4736-9

3. fail-safe fiber sensor which either produces a calibrated output signal or fails to a state of no light output.

In addition, microbend sensors have been used in hostile environment applications such as in high-temperature zones and explosion hazard areas.

## 8.2 BACKGROUND AND PHYSICS OF MICROBENDER TECHNOLOGY

What is microbending? In the simplest terms, the mechanical perturbation of a multimode fiber waveguide causes a redistribution of light power among the many modes in the fiber. The more severe the mechanical perturbation or bending, the more light is coupled to radiation modes and is lost. Thus, the important characteristics of a microbend fiber optic sensor (microbender) are that it uses a multimode optical fiber, it is a light intensity sensor, and the light intensity decreases with mechanical bending. Most of these characteristics are problems for the fiber cable designer, but Fields *et al.* [1] came up with a way to enhance the microbending effect by squeezing the fiber between a set of corrugated plates called deformer plates, or tooth blocks. Their configuration is shown in Fig. 8.1. They found that by tuning the mechanical bend frequency, the microbending loss could be increased by several orders of magnitude. Figure 8.2 shows plots of sensor transmission *vs.* tooth spacing for different tooth block displacements, *D*, for both graded- and step-index fiber. The graded-index fiber exhibits a resonant response and the step-index fiber a threshold response.

A microbender is called an intrinsic fiber optic sensor because light is not permitted to exit from the fiber into free space. The microbender is a displacement sensor. When the separation between the tooth blocks changes (Fig. 8.1), the sinusoidal amplitude of the clamped fiber changes accordingly. The light transmission through the clamped region is a very sensitive function of the sinusoidal amplitude of the bent fiber.

Fields [2] analyzed all of these results and arrived at the basic set of equations used to predict microbend sensor performance.

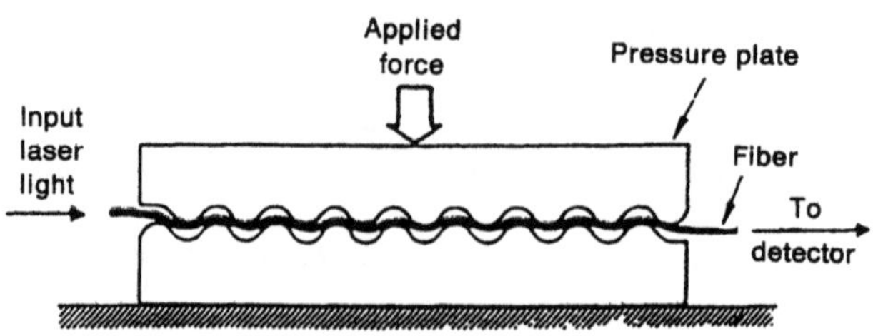


**Fig. 8.1** Original diagram of a microbend fiber optic sensor taken from [1].

The power law refractive index profile is

$$n^2(r) = n_0^2[1 - 2\Delta(r/a)^\alpha]. \tag{8.1}$$

The propagation constant difference is

$$\delta\beta_m = \frac{2}{a}\left(\frac{\alpha\Delta}{\alpha+2}\right)^{1/2}\left(\frac{m}{M}\right)^{(\alpha-2)/(\alpha+2)} \tag{8.2}$$

For graded-index fibers, $n(r)$ is the refractive index, which is a function of distance, $r$, between the center of the core and the cladding, $\Delta$ is the normalized index difference between core and cladding, $a$ is the fiber radius, $\beta$ is the propagation constant for the $m$th mode, $m$ is the mode number and $M$ is the total number of modes. For $\alpha = 2$, the refractive index profile is parabolic and the difference $\delta\beta_m$ between propagation constants for adjacent modes is a constant. This allows calculation of the period of the tooth spacing $\Lambda_c$ on the tooth blocks to achieve the resonance condition where the microbend loss is sharply peaked (see Fig. 8.2):

$$\delta\beta_c = 2\pi/\Lambda_c \tag{8.3}$$

$$\Lambda_c = 2\pi a/(2\Delta)^{1/2}. \tag{8.4}$$

For step-index fibers $\alpha = \infty$ and there is no resonance condition because the difference $\delta\beta_m$ between adjacent mode pairs is not a constant. However, it is possible to calculate a threshold value for the period of the tooth spacing, and for $\Lambda$ less than the threshold, the microbend loss increases.

$$\delta\beta_m = (2\Delta^{1/2}/a)(m/M) \tag{8.5}$$

$$\Lambda \leqslant \Lambda_s = \pi a/\Delta^{1/2}. \tag{8.6}$$

Lagakos *et al.* [7] confirmed that in a microbender, light is lost from the fiber core via coupling between guided modes and radiation modes. In a multimode fiber, the higher-order modes are those modes that are most easily coupled out of the fiber at small bends. Work done by Horsthuis and Fluitman [8], Diemeer and Trommel [9] and Lagakos and Bucaro [10] provides the microbender designer with information on how to select the tooth spacing $\Lambda$ on the microbender blocks to achieve the best sensitivity.

Other workers, including Mavadaat [11], have done ray analyses to predict microbend sensor performance. This work was done in 1984 and produced results that agreed well with the modal analyses. However, both the modal and ray models have had difficulty in predicting the light loss *vs.* displacement of the deformer plates, as well as the saturation effects that occur at large displacements when the core light is substantially depleted. Analyses are still being performed on microbend sensors to improve predictions of light loss *vs.* displacement of the deformer plates. Recent work combines finite element modeling of the sensor mechanics with numerical model soft beam propagation [12], and wave

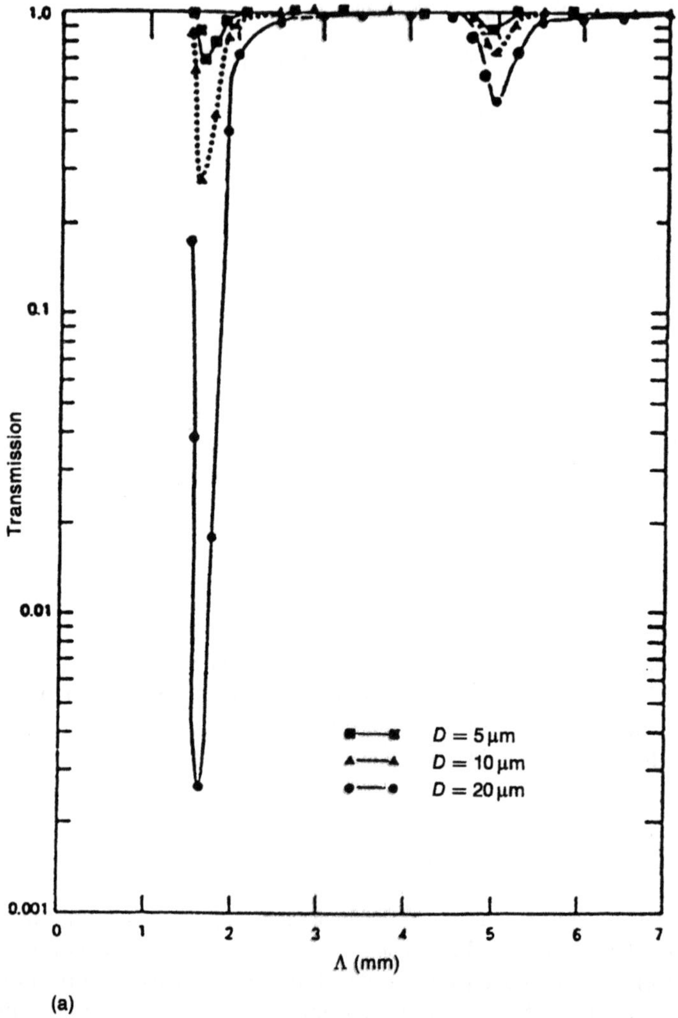


**Fig. 8.2** Transmission *vs.* tooth spacing (Λ) for various tooth block displacements (*D*) (from [1]). (a) Parabolic-profile fiber. (b) Step-profile fiber.

optics calculations for all the modes in bent multimode fibers [13] hold great promise for understanding effects in microbend sensors such as changes in sensitivity with temperature.

In the clamped region of the microbender, it is important to ensure that light coupled out of the core does not re-enter. Yao and Asawa [14] have shown how radiation modes in the cladding can re-enter the core and reduce the sensitivity of the microbender. This work was presented at the first Optical Fiber Sensors (OFS) Conference held in London in 1983. The most straightforward way to

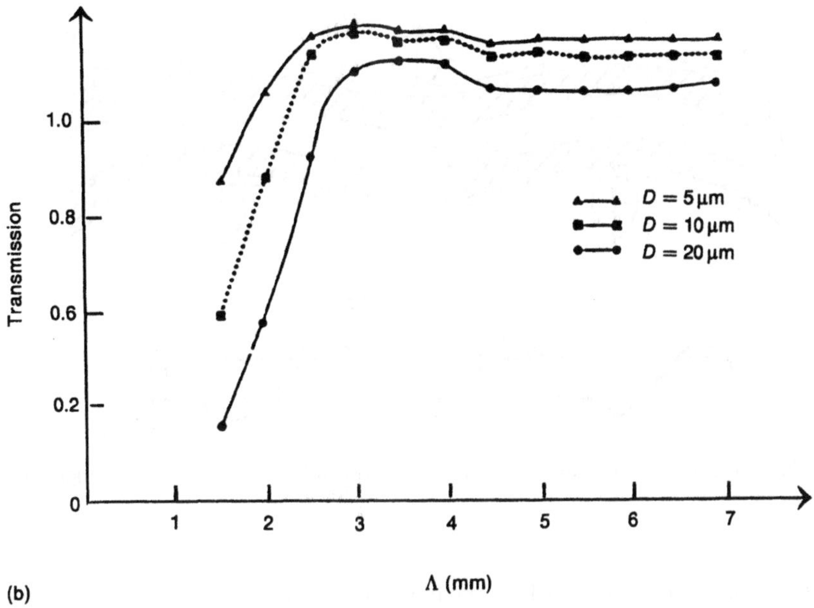


**Fig. 8.2** Continued.

remove light from the cladding is to use an optical fiber with an absorptive buffer coating to perform radiation mode stripping along the sensing region.

## 8.3 EARLY APPLICATIONS

Two examples of early applications of microbend sensors are described in this section. At the third OFS conference held in San Diego in 1985, Harmer [15] presented the results of a study on the effects of fiber jacket materials on the performance of microbenders. The intention of the study was to optimize the performance of a distributed microbender over the entire length of the jacketed fiber. This was one of the first reports on the development of distributed fiber optic sensors. He found that fibers with rubber and teflon jackets exhibited microbend sensitivity in a similar way to fibers with no jacket.

Harmer's work led to the commercialization of a distributed microbend fiber optic sensor by Herga Electric. In this sensor, described by Oscroft [16] and illustrated in Fig. 8.3, the fiber is deformed by a spiral deformer such as a stiff spring, where the period of the spiral matches the optimum fiber bend period (see equation 8.4). The Herga distributed sensor has been used in a number of commercial applications including tactile sensing in the bumpers of automatic guided vehicles to provide automatic shut-off.

A second interesting application of microbenders was investigated by Jenstrom and Chen [17] at Purdue University. Two parallel arrays of optical fibers at right

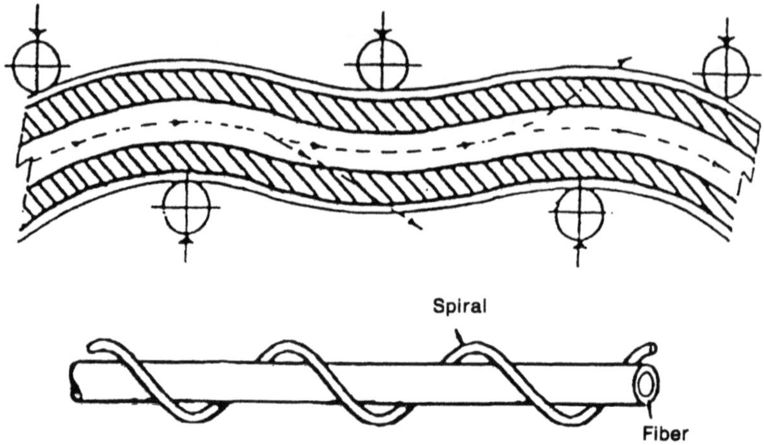


**Fig. 8.3** Diagrams of the Herga distributed microbend sensor, the first commercialized microbender.

angles form an $n \times n$ array where the fiber spacing equals the optimum deformer period $\Lambda$. The parallel array of fibers below acts to deform each of the fibers laid across in the right angle parallel array above, so that an array of $m \times n$ microbend sensors is created. Each fiber has its own light source and detector. The problem with decoding the light intensity changes is that with a $5 \times 6$ fiber array, for example, there are 11 known variables, i.e. the light intensity changes through each fiber, but a total of 30 unknown light intensities at each overlap joint in the $5 \times 6$ array. Jenstrom and Chen arrived at a very clever algorithm to decode the 11 light intensity changes from each fiber to obtain the pressure distribution and contour over the entire grid.

An important result from the design of this array is the linearity and low hysteresis of each microbend sensor in the array and the potential low cost of the arrays, all of which would be ideal for robotic tactile sensing.

## 8.4 MICROBENDER ENGINEERING

### 8.4.1 General

In 1987, Lagakos *et al.* [18] published an excellent analysis on the operation of microbend sensors which built on the early work of Fields [2]. Lagakos *et al.* derived some very useful formulas for microbender modeling and design. The change in light transmission, $\Delta T$, through a microbend sensor can be written as

$$\Delta T = \left(\frac{\Delta T}{\Delta X}\right) \Delta F \left(k_{\mathrm{f}} + \frac{A_{\mathrm{s}} Y_{\mathrm{s}}}{l_{\mathrm{s}}}\right)^{-1}, \tag{8.7}$$

where $\Delta X$ is the displacement of the deformer plates, $\Delta T/\Delta X$ is the sensitivity (slope of the transmission *vs.* displacement response curve for a microbender), $\Delta F$ is the applied force, $A_s$, $Y_s$ and $l_s$ are the area, Young's modulus and the thickness of the spacer material between the deformer plates, respectively, and $k_f$ is the effective spring constant of the optical fiber clamped between the deformer plates:

$$k_f^{-1} = \frac{\Lambda^3}{3\pi Y d^4 \eta}. \tag{8.8}$$

The effective spring constant for the assembly is an important parameter for the designer, and is expressed in terms of $\Lambda$, the deformer tooth spacing, cubed, Young's modulus, $Y$, for the glass, the fiber diameter $d$, and the number of bends, $\eta$.

The critical parameter from the design standpoint is the sensitivity $\Delta T/\Delta X$ in equation (8.7). This parameter depends on the properties of the optical fiber, including core and clad refractive indices, the change in these indices with temperature, the type of fiber buffer coating, and the optical power distribution among the many modes in a multimode fiber. The engineering of microbenders for specific applications requires a detailed understanding of how all of these factors affect microbender sensitivity, and how to trade off variations of these factors in order to optimize $\Delta T/\Delta X$ for the specific application.

### 8.4.2 Microbender advantages and disadvantages

Shown in Fig. 8.4 is a microbender configured as a weldable strain gage. It is useful to examine this simple microbender in some detail in order to understand

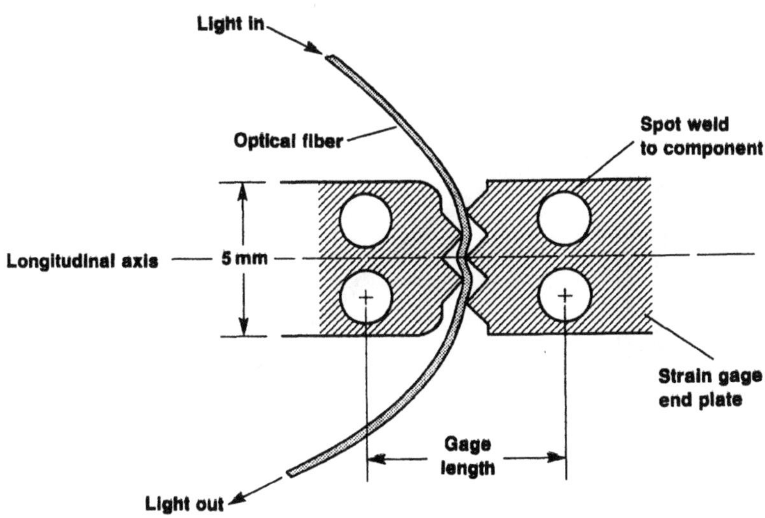


**Fig. 8.4** Fiber optic microbend sensor configured for longitudinal strain measurement.

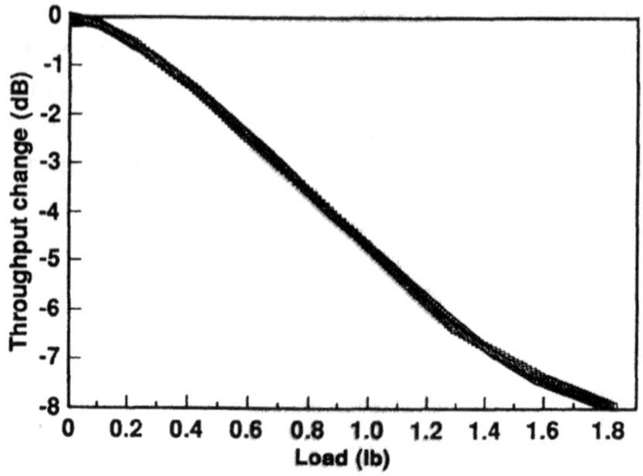


**Fig. 8.5** Measured throughput (log transmission) *vs*. load.

the specific performance advantages and shortcomings of microbenders. The deformer plates in this application are welded in place, and the period of the tooth spacing $\Lambda$ is 1.5 mm. The fiber is step-index, polyamide-buffered 100/140/170. For this specific configuration, the change in the log of transmission *vs.* applied force is given in the plot in Fig. 8.5. This plot represents hundreds of measured data points, as does the plot in Fig. 8.6 which shows the measured log of transmission *vs.* displacement of the deformer plates as the plates are moved closer together. The characteristics of this typical microbend sensor response function are worth noting. There are three distinct regions of sensitivity. The first nonlinear region (0–15 μm displacement) where the deformer spacing is approximately the fiber diameter, the intermediate linear operating region (15–50 μm displacement), and the final nonlinear region beyond 50 μm, where depletion of light from available modes is observed. The mechanical properties

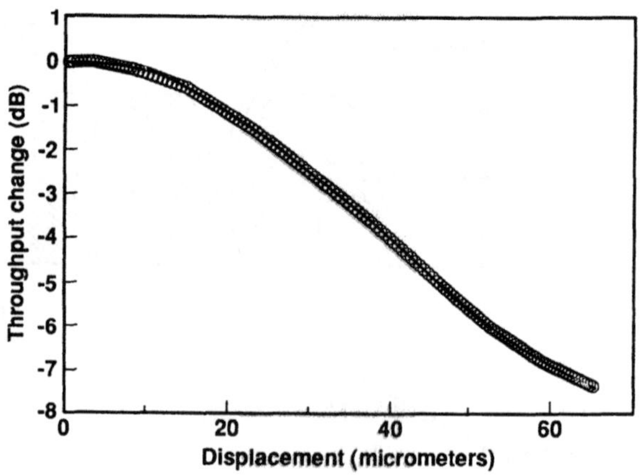


**Fig. 8.6** Measured throughput (log transmission) *vs*. displacement.

of the fiber buffer coating determine the shape of the first region and whether or not hysteresis is present. The slope of the linear region determines the sensitivity, which is $0.15\,\mathrm{dB}\,\mu\mathrm{m}^{-1}$, and the location of the knee in the curve ($50\,\mu$ in this example) sets the high end of the dynamic range.

Although not as sensitive as interferometric sensors, microbenders have respectable sensitivity to small displacements of the tooth blocks. Displacement sensitivities of the order of $10^{-10}$ m per $\sqrt{\mathrm{Hz}}$ have been reported [18, 19]. The dynamic range of microbenders is remarkable, and one of the major advantages of these sensors. A practical mode of operation is to use a photodiode detector to convert the light intensity to a photocurrent, and then process the photocurrent to obtain a microbender response function which is the log of transmission *vs.* tooth block displacement (Fig. 8.6). In this mode, the microbender response to small displacements exhibits a linear (1%) dynamic range greater than $10^4$. This dynamic displacement range is achievable for d.c. measurands if 0.001 dB changes in light power can be repeatedly detected. This performance is possible using present-day silicon photodiodes and high-quality instrumentation amplifiers.

Microbenders are most useful as contact sensors for measurement of small displacements in mechanical systems. In general, these applications include those where the mechanical components include two or more springs, either in series or in parallel. In order to make effective use of microbenders in such measurement applications, it is important to understand the limitations of the sensors and how to design around these limitations. The critical consideration from a mechanical standpoint is the fiber spring constant (see equation 8.8). The fiber clamped in the tooth blocks acts as a spring. The spring constant (stiffness) varies with tooth spacing, number of teeth and fiber diameter, but is generally in the range 500–2000 pounds per inch. To ensure linear and repeatable performance, it is important that the microbender stiffness be small compared with the stiffness of other components in the mechanical system.

Microbend sensors share disadvantages common to other fiber optic intensity sensors. The primary disadvantage, which has many manifestations, is its sensitivity to optical power level. In a fiber optic system, changes in optical power level can result from changes in the output of the optical source, changes in the coupling between fiber connectors, changes in transmission along the fiber link, and changes in transmission or coupling through other fiber optic components in series with the microbender. There can also be problems with fiber strength, and the effects of the fiber buffer coating on microbender accuracy and repeatability. Progress has been made in all of these areas over the last few years, and each area is briefly discussed in the next few paragraphs.

### 8.4.3 Self-referencing

The development of microbender signal processing methods to compensate and correct for all of the potential changes in optical power level has been the object of extensive work over the last few years. Various schemes have been

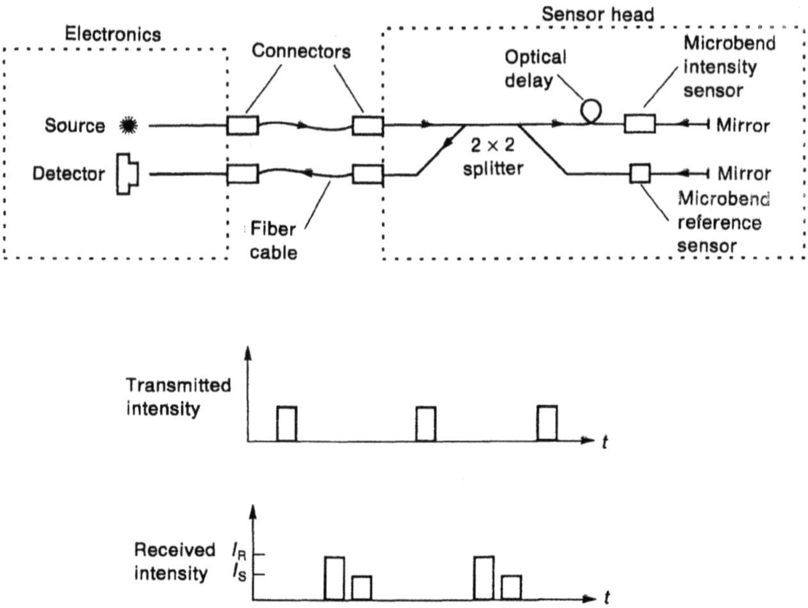


**Fig. 8.7** Time-division normalization.

tried to self-reference fiber optic sensors, and several of these self-referencing methods work well with microbenders. [20–22].

Time-division normalization is a self-referencing method that has been successfully used with microbend sensors to reduce significantly the error contribution of light intensity changes caused by cable bending, light source fluctuations and connector losses. Spillman and Lord [20] and Lindsay and Morton [21] have developed similar types of time-referencing methods. In the example shown in Fig. 8.7, time delay is introduced into the sensing arm of a 2 × 2 splitter so that light pulses in the sensing and reference arms arrive back at the splitter at different times. The ratio of the two pulses is calculated in the receiver. Spurious intensity changes which affect both light signals in the same proportion cancel in the ratio. An error plot (Fig. 8.8) shows the effect that a 6-dB power change has on the output signal. The error introduced is about one-quarter of 1% of full scale.

A similar approach based on wavelength normalization instead of time has been used by other workers for intensity self-referencing. In the example shown in Fig. 8.9, a broadband light source is divided into two wavelength channels – a sensor and a reference channel. The two channels are recombined, and separated again at the receiver. The signals are processed in two steps to eliminate light-level changes in the fiber link caused by cable bends or connector losses that result in proportional signal modulation in both channels. First, a reference light intensity signal from each LED source in the electro-optic transmitter is used to compensate the corresponding signal channel from the sensor.

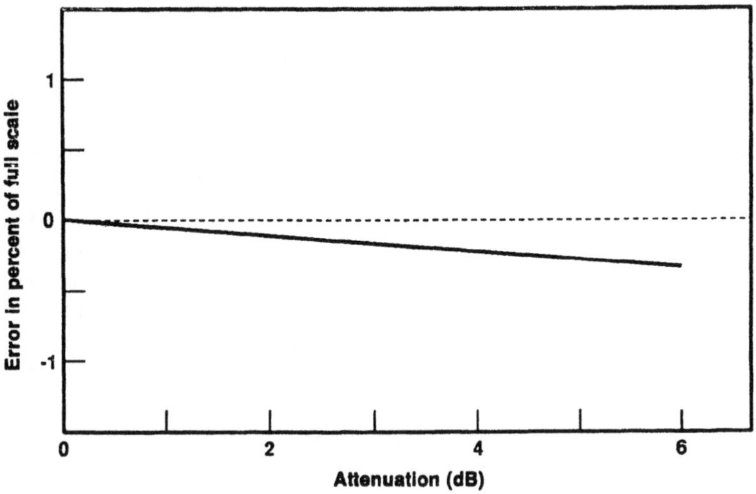


**Fig. 8.8** Error *vs.* power change measured with a time-division normalization processor.

The compensation may be performed using bi-cell photodetectors and log ratio current amplifiers. This operation on each sensor wavelength channel removes optical signal variations caused by light-source intensity fluctuations. Next, the resultant signal voltages are subtracted in a difference amplifier. This operation removes optical signal variations caused by cable and connector losses, which have the same effect on both wavelength channels (to first order). Ideally then, all resulting changes in the difference amplifier output are proportional only to measurand-induced changes in the microbenders.

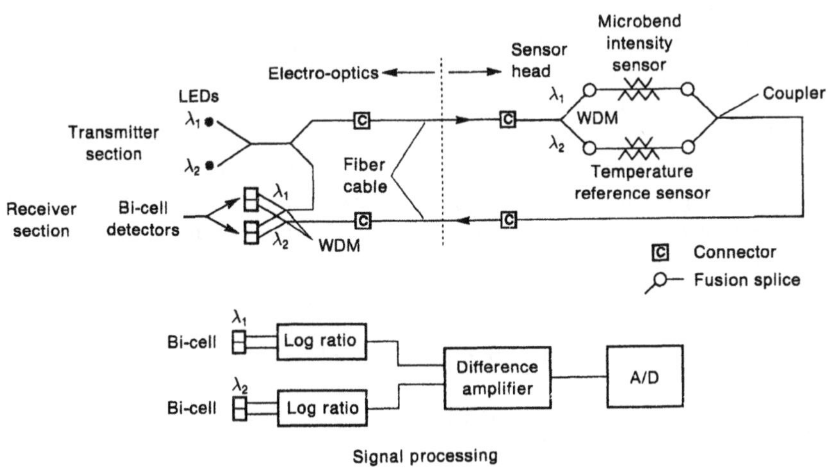


**Fig. 8.9** Wavelength division normalization and signal processing. WDM, wavelength division multiplexing.

### 8.4.4 Modal sensitivity

A significant disadvantage unique to microbenders is the result of the very phenomenon which makes this sensor work. The modal sensitivity of the microbender also makes it a modal filter. Thus, any modal filtering action by other components in the multimode fiber optic link is a potential error source. The fiber itself is a modal filter as well as fiber connectors and power splitters. Self-referencing methods help to correct for the modal filtering action of the fiber optic link and the connectors. Careful system choices (e.g. waveguide coupler *vs.* fused biconic coupler) can help avoid unwanted modal filtering by components, and various proprietary methods can be used to ensure modal mixing of the light input/output to/from the microbender. One method to reduce modal sensitivity is to perform mode mixing at the microbender input and output. High numerical aperture (NA)/low NA fiber junctions may be used for this purpose [18]. The trade-off with this approach is that the microbender becomes a high insertion loss component, 20 dB being typical. This high insertion loss severely limits the number of microbend sensors which can be multiplexed. Typically, under these conditions, only two or three high-performance microbend sensors can be multiplexed onto a single fiber.

### 8.4.5 Fiber strength and buffer coating

In addressing the fiber strength issue, remember that deformer plates clamp the fiber. Thus, large stresses can be produced in the fiber. If the plates are brought too close together, the fiber will break. An empirical design rule is to maintain a ratio of maximum applied stress to fiber break stress of less than 1:4. Under these conditions, the fiber life is effectively unlimited. (One attractive feature for many applications that results from the clamped fiber microbender is the positive 'no-light' indication of a broken fiber.) The 1:4 stress ratio guideline was confirmed by Miers *et al.* [23] over 10 years ago and was based on extensive cyclic life-test data for microbenders. Lifetimes exceeding 10 million cycles of the deformer plates can be expected for modest fiber stresses of 100 000 psi.

Since the microbend teeth push into the buffer coating, it is important to understand the mechanical, physical and chemical interactions among the deformer plate material, the buffer coating material and the glass optical fiber. Empirical data has been gathered, much of it proprietary. The important issues with any buffer coating material are whether or not it introduces sensor zero shift and span shift. All buffer coating materials introduce these errors to some extent, but the errors can be made negligible by doing a cyclic shake-down of the microbender over the displacement range of the deformer plates. The desirable characteristics for a fiber buffer coating are listed in Table 8.1.

**Table 8.1** Desirable buffer coating characteristics

| |
|---|
| Mechanically stable |
| Wear resistant |
| Light absorber |

**Table 8.2** Application areas for microbenders

| | |
|---|---|
| Aerospace | Medical |
| Automotive | Mining |
| Chemical processes | Power generation |
| Environmental | Robotics |
| Maglev | Structural/civil |
| Manufacturing | Undersea/offshore |

## 8.5 APPLICATIONS SUMMARY

Over the years, microbenders have been configured for the measurement of many different parameters, including pressure [24–27], temperature [28, 29], acceleration [30, 31], flow [32], local strain [33, 34] and speed [35]. Microbender arrays have been used in tactile sensing systems [17, 36] and in distributed sensing systems for temperature [37], strain [38, 39] and structural monitoring [40, 41]. Table 8.2 lists areas of application where microbenders have now at least been designed. In some of these areas, prototypes have been built and tested, and in a few of the areas, commercial products have been or are currently offered.

For detailed discussions on many of these applications the reader may refer to several excellent references [42–44]. It is important to note that in these applications, microbenders have been shown to be stable and repeatable [45, 46]. Accelerated life-test data [23] have shown that the lifetime of the sensing fiber can exceed $10^{10}$ cycles, and that wear of the fiber buffer coating over the microbend lifetime is not a performance limitation.

Table 8.3 summarizes the performance characteristics of microbenders based on demonstration of prototypes and commercial devices up to the present time. The advantages and disadvantages are listed, along with a subjective assessment of the level of electronics complexity needed to do the signal processing functions in order to achieve high levels of performance.

'Easy-to-hold fiber' refers to the mechanically clamped fiber in the microbender configuration. No glass-to-metal adhesives are needed, which is an important feature necessary to achieve repeatable performance.

'Intrinsic' implies resistance to fouling, which is a common problem with fiber optic sensors that have exposed fiber ends. The fail-safe feature refers to the fact that a broken optical fiber fails to a 'no-light' condition.

Thermal sensitivity may be improved to $<5 \times 10^{-5}$ per °C with an independent temperature measurement, but this approach is not always practical.

**Table 8.3** Microbender performance summary

| Advantages | Neutral | Disadvantages |
|---|---|---|
| Repeatability<br>Easy-to-hold fiber<br>Intrinsic<br>Fail-safe | Moderate electronics complexity | Thermal sensitivity about $10^{-4}$ per °C<br>Insertion loss |

A 20-dB insertion loss for a fully self-referenced and modally conditioned microbender precludes multiplexing of more than just a few sensors on a single fiber.

Engineering trade-offs associated with the required operating temperature range *vs.* specified dynamic range *vs.* measurement accuracy set the microbender design constraints, which in turn determine the influence that modal filtering has on the overall achievable error limits. What this means in practice is that presently achievable microbender performance is about ±0.05% absolute accuracy at constant temperature, about ±1% relative accuracy with temperature compensation (0–100°C compensated range using a reference microbender as a compensator), and about ±0.5% absolute accuracy over the temperature range using an independent temperature measurement [46].

## 8.6 CONCLUSIONS

Microbend fiber optic sensors have found many applications in metrology, and have been used in several commercial products. Significant improvements in the performance of microbenders have been made in parallel with gains in practical understanding and with the introduction of effective self-referencing methods to correct for changes in optical power level.

Continued improvements in the performance of microbenders may be expected as the modal sensitivity of fiber optic components is reduced, and as improved methods of modal mixing and temperature compensation are developed. Even at the present time, however, the overall performance of microbenders is more than adequate for many industrial and military applications, especially in hostile environments. Furthermore, microbenders are readily adapted to distributed sensing. The relatively low cost of the microbender compared with other types of fiber optic sensors ensures that it will be used in many applications and systems over the next few years.

## REFERENCES

1. Fields, J. N. *et al.* (1980) Fiber optic pressure sensor. *J. Acoust. Soc. Am.*, **67**, 816–818.
2. Fields, J. N. (1980) Attenuation of a parabolic-index fiber with periodic bends. *Appl. Phys. Lett.*, **36**, 799–801.
3. Nelson, D. F. *et al.* Vibration-induced modulation of fiberguide transmission. *Proc. Topical Meeting on Optical Fiber Transmission*, TU-E7-1 to TU-E7-4.
4. Fields, J. N. and Cole, J. H. (1980) Fiber microbend acoustic sensor. *Appl. Opt.*, **19**, 3265–3267.
5. Lagakos, N. *et al.* (1981) Microbending fiber optic sensor design optimization. *Proc. CLEO'81*, p. 100.
6. Lagakos, N. *et al.* (1982) Microbend fiber optic sensor as extended hydrophone. *IEEE J. Quantum Electron.*, **QE-18**, 1633–1638.
7. Lagakos, N. *et al.* (1981) Multimode optical fiber displacement sensor. *Appl. Opt.*, **20**, 167–168.

8. Horsthuis, W. H. G. and Fluitman, J. H. J. (1982) Sensitivity dependence on number of bends in a fiber optic microbend pressure sensor. *NTG-Fachberichte*, **79**, 147–152. (1983) Development of fibre optic microbend sensors. *Sensors Actuators*, **3**, 99–110.
9. Diemeer, M. B. J. and Trommel, E. S. (1984) Fiber optic microbend sensors: sensitivity as a function of distortion wavelength. *Opt. Lett.*, **9**, 260–262.
10. Lagakos, N. and Bucaro, J. A. (1987) Optimizing fiber optic microbend sensor. *Proc. SPIE*, **718**, 12–20.
11. Mavadaat, R. (1984) Ray analysis of microbend fibre sensors. *Sensors Actuators*, **6**, 289–295.
12. Hastings, M. C. and Nippa, D. F. (1994) Integration of the finite element and beam propagation methods to determine performance of microbend sensors. *Proc. OFS*, **10**, 376–379.
13. Wanser, K. H. *et al.* (1994) Novel fiber devices and sensors based on multimode fiber Bragg gratings. *Proc. OFS*, **10**, 265–268.
14. Yao, S. K. and Asawa, C. K. (1983) Microbending fiber optic sensing. *Proc. SPIE*, **412**, 9–13.
15. Harmer, A. L. (1985) Distributed microbending sensor. *Proc. OFS*, **3**, 126–128.
16. Oscroft, G. (1987) Intrinsic fiber optic sensors. *Proc. SPIE*, **734**, 207–213.
17. Jenstrom, D. T. and Chen, C. L. (1989) A fiber optic microbend tactile sensor array. *Sensors Actuators*, **20**, 239–248.
18. Lagakos, N. *et al.* (1987) Microbend fiber optic sensor. *Appl. Opt*, **26**, 2171–2180.
19. Berthold, J. W. (1984) High temperature fiber optic pressure sensor. *Proc. ISA*, **30**, 85–94.
20. Spillman, Jr., W. B. and Lord, J. R. (1987) Self-referencing multiplexing technique for intensity modulating fiber optic sensors. *Proc. SPIE*, **718**, 182–191.
21. Lindsay, T. A. and Morton, J. D. (1989) Standard fiber optic interface for aerospace applications: time domain intensity normalization. *Proc. SPIE*, **989**, 45–55.
22. Wlodarczyk, M. T. (1991) Environmentally insensitive commercial pressure sensor. *Proc. SPIE*, **1368**, 121–131.
23. Miers, D. R. *et al.* (1986) Life tests of optical fibers in a microbend configuration. *ACerS, 88th Annual Meeting*, Glass Division, Chicago, IL.
24. Krohn, D. A. *et al.* (1983) Understanding fiber optics for automated control. Part II. Sensing techniques, applications, and economics. *Plant Eng.*, **37**, 56–58.
25. Spenner, K. (1985) Microbending pressure and displacement sensor. *Proc. OFS*, **3**, 146–148.
26. Berthold, J. W. *et al.* (1987) Design and characterization of a high temperature fiber optic pressure transducer. *J. Lightwave Technol.*, **LT-5**, 870–876.
27. Reed, S. E. *et al.* (1993) Fiber optic total pressure transducer for aircraft applications. *Proc. SPIE*, **2070**, 17–23.
28. Lumholt, O. *et al.* (1991) Simple fiber optic low-temperature sensor that uses microbending loss. *Opt. Lett.*, **16**, 1355–1357.
29. Cutolo, A. and Gallo, M. (1991) Microbending optoelectronic sensor for on-line temperature measurements in high-power electrical systems. *Eur. Trans. Elec. Power Eng.*, **1**, 281–287.
30. Freal, J. B. *et al.* (1987) A microbend accelerometer for borehole deployment. *J. Lightwave Technol.*, **LT-5**, 993–996.
31. Miers, D. R. *et al.* (1988) Design and characterization of fiber optic accelerometers. *Proc. SPIE*, **838**, 314–317.

32. Sovik, R. E. (1985) Implementation of fiber optics in a vortex shedding flowmeter. *Proc. ISA Digitech '85*, pp. 207–211.

33. Reed, S. E. and Berthold, J. W. (1986) Development of a microbend strain gage. *SEM Fall Conference on Experimental Mechanics*, Keystone, CO. See also US Patent 5 020 379.

34. Weiss, J. D. (1989) Fiber optic strain gage. *J. Lightwave Technol.*, **7**, 1308–1318.

35. Varshneya, D. *et al.* (1991) Fiber optic speed sensor for advanced gas turbine engine control. *Proc. SPIE*, **1367**, 181–191.

36. Schoenwald, J. S. and Martin, J. F. (1984) Fiber optic tactile sensor for robot grippers. *Soc. Manuf. Eng.*, MS84-1041.

37. Falco, L. and Debergh, P. (1989) Birmorphous distributed transducer for temperature threshold sensor. *Proc. SPIE*, **1011**, 166–172.

38. Asawa, C. K. *et al.* (1982) High sensitivity fibre optic strain sensors for measuring structural distortions. *Electron. Lett.*, **18**, 362–364.

39. Rogers, A. J. (1991) Distributed optical fiber sensing. *Proc. SPIE*, **1508**, 2–24.

40. Udd, E. *et al.* (1990) Microbending fiber optic sensors for smart structures. *Proc. SPIE*, **1170**, 478–482.

41. Wanser, K. H. *et al.* (1993) Distributed fiber optic sensors for civil structures using OTDR. *CA State Fullerton Conference Monograph 9308 W3*, pp. 303–327.

42. Culshaw, B. and Dakin, J. (eds) (1989) *Optical Fiber Sensors: Systems and Applications*. Artech House, Norwood, MA.

43. Udd, E. (1991) *Fiber Optic Sensors: An Introduction for Engineers and Scientists*. Wiley, New York.

44. Zuech, N. (1992) *Handbook of Intelligent Sensors for Industrial Automation*. Addison-Wesley, Reading, MA.

45. Berthold, J. W. (1991) Field test results on fiber optic pressure transmitter system. *Proc. SPIE*, **1584**, 39–47.

46. Berthold, J. W. *et al.* (1994) Flight test results from FOCSI fiber optic total pressure transducer. *Proc. SPIE*, **2295**, 216–222.

# 9

# Intrinsic position sensing using optical fiber and coherence domain polarimetry

S. Chen and B. T. Meggitt

## 9.1 INTRODUCTION

Detection of the physical position of an object is one of the most important sensing applications in industry and engineering. Not only is the gauging of physical dimensions with such position sensor systems important in its own right, but the monitoring of many operational and control parameters in engineering applications can also be converted into position sensing by use of various transducer systems.

Currently, the position sensing of short distances of up to 1 mm is dominated by capacitance measurement-based systems with reported resolutions in the sub-nanometer region. In the longer distance range, most products now available are based on inductive coupling methods, e.g. the linearly variable differential transformer (LVDT). In many other applications, optical position encoding/decoding is desirable where the property of electro-magnetic induction (EMI) immunity is of particular concern. The international standard for length measurement is based on optical interferometry using frequency-stabilized lasers. Many position-sensing systems use the same operating principles, where a bulk mirror or retroreflector is attached to the object under measurement, and its position is determined by measuring the imbalance or optical path difference (OPD) of the interferometer. Although such systems are capable of achieving a dynamic range better than $10^9$, they are generally too expensive and fragile to be employed in most industrial environments. Systems with optical grating rulers and Moiré fringe read-out schemes have been used in industrial environments, but like conventional interferometric systems, the use of bulk optic components, such as mirrors, lenses and gratings is prone to unavoidable contamination build-up. This fact severely limits their performance capability and sensor lifetime. Optical-fiber-based and in-fiber position sensing techniques may be considered to perform position measurements, even intrinsically within a piece of highly birefringent optical fiber. Such a scheme takes full advantage of the linear,

*Optical Fiber Sensor Technology*, Vol. 3. Edited by K. T. V. Grattan and B. T. Meggitt.
Published in 1998 by Kluwer Academic Publishers, London. ISBN 978-1-4419-4736-9

one-dimensional, dielectric nature of the optical fiber medium, and like all optical fiber sensor methods offers immunity to electromagnetic interference as well as environmental contamination, showing a significant operating advantage over previous (extrinsic-type) position-sensing devices.

In addition, owing to the flexibility of optical fibers, this type of intrinsic technique can be used to monitor positions along routes of any topography over a wide spread of dynamic ranges. This feature means that the approach suits a variety of position-sensing application areas, including rotary angular measurements, as well as linear displacement measurement, and offers an alternative to many of the extrinsic optical position-sensing systems discussed in previous chapters.

## 9.2 PRINCIPLES OF OPERATION

### 9.2.1 Optical fiber white-light interferometry

The fundamental operating principles of this intrinsic fiber sensor approach is based on optical fiber white-light interferometry [1]. As shown in Fig. 9.1, an optical fiber sensor system using white-light interferometry normally consists of two interferometers linked by an optical fiber. One is a sensing interferometer, of which the OPD responds to the measurand, and the other is a processing interferometer in which the OPD is known and can be scanned over a certain range. A low-coherence optical source is used in the system, which has a coherence length much shorter than the OPDs in either the processing or sensing interferometers, and these two OPDs are presented as $OPD_p$ and $OPD_s$ in Fig. 9.1. Under this condition, the interference fringes will only appear when $OPD_p \approx 0$ and $|OPD_p| \approx |OPD_s|$, as shown in Fig. 9.2, where the precise position is indicated by the bright fringe of maximum visibility at the center of the fringe signature. The interference order of this fringe is zero, and it is thus termed the 'zero-order fringe' or 'central fringe'. If this particular indicator can be detected by the system, the OPD of the sensing interferometer can be measured by the known OPD of the processing interferometer.

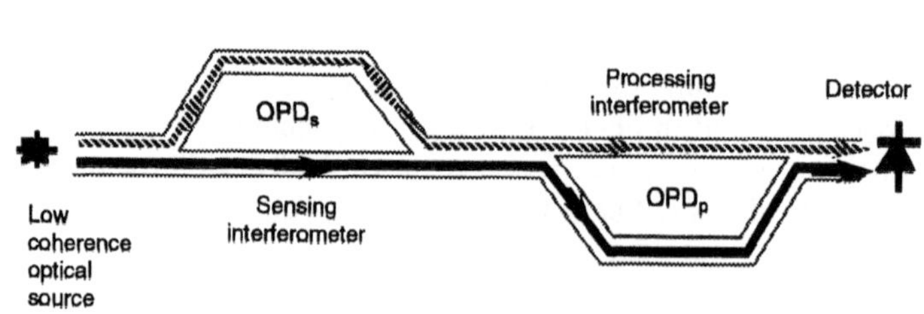


**Fig. 9.1** Theoretical model of an optical fiber sensor system employing white-light interferometry.

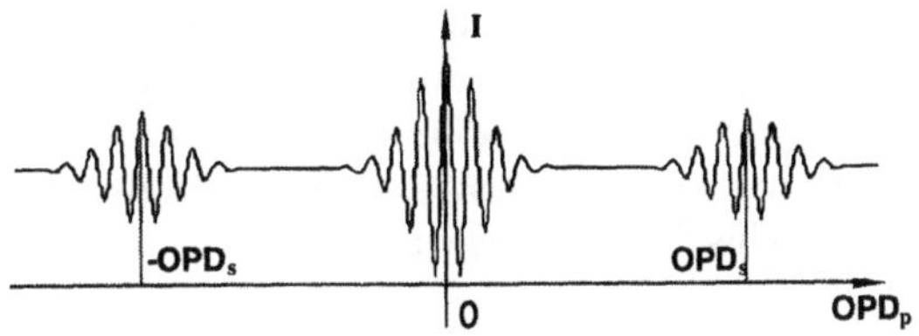


**Fig. 9.2** Theoretical output fringe pattern of an optical fiber sensor system employing white-light interferometry.

### 9.2.2 Optical coherence domain polarimetry

Optical coherence domain polarimetry (OCDP) [2] is a special white-light interferometric system incorporating highly birefringent fiber. As shown in Fig. 9.3, the system presented uses a length of optical fiber into which is launched polarized light from a low-coherence source such as superluminescent diode (SLD) or multimode laser diode. The fiber is of the high birefringence type, which possesses two polarization modes (axes) orthogonal to each other. The input polarizer is so oriented that all the optical power is initially launched into only one of the polarization axes, say the fast axis, in the highly birefringent fiber. The sensor is configured around a set (straight) length of highly birefringent fiber along which a movable wiper exerts a pressure force at its point of contact. The wiper arm is attached to the position measurement arm of the sensor, and its point of contact with the fiber will vary with the movement of the position arm between a minimum and a maximum travel motion, in a similar fashion to the classic electrical potentiometer structure. By arranging the correct orientation of the optical axes of the highly birefringent fiber, the action of the point pressure

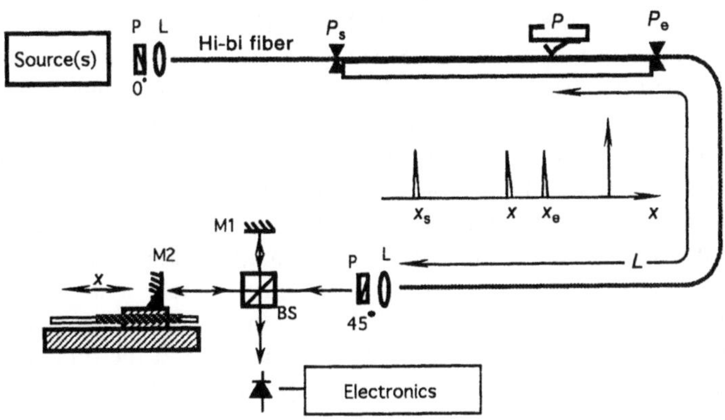


**Fig. 9.3** Schematic diagram of the intrinsic optical fiber position monitoring system. $P_s$ and $P_e$, end markers; $P$, variable position wiper; M1 and M2, J2 mirrors; BS, beam splitter; P(0) and P(45°), polarizers.

contact of the wiper arm will be to induce some of the light in the fast axis to couple across to the unpopulated orthogonal polarization mode, the slow axis. There is a difference in refractive index between the two polarization modes, and therefore an OPD will exist between the wave packets emerging from the two orthogonal modes. This OPD, $x_w$, is written as

$$x_w = BL_w, \tag{9.1}$$

where $B$ is the birefringence of the fiber, and $L_w$ is the length of the fiber from the contact point to the exit end. It can be seen that this OPD will depend upon the position of the wiper arm relative to the fiber exit end and will be proportional to it. Hence, by making a measurement of the OPD between the two modes, the position of the wiper arm can be determined. From the previous discussion, it is clear that this OPD can be measured using optical fiber white-light interferometry. By using a Michelson interferometer at the fiber output as a processing interferometer, it is possible to adjust one arm of the interferometer away from its balance point until an equal, but opposite, temporal delay is introduced between the two modes of the beam. The degree of imbalance of this processing interferometer, $x_p$, is a direct measure of the wiper position. This can be expressed as

$$L_w = 2x_p/B. \tag{9.2}$$

By using a low-coherence optical source, the position where the coupled and remaining fractions of light are brought back into exact temporal coincidence could be detected through the location of a central (zero-order) fringe within the interferometric fringe signature produced.

The birefringence of the highly birefringent fibers is normally in the range 2–$8 \times 10^{-4}$. This means that a precision positioning device with a small scanning range can be used to detect the wiper position over a much larger range.

Experiments undertaken to verify the concept of the position-sensing system described above [3] used a set-up similar to the one shown in Fig. 9.3. A single low-coherence optical source was first used, which was a multimode laser diode operating below its threshold, yielding a coherence length of 30 fringes at 750 nm. A 20-cm length of York HB750 bow-tie-type highly birefringent fiber was laid in a straight 'U' slot on a metal plate. The fiber was carefully oriented and glued to the plate in the slot, the depth of which was about half of the fiber diameter. The wiper, which was mounted on a translation stage, exerted a constant pressure (~50 g) onto the fiber through a pre-loaded spring. The wiper and two markers each introduced approximately 8% of the optical power into the orthogonal axis. Figure 9.4 shows the absolute wiper displacement along the fiber *vs.* the position read-out from the stage in the processing interferometer. A scanning range of ~70 μm in the processing interferometer covered a positioning range of 20 cm along the fiber. The scanning frequency was 10 Hz. Using these data and equation 9.3, the birefringence of the fiber used in the experiment was measured as $6.75 \times 10^{-4}$, yielding a measured beat length ($L_p$) of 1.2 mm.

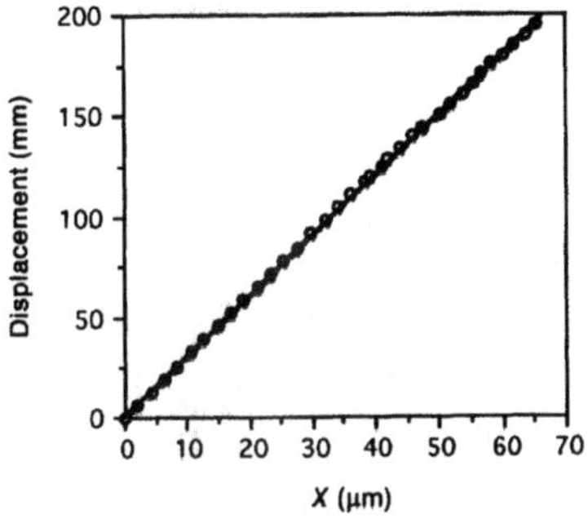


**Fig. 9.4** Wiper displacement *vs.* the imbalance of the processing interferometer measured in experiment.

## 9.3 RESOLUTION ENHANCEMENT

The absolute accuracy of the wiper position measurement depends on the accuracy of the translation stage in the processing interferometer, which gives the reading of its imbalance. Further, it also depends on the ability of the system to locate exactly the position where the coupled and remaining fractions of light inside the highly birefringent fiber are brought back into total temporal coincidence. As discussed in previous sections, this position is marked by the precise position of the zero-order fringe at the center of the fringe signature [4].

Since most low-coherence optical sources have a Gaussian profile in the spectral domain [5], the normalized output, $I(x)$, of such a system can be written as [4]

$$I(x) = \exp\left[-\left(\frac{2(x - x_s)}{\lambda C}\right)^2\right] \cos[2\pi(x - x_s)/\lambda], \tag{9.3}$$

where $\lambda$ is the central wavelength of the optical source, $x$ is the known OPD in the processing interferometer, $x_s$ is the OPD in the sensing interferometer and hence is the one to be measured, and $C$ is the coherence length of the source. For convenience, the coherence length, $C$, in this chapter is defined as the range of OPD measured by the number of interference fringes counted within which the fringe visibility is no less than $1/e$ of its maximum at the center of the fringe signature. In this way, the coherence length of a source will be independent of its wavelength. When $x$ is scanned near the region $x = x_s$, the computer-simulated output pattern of a white-light interferometric system with $C = 30$ is shown in Fig. 9.5, which agrees with the real output of such a system obtained by experiment, as shown in Fig. 9.6. Here, the low-coherence optical source was constructed using a multimode laser diode operating below its lasing threshold with a wavelength centered at 750 nm.

Figures 9.5 and 9.6 show that the fringe signature is very flat around this zero-order fringe in the central region. This makes the identification of the central fringe by its visibility alone very difficult. From equation 9.3, the amplitude

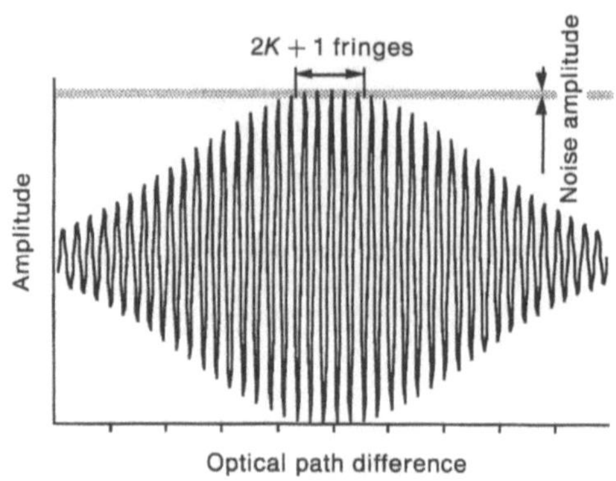


**Fig. 9.5** A computer-simulated fringe pattern of a white-light interferometric system using a single low-coherent source with a coherence length of 30 wavelengths.

difference, $\Delta I$, between the central fringe (zero-order) and its adjacent fringe can be written as

$$\Delta I = I(x_w) - I(x_w \pm \lambda) = 1 - \exp(-4/C^2). \tag{9.4}$$

When the amplitude of intensity noise is larger than $\Delta I$, i.e.

$$SNR \leq -20 \log (\Delta I/2), \tag{9.5}$$

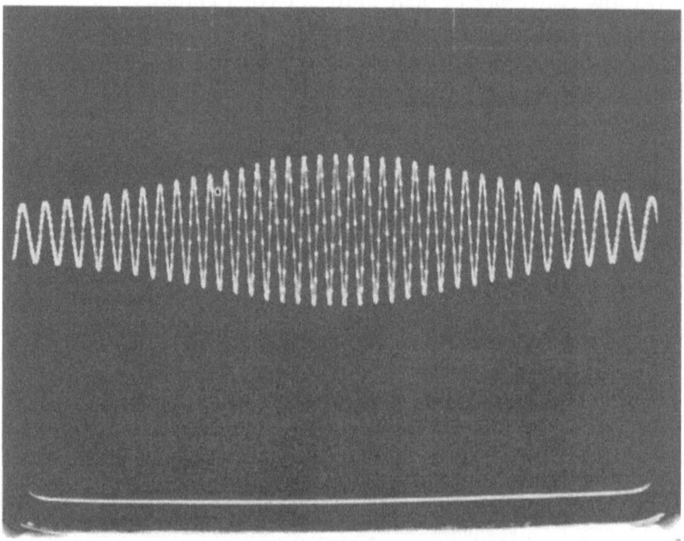

**Fig. 9.6** A real interference fringe pattern obtained from an experimental white-light interferometric system using a single 780 nm laser diode with the driving current below the lasing threshold.

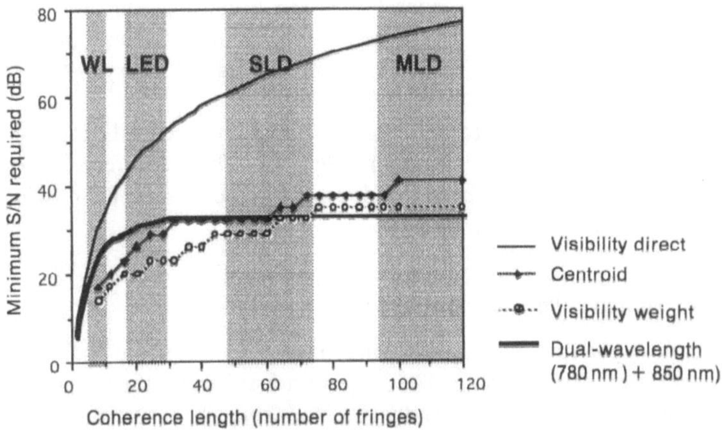


**Fig. 9.7** The minimum system signal-to-noise ratio required to identify the central fringe using different techniques. WL, real white-light lamp; LED, light-emitting diode; SLD, superluminescent diode; MLD, multimode laser diode.

where *SNR* is the system signal-to-noise ratio, the system is not able to distinguish the central fringe from adjacent fringes in the central region directly by its visibility. This required *SNR* for sources with different coherence lengths is computed and presented as the thin solid curve in Fig. 9.7. The shadowed bands in the figure represent the coherence-length ranges of available low-coherence sources. Since the SLD is the most practical low-coherence source for single-mode optical fiber sensor systems, the required minimum SNR is 60–70 dB, which is far too high for most sensor systems.

In the work done by Chen *et al.* [3], an SNR of 38 dB has been achieved. Figure 9.8 shows the statistic distribution of 50 measurements taken on a stationary wiper. These results have a standard deviation of 5.2 mm for the measured

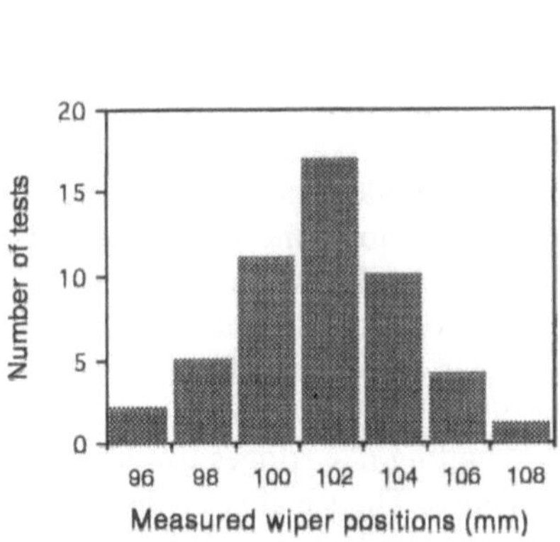


**Fig. 9.8** Statistical distribution of 50 position measurement results with a single low-coherent source.

wiper position along the fiber. In general, identification of the central fringe by its visibility can only achieve a position detection resolution at the level of 5–10 mm. This resolution is respectable if the measurement range of the system is tens or hundreds of meters.

For applications where a better resolution is needed, the central fringe has to be identified precisely even under an unfavorable SNR. Three different fringe-order identification techniques have been developed to solve this problem, which are described in the following sections.

### 9.3.1 Centroid algorithm technique

In the centroid algorithm technique [4], the output fringe pattern is digitized and sent to a computer for processing. The algorithm first locates the fringe signature using a simple threshold, and the centroid of the fringe signature is calculated as

$$c = \sum I_i x_i / \sum I_i, \tag{9.6}$$

where subscript $i$ represents the pixel numbers in the data array obtained from the digitized fringe pattern, and $I_i$ and $x_i$ are the output intensity and $\mathrm{OPD_p}$ at pixel $i$, respectively. The fringe which contains the centroid is identified as the central fringe.

Computer simulations were carried out to assess the performance of the centroid algorithm technique. Random numbers of certain amplitude were added to computer-generated fringe signatures similar to the one in Fig. 9.5 to simulate certain levels of noise. The designed centroid algorithm was then applied, and a 'hit' indicates that the calculated centroid is within the central fringe. The level of simulated noise was increased until the hit rate was just below 99.99%. Then the associated SNR is regarded as the minimum SNR required for the central fringe identification using the centroid algorithm technique. This is shown by the curve with triangular dot markers in Fig. 9.7, which is much lower than that required for direct identification.

### 9.3.2 Shoulder weights algorithm

The technique is based on the observation that although the fringe signature is flat in the central region, its 'shoulder' regions, i.e. $\pm C/4$ fringe away from the central fringe, have the steepest slope and the whole signature is symmetric. In this technique, the fringe pattern also has to be digitized. The algorithm makes the first approximation of the central fringe directly by the position of maximum visibility, $c_0$, and then it starts an iterative approach toward the real central fringe, as shown in Fig. 9.9. First, the current approximation of the shoulder positions are located by counting $C/4$ fringes towards the left and right sides of the approximate central fringe. The weights of the left and right shoulders, $W_\mathrm{l}$ and $W_\mathrm{r}$, respectively, are calculated as

$$W_{\mathrm{l,r}} = \sum I_i^2. \tag{9.7}$$

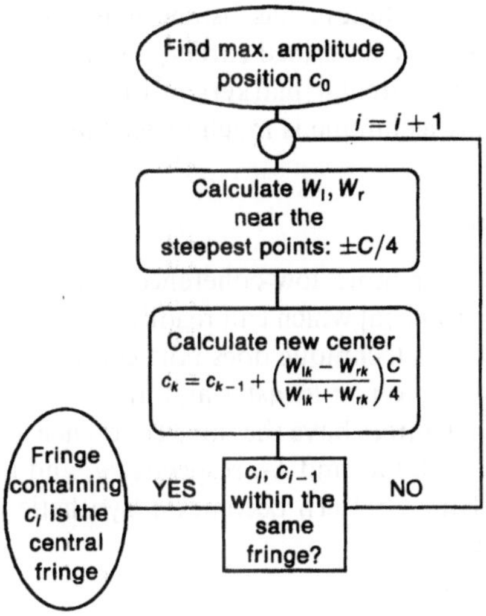


**Fig. 9.9** Flow chart of the shoulder weights algorithm.

The sums are carried out to values at the pixels centered at the left and right shoulder points, and each covers the same width of two to three fringes. As shown in Fig. 9.10, if the previous estimation is not the real central fringe, off to the left, for example, the calculated left shoulder weight $W_l$ will be larger than that on the right, $W_r$. Then an improved estimation of the central fringe position is calculated as

$$c_k = c_{k-1} + \left(\frac{W_{lk} - W_{rk}}{W_{lk} + W_{rk}}\right)\frac{C}{4}, \tag{9.8}$$

where subscript $k$ is the number of the iteration. If $c_k$ and $c_{k-1}$ locate within the same fringe, then this fringe is identified as the central fringe. Otherwise, the newly calculated $c_k$ is regarded as the current approximation of the central fringe position and another iteration begins.

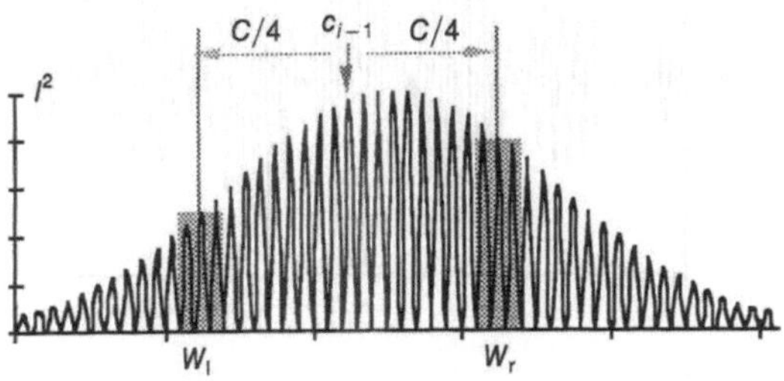


**Fig. 9.10** An illustration of the use of shoulder weights to improve the estimation of the central fringe position.

The performance of this technique has also been assessed by computer simulation using the same procedure described earlier, and its minimum required SNR is shown by the curve with circular markers in Fig. 9.7. Computer simulations have also shown that this technique is much faster than the centroid algorithm.

### 9.3.3 Compound source

The combination of two or more low-coherence sources is used to produce a more dominant central fringe [6] which can readily be identified with simple analogue methods. Because this technique does not require a digitization procedure and related computation power, it is faster and more economical to use.

Assuming that the two sources have the same coherence length, $C$, but different central wavelengths, given by $\lambda_1$ and $\lambda_2$, respectively, and neglecting the dispersion effect, the output fringe pattern becomes a 'synthetic wave', which can be expressed in intensity terms as

$$\begin{aligned} I(x) &= [I_1(x) + I_2(x)]/2 \\ &= \exp\left[-\left(\frac{2(x - x_w)}{\lambda C}\right)^2\right]\cos[2\pi(x - x_w)/\lambda_{sy}]\cos[2\pi(x - x_w)/\lambda], \end{aligned} \tag{9.9}$$

where $\lambda = (\lambda_1 + \lambda_2)/2$, $-\lambda = \lambda_1 - \lambda_2$ and $\lambda_{sy}$ is the 'synthetic wavelength', which can be written as $\lambda_{sy} = 2\lambda_1\lambda_2/\Delta\lambda$. The amplitude difference between the central fringe and the adjacent fringe, $\Delta I'$, becomes

$$\begin{aligned} \Delta I' &= I(x_s) - I(x_s \pm \lambda) \\ &\approx 1 - \exp[-(4/C^2)]\cos[\pi\Delta\lambda/\lambda]. \end{aligned} \tag{9.10}$$

The fringe signature profile at the central region is clearly sharpened by the factor presented by the cosine function in the above equation, so that the central fringe can be identified directly from its amplitude. Fig. 9.11 shows a computed fringe

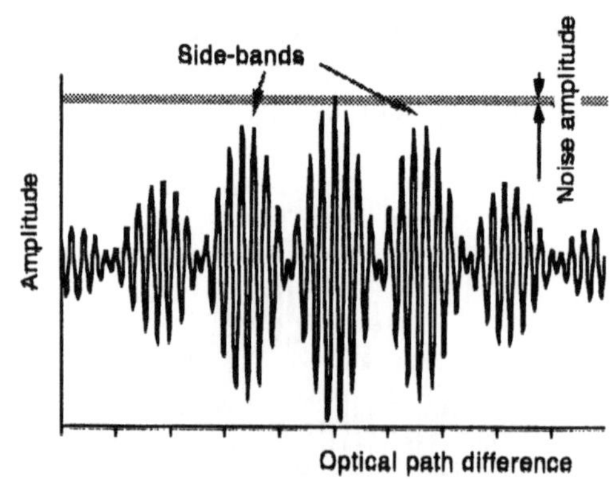


**Fig. 9.11** A computer-simulated fringe signature of a white-light interferometric system using a combination of two sources with coherence lengths of 30 fringes and operating at central wavelengths of 750 nm and 840 nm.

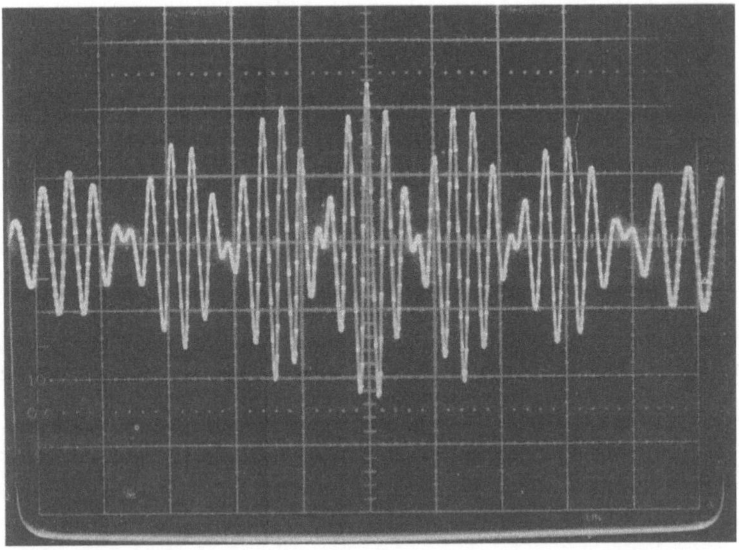

**Fig. 9.12** A real fringe signature from an experimental white-light interferometric system using a combination of two laser diodes centered at 750 nm and 840 nm, and biased below their lasing thresholds.

signature of a white-light interferometric system using a combination of two sources with coherence lengths of 30 fringes and operating at central wavelengths of 750 nm and 840 nm. The simulation fits well with the real output signal, shown in Fig. 9.12, of an experimental system using two low-coherence sources with similar coherence lengths at these central wavelengths. These figures show that the central fringe became very dominant in its amplitude and can easily be located using simple analogue methods such as with a pre-set threshold. The thick solid curve in Fig. 9.7 indicates the minimum required SNR using the dual wavelength method.

When the two optical sources used have longer coherence lengths, the maximum amplitude of fringes in the harmonic side-bands of the synthetic wave (as shown in Figs. 9.11 and 9.12) can be very large, causing new problems for the identification of the central fringe. In such circumstances, additional optical sources can be used with their wavelengths equally dividing the spectral gap between the two original sources, and as a result the harmonic side-bands in the fringe signature can be suppressed. However, the difference between the amplitude of the central fringe and that of its two adjacent fringes will not increase. Figure 9.13 shows a computed fringe signature for the combination of four sources at 750 nm, 780 nm, 810 nm and 840 nm.

The central fringe identification techniques described earlier have been successfully demonstrated experimentally [4, 6]. After the identification of the central fringe, the resolution of the system can be further enhanced by measuring the

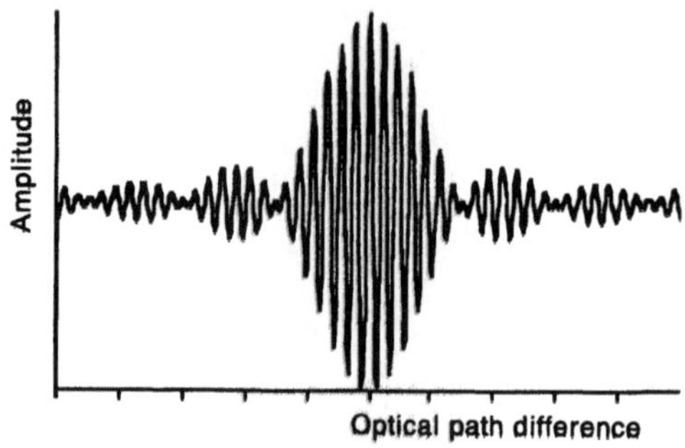


**Fig. 9.13** A computer-simulated fringe signature for the combination of four sources at 750 nm, 780 nm, 810 nm and 840 nm, all with the same coherence length of 30 wavelengths.

phase of the central fringe within a $2\pi$ unambiguous range using the wide range of phase-detection methods available [7, 8]. Depending on the SNR of the system, a resolution of up to 1/100 fringe can be achieved.

## 9.4 TEMPERATURE COMPENSATION SCHEME

Since the birefringence of the fiber and the imbalance of the processing interferometer are all sensitive to temperature variations [9], from equation 9.1, the directly measured wiper position, $L_w$, will be subject to the effect of temperature change. In order to solve this problem, a scheme of measuring the 'relative position within the stroke' has been developed. As shown in Fig. 9.3, there are two fixed markers, $P_s$ and $P_e$, at the start and the end positions of the wiper travel range. Each of them couples a small amount of optical power into the slow axis. It has been demonstrated that as long as the amount of optical power coupled across is much smaller than that which remains in the fast, the cross-coupling effect between the three points can be neglected [10]. Their positions, $L_s$ and $L_e$, can therefore be detected by the corresponding OPDs using the processing interferometer in the same way as the movable wiper, that is

$$L_s = 2x_s/B \quad \text{and} \quad L_e = 2x_e/B, \tag{9.11}$$

where $x_s$ and $x_e$ are the imbalances of the processing interferometer corresponding to $L_s$ and $L_e$, respectively. The 'relative' wiper position, $p$, within its movable range may be defined as

$$p = (L_w - L_s)/(L_e - L_s). \tag{9.12}$$

Using equations 9.2, 9.11 and 9.12, $p$ can be rewritten to become

$$p = (x_w - x_s)/(x_e - x_s). \tag{9.13}$$

No 'B' term is present in the above equation, and in addition, the first-order temperature effects on $x_w, x_s$ and $x_e$ are canceled. However, the above discussion is based on the presumption that the temperature is uniform in the processing

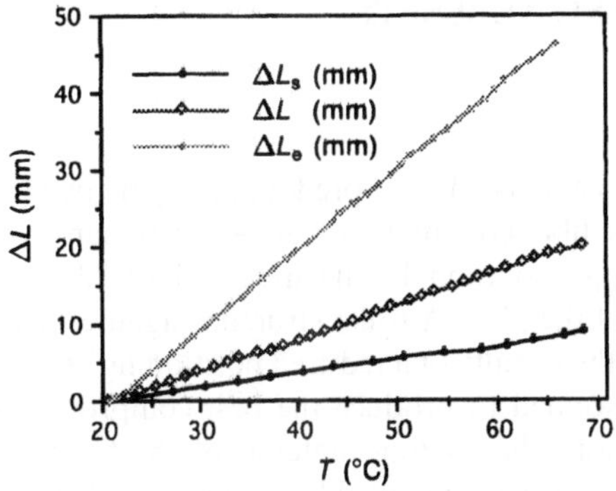


**Fig. 9.14** Temperature effects on measured absolute positions.

interferometer and across the whole length of the sensing section of the highly birefringent fiber from $P_s$ to $P_e$. This is a close approximation for applications where the stroke length is moderate. Therefore, by detecting the absolute position of the wiper, as well as that of the two end-markers, and then calculating the relative position of the wiper within its stroke, the temperature effect on the measurement will largely be eliminated.

The temperature stabilization scheme was also assessed in the preliminary experiment mentioned above [3]. The wiper and the section of highly birefringent fiber between the two markers was placed in an oven which provided a controlled temperature variation from 18°C to 70°C. The wiper was positioned at approximately one-third of its complete range and remained stationary during the temperature excursion. Figure 9.14 shows the measured absolute positions of the wiper and the two markers, all of which changed following the temperature variation. The rates of these changes depended on their relevant positions along the fiber. The 'relative position' of the wiper, however, remained constant within 1% throughout the approximately 50°C temperature change, as shown in Fig. 9.15.

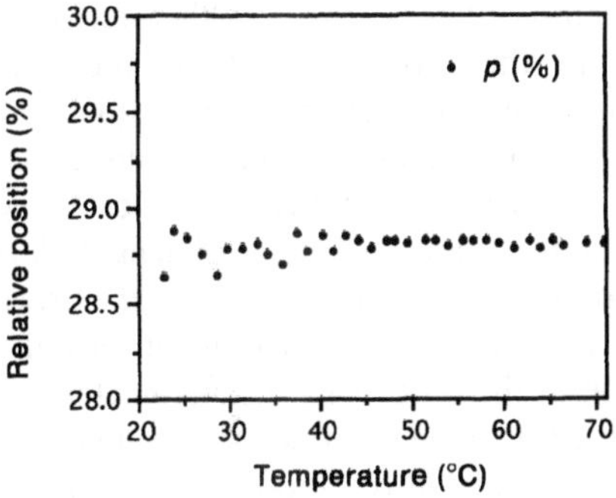


**Fig. 9.15** Temperature stability achieved in measured relative positions.

## 9.5 OTHER TECHNICAL ISSUES

### 9.5.1 Durability

The durability of the system is affected predominantly by two factors: the shape of the wiper head and the fiber coating material. A sharp, hard wiper head may provide the best positioning resolution, but it can scratch the fiber coating and severely reduce the durability of the fiber. A roller structure significantly improves the fiber durability, but suffers the penalty of a reduced positioning resolution. A 'soft' roller with miniature teeth seemed to produce the best compromise in the experiment reported [3]. In addition, the coating material of the fiber has to be extremely hard-wearing, thus presenting some problems in practical systems.

### 9.5.2 Orientation of fiber birefringent axes

Ideally, the high-birefringence fiber used in the system should be laid with its birefringent axes at 45° towards the direction of the pressure exerted by the wiper, because at the orientation, sufficient cross-coupling can be produced with maximum pressure. However, in many applications it is not practical to maintain this orientation over a long distance. In addition, since the coupling ratio at this orientation is very sensitive to the pressure, the amplitude of the output signal varies over a very wide range when the wiper is traveling along the fiber. This makes signal processing very difficult. It is therefore necessary to make the coupling ratio less sensitive to the pressure inserted. This can be achieved by using a relatively soft coating and by laying the fiber into a 60° 'V' groove.

## 9.6 DISCUSSION

As already mentioned, the applications of the technology described can roughly be classified into two categories. In the first category, the position range is relatively short along the length of the sensing fiber, i.e. in the range up to 50 cm. A piezoelectric transducer can be used as the scanning actuator in the processing interferometer, with a capacitance-based sensor as the precision position sensor. The nanometer resolution in the processing interferometer coupled with central fringe identification and phase measurement techniques can allow such a set-up to achieve a sub-millimeter positioning resolution along the fiber. A scan rate of up to 50 Hz can also be achieved. In the second category, the position range along the fiber can be up to a thousand meters in length. A motorized mechanical stage may be employed as the scanning actuator in the processing interferometer, with an optical rotary encoder as the displacement sensing device. Owing to the relatively low resolution in the processing interferometer, there is little advantage in employing special central fringe identification techniques. The central fringe is estimated by its visibility, and the resolution will be in the region of 5–10 mm. The measurement speed will also be relatively slow. In either case, the accuracy

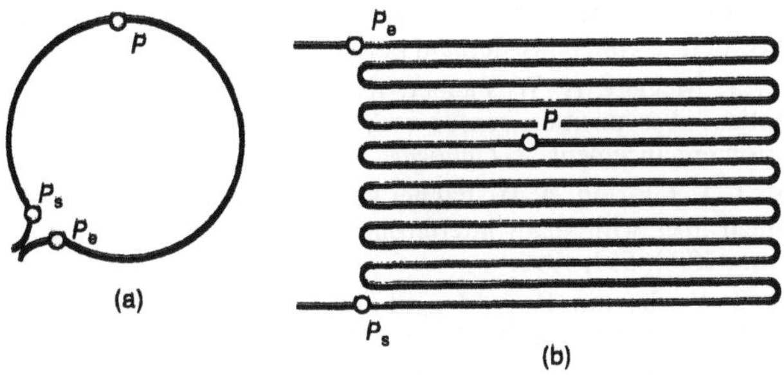


**Fig. 9.16** Two examples of different sensing fiber routes to facilitate (a) angular and (b) two-dimensional position detection.

of the present system is always going to be moderate due to the temperature instability of the highly birefringent optical fibers.

The optical fiber position monitoring system described has many distinct features. Firstly, it is an intrinsic sensor system in which the light remains within the optical fiber in the sensing area. Secondly, a scheme of relative position-detection is used to greatly reduce the effect of temperature on the measurement result. Thirdly, the system employs a simple dual wavelength fringe-order identification technique to enhance the resolution of the position measurement to the region within one beat-length of the highly birefringent fiber used. Finally, one additional attractive feature of the system presented is that the structure of the sensor head is compatible with a conventional potentiometer-type position sensor.

In summary, this position-sensing approach has the following major advantages:

1 it is robust and compact in structure;
2. the sensing interferometer is completely passive and has a compact, intrinsic structure, while the processing interferometer can be placed away from the hostile environment in the sensing area;
3. it is immune to electromagnetic interference effects (which is an inherent advantage of all optical-based sensors);
4. it is versatile in its potential applications.

Because of the flexibility of optical fibers, the sensing high-birefringence fiber can be laid to follow almost any topography in any shape or form. This feature allows the technology to be used in a wide range of applications such as angular position monitoring or even monitoring the position in two- and three-dimensional space, as shown in Fig. 9.16. It has been reported that the same principle has been used to detect and assess impact damage in aerospace structures [11].

## REFERENCES

1. S. Chen, B. T. Meggitt and A. J. Rogers, Electronically scanned white-light interferometry with enhanced dynamic range, *Electron. Lett.*, **26**(24), 1663–1665, 1990.

2. S. Chen, I. P. Giles and M. Fahadiroushan, A quasi-distributed pressure sensor using intensity type optical coherence domain polarimetry, *Opt. Lett.*, **16**(5), 342–344, 1991.
3. S. Chen, B. T. Meggitt, A. W. Palmer, K. T. V. Grattan and R. A. Pinnock, An intrinsic optical-fiber position sensor with schemes for temperature compensation and resolution enhancement, *J. Lightwave Technol.*, **15**(2), 1997.
4. S. Chen, K. T. V. Grattan, A. W. Palmer and B. T. Meggitt, Fringe-order identification in optical fiber white-light interferometry using a centroid algorithm method, *Electron Lett.*, **28**(6), 553, 1992.
5. Y. Ning, K. T. V. Grattan, B. T. Meggitt and A. W. Palmer, Characteristics of laser diodes for interferometric use, *Appl. Opt.*, **28**(17), 3657–3661, 1989.
6. S. Chen, A. W. Palmer, B. T. Meggitt and K. T. V. Grattan, Instantaneous direct identification of central fringe using dual sources with widely spaced wavelengths, *Electron. Lett.*, **29**(4), 334–335, 1993.
7. S. Chen, A. W. Palmer, K. T. V. Grattan and B. T. Meggitt, Digital signal processing techniques for electronically scanned optical-fiber white-light interferometry, *Appl. Opt.*, **31**(28), 6003–6010, 1992.
8. V. Gusmeroli and M. Martinelli, High-resolution distance measurement by dynamic spectral filtering of a superluminescent source, *Opt. Lett.*, **16**(15), 1195–1197, 1991.
9. W. Eickhoff, Temperature sensing by mode–mode interference in birefringence optical fibers, *Opt. Lett.*, **6**(4), 204–206, 1981.
10. S. Chen and I. P. Giles, Optical coherence domain polarimetry: Intensity and interferometric type for quasi-distributed optical fibre sensors, *SPIE Proceedings*, **1370** (for OE/Fiber'90 in San Jose, CA).
11. I. P. Giles, P. Lloyd, C. Doran and M. Mondanos, Optical fibre based damage detection for aerospace structures, *IEE Colloquium on Optical Techniques for Smart Structures and Structural Monitoring*, London, 1997.

# 10

# Commercial activity in optical fiber sensors

Z. Y. Zhang and K. T. V. Grattan

## 10.1 INTRODUCTION

### 10.1.1 Background to the work

In a dynamic field such as that of optical fiber sensing (OFS) and measurement, it is particularly important for the researcher, instrumentation developer, manufacturer, supplier and end user to be fully aware of the existence of, and potential for the use of, the rapidly developing technology that is available. It is effectively impossible to attend all the relevant meetings, conferences, trade shows and exhibitions, and at the same time conduct normal business, and synopses of data and developments are particularly useful. Thus their essential value is seen as providing a summary of the information required on developments and trends, and presenting the data in an accessible and quickly digestible form.

A range of reports on, and most significantly several surveys of, commercial optical fiber sensor technology have been carried out in recent years. Most have been produced either for specific clients, or to be marketed to major industries, to inform them about the trends seen in the field. Such surveys are rarely available for less than several thousand dollars, and only the most general details are published widely, often by way of advertising the full material itself. Of those that have been more widely circulated, that carried out in 1988 by ERA Technology [1] was a report to the Optical Sensors Collaborative Association (OSCA) in the UK on a world-wide survey of manufacturers in the field. In 1994, this work was extended in a further document by Crossley [2], adding to the data available through a report based on interviews (mostly in the USA) and a questionnaire to major manufacturers.

It is thus timely to provide an update of the information base yet again, with the goal of reporting the data created not from the responses to questionnaires or interviews, but from information based upon the rapidly developing availability of data by electronic means, through e-mail and the Internet. This has been seen, almost inevitably, as the most effective means of distributing information to the technically active community, and manufacturers and agencies, both large and small, can make extensive use of it. This has been the major change in recent years, since the previous surveys [1,2] were conducted when such techniques were in their

*Optical Fiber Sensor Technology*, Vol. 3. Edited by K. T. V. Grattan and B. T. Meggitt.
Published in 1998 by Kluwer Academic Publishers, London. ISBN 978-1-4419-4736-9

infancy. Its use removes some of the subjective factors relating to questionnaires and the responses received (or not, as the case may be), and the accuracy of the data thus incorporated. Also, the potential for influence, if the interviewing method is on the basis of being face-to-face, is thus avoided. A further factor is the convenience and comprehensive approach of this means of acquiring data, analyzing it and producing a synopsis, making an electronic survey the most accessible means by which such effective and meaningful data can be produced.

Thus, it was decided to focus this chapter upon patent information, which is representative of both real commercial potential and exploitation, and is also more readily quantifiable than much of the more research-oriented information available in the open scientific literature and discussed extensively in previous chapters. The former is seen as being much closer to, or actually on, the market, rather than having the typical development period of perhaps up to 10 years that could be expected for the exploitation of more recent research data.

The aims and objectives of the analysis carried out were:

1. to survey available data from a range of electronic sources, from manufacturers, patent sources and researchers in the broad field of optical fiber sensor technology;
2. to access and categorize this information in a meaningful form in relation to measurands, geographical factors and trends;
3. to summarize the data obtained in a way that is complementary to, and builds upon, previous work [1, 2]. (To that end, as far as possible, similar categorization criteria and headings were used, and data are presented in summary and tabular form for ready access and comparison.)

In this chapter, Section 10.1 discusses the methodology applied and the categorization criteria incorporated, and Section 10.2 gives the results of the OFS product survey, together with summarized data on sensor suppliers and products, and comments on the survey results. Section 10.3 deals with patent data from the prime market, the USA, illustrating the technological development mirrored by the patents issued, details of the geographical distribution of OFS patent assignees, and measurand distribution details of OFS patents, together with a full listing of relevant OFS patents over the years 1994–1997. The tables in Section 10.4 detail the range of technical specifications for available sensors. A brief discussion of trends and a summary of the major themes of the findings is included in the final section.

### 10.1.2 Methodology applied in this study

This report was based mainly on surveys carried out using various electronic media. The methodology adopted is described below.

#### *(a) Collection of data for commercially available OFS products and services*

Various net search machines, accessible from the World Wide Web server through Netscape Navigator, were used to conduct searches for existing web sites for companies surveyed previously [1, 2] (in 1994), including any other companies which either were overlooked in the last survey or are new players in the OFS field. The former was based on the company names listed in the last survey [2], and the latter was conducted using appropriate key words. Data for OFS products and services were either retrieved from these sites or requested by electronic mail.

### *(b) Browsing US Patents on OFS*

Currently, there are two major World Wide Web sites which provide freely searchable databases of information from US patents. One is at the US Patent and Trademark Office (USPTO) at <http://patents.uspto.gov>, and its US Patent Bibliographic Database contains front-page information from US patents issued from 1 January 1976 to 1994. The other is the IBM Patent Server at <http://patent.womplex.ibm.com>, and it provides access to the bibliographic data and text of all claims of patents issued by USPTO from 1974 to the present, plus some patents issued during 1971 to 1973.

The IBM Patent Server was employed here to search for OFS-related patents issued by USPTO from January 1980 to the end of 1997. Complete scanned images are available from the server for patents issued from 1974 to 1997. Such a service is not yet available from the USPTO server.

The IBM server currently supports simple keyword, phrase and patent number searches, as well as the use of Boolean operators. The following fields are searchable: patent number, title, abstract, all claims, assignee, inventors and attorney/agent. The patent references are hyperlinked, allowing users to locate and view prior art by navigating through all the cited patents. The link is bi-directional. Users can access not only the patents referenced by the examiners, but also later patents that reference the existing document.

All of the above functions of the server were used in the search, and OFS-related US patents and relevant information were retrieved and stored for further classification.

### *(c) Primary analysis of data collected*

Primary analysis of data gathered in Sections (a) and (b) was carried out to list and classify the existing OFS technologies or application areas, including those overlooked by the survey in 1994 or emerging subsequently. To create a comprehensive collection of data, if any OFS technology area was identified which had not been expected in the prior search, the previous stages had to be repeated under redefined search items.

### *(d) Compilation of data into a coherent form*

The classification system was finalized and the statistical analysis of the data was developed to show the status of current OFS developments and the trends of future developments.

**Table 10.1** Categories of sensor covered in the survey

| | |
|---|---|
| **Temperature sensors** | |
| Birefringence | Liquid core fiber |
| Distributed method | Microbend |
| Fiber amplifier | Photoconductivity |
| Fluorescence/luminescence | Pyrometer/blackbody |
| Fotonic | Reflectance |
| In-fiber Bragg grating | Thermal switch |
| Infrared radiation | Transmission/absorption |
| Interferometer | |
| **Pressure sensors** | |
| Birefringence | Ruby fluorescence |
| Diaphragm + fotonic | Shock-induced radiation |
| Hollow fiber | Tapered fiber |
| Microbend | |
| **Displacement sensors** | |
| Dual wavelength method | Magneto-optic |
| Fizeau cavity | Microbend |
| Fotonic | Modal noise |
| FP interferometer | Prestressed fiber |
| **Liquid (fluid) level and presence sensors** | |
| Evanescent wave | Prismatic methods |
| Pressure | TIR |
| **Gas sensors** | |
| Absorption | Fluorescence/luminescence |
| Evanescent wave | Surface plasmon resonance |
| **Chemical sensors** | |
| Acidity | Osmometer |
| Chromophone | pH |
| Dye-containing polymer | Raman spectroscopy |
| Dyed thin film | Refractometry |
| Evanescent wave spectroscopy | SERS[a] |
| Fiber laser spectroscopy | Spectrometer |
| Fluorescence/luminescence | Spectrophotometer |
| Hydrocarbon | Surface plasmon resonance |
| **Biosensors** | |
| Blood analyte | Endoscopic temperature/pressure |
| Blood flow | Fluorescence/evanescence wave immunoassay |
| Blood oxygenation | Ligands |
| Blood perfusion | Tissue diagnosis |
| Blood pressure | Tissue penetration monitoring |
| Dental anatomical structure | |
| **Strain/stress/structure monitoring** | |
| **General purpose** | |
| Auxiliary technologies | Interferometry |
| Coherent detection | Multimechanical parameter |
| Distributed measurement | OTDR |
| Evanescent wave | Polarimetry |
| Fiber laser/amplifier sensor | Raman scattering |
| Fiber modulation | Spectroscopy |
| Fluorescence | Special fiber |
| In-fiber Bragg grating | Two-mode fiber sensor |
| Bragg grating | |

[a] Surface enhanced resonance Raman spectroscopy.

**Table 10.1** (*Continued*)

| | |
|---|---|
| **Gyroscope** | |
| **Electric and magnetic sensors** | |
| Electric current/voltage | RF field |
| Electromagnetic field | Sagnac effect |
| **Miscellaneous sensors** | |
| Acceleration | Mach number |
| Acoustic and ultrasonic sensors | Microscope |
| Corona/spark detection | Moisture |
| Corrosion | Object discrimination |
| Diffuse reflectance | Object presence |
| Doppler velocimetry | Patterns |
| Emissivity | Polymer molding |
| Fault detection | Position |
| Fiber kink detection | Quadrant |
| Film thickness | Radiation |
| Flow rate/continuity | Rangefinder |
| Fluid interface/constituent | Refractive index |
| Force/load | Scatterometer |
| Gravity | Security key |
| Hydrophone | Sheet material monitoring |
| Image sensor array | Speed/velocity |
| Intrusion/disturbance | Surface roughness/flatness |
| Irradiation dosage | Vibration |

### 10.1.3 Categorization used in the survey

The list of categories used, and the associated physical techniques involved arising from the survey, are shown in Table 10.1. This indicates the very wide range of techniques and measurands applicable to OFS methods.

## 10.2 RESULTS OF THE OFS PRODUCT SURVEY

### 10.2.1 OFS suppliers

An earlier report [1], which was completed in May 1988, identified over 213 companies as being potential manufacturers of OFS, and information was obtained from over 100 of these. This work was updated by Crossley [2] in 1994, when the total number of companies identified as being potential OFS suppliers rose to around 340. At that stage, the majority of these potential manufacturers were sent a questionnaire seeking information on the company and on OFS products at any stage of development, and 94 OFS suppliers were located.

During the intervening years several more companies have entered the OFS manufacturing field; for example, the entry of BICC, the $6 billion international cable and construction company, into the optical fiber DTS (distributed temperature sensor) market through a UK subsidiary, BICC Thermoheat [3], to compete with York Sensors, which is growing at about 25% a year. A number of new OFS products have also entered the marketplace. Some examples are a range of OFS products trademarked as SHAPE SENSORS, SHAPE TAPE, and manufactured

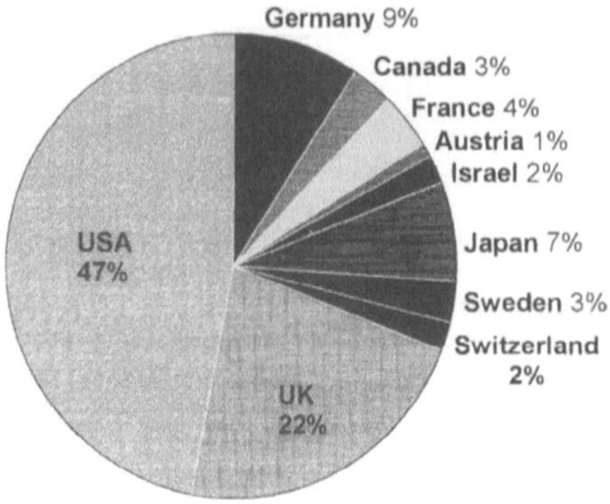


**Fig. 10.1** Geographical distribution of FOS manufacturers surveyed.

by Measurand Inc. for measurements ranging from curvature, displacement, pressure, gas and liquid flow to acceleration. Measurand was incorporated in 1993 to commercialize a patented OFS technology, and now has a production capacity of 10 000 units per year.

To date, a total of around 350 companies have been identified as potential OFS manufacturers or suppliers, and 101 of these companies were confirmed as OFS manufacturers. The geographical distribution of these OFS manufacturers is shown in Fig. 10.1, showing that, as expected, the USA is the dominant player in the OFS market under this criterion.

### 10.2.2 OFS sensor products

The distribution of the OFS products offered by the 101 manufacturers located is illustrated in Fig. 10.2, where sensors are categorized by their measurands as listed in Table 10.1. Of the 164 sensors whose details appear here, temperature (30), displacement (28) and pressure (12) sensors account for about 43%, with gas (12) and chemical (22) sensors, mostly based on absorption spectroscopy techniques, and liquid-level sensors (10) representing another 26%. The remainder amount to 31%. This shows no significant difference from the results of previous work [2] in spite of the updating of the data used. Further details on the specifications reported for these sensors are given in Section 10.4, where they are categorized by a range of measurands.

Such sensors may also be crudely categorized in terms of the exploitation of the optical fiber into the familiar intrinsic and extrinsic categories. In most cases, the sensors are extrinsic and the fiber is simply a light-guide, often simply facilitating the remote operation of a conventional optical sensor. This extrinsic category includes the majority of the displacement sensors (remote photo-electric sensors), as well as nearly a third of the temperature sensors (remote pyrometers) and the majority of the gas/chemical sensors which are based on remote absorption spectroscopy. In the main, these are sensors produced by well-established

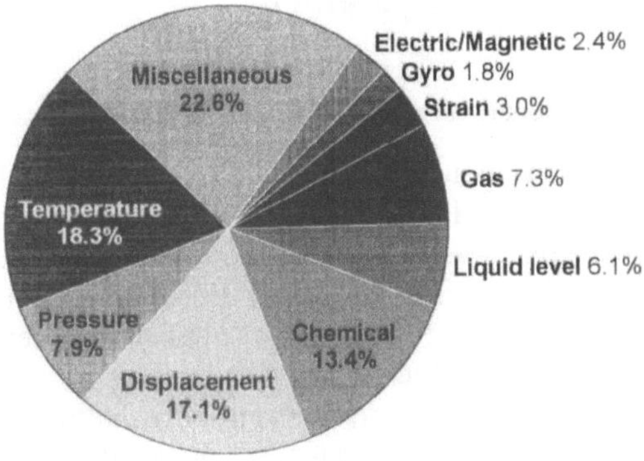


**Fig. 10.2** Measurand distribution of FOS products surveyed.

manufacturers via the modification of existing nonfiber-optic products, and market acceptance is generally good.

### 10.2.3 Comments on the product survey results

It has been acknowledged in previous OFS survey results that the questionnaire is not a particularly efficient means of gathering data. The report in 1988 [1] suggested that a return rate of approximately 50% might be expected, and this would normally be considered a very high figure for a survey of this type. In the survey prior to the 1994 report [2], of the 300+ questionnaires mailed, only 21 replies were received, of which seven indicated that the company was not involved in OFS manufacture, and communications by telephone and fax had to be used, which was less useful for gathering information from non-English-speaking countries. This clearly shows the comparative inefficiency and relatively subjective nature of this type of survey.

In the present survey, the Internet was used for data gathering, but somewhat surprisingly for a 'hi-tech' industry (at the end of 1997), of the 101 OFS manufacturers located, only nine have put their OFS product information on web sites. This has inevitably meant some distortion of the data, and it can be seen that the UK share of the geographical distribution shown in Fig. 10.1 is greater than could reasonably be expected.

The product survey result can only reflect the current status of the OFS market, and cannot effectively be used as a prediction of the trend of developments. Not many manufacturers would be willing to disclose their plans for OFS development in a commercially sensitive field. Nevertheless, the development of OFS products depends to a large extent on the most effective exploitation of the technologies and the associated market drivers. One better way to monitor the technology development and to attempt to foresee the trends of the OFS market is to review patents issued in the OFS field over a period of several years, and this is used as the prime data source for this survey.

## 10.3 US PATENTS IN THE OFS FIELD

### 10.3.1 Technology development mirrored by patents issued (1980–1997)

A total of 1310 US patents, issued between 1980 to the end of 1997, have been located and identified as being in the field of optical fiber sensors. The numbers of such patents issued over these years are illustrated in Fig. 10.3, on an annual basis. This shows the general trend of a clear increase in the number of OFS patents issued each year from 1980 to 1994 (by almost an order of magnitude), followed by a noticeable decline thereafter.

Statistics gleaned from the patent information show that the average length of time from a patent application being filed to it being issued is about 2 years. Thus, the number of successful OFS patent applications filed each year is estimated to be as depicted by the thick line in Fig. 10.3. This indicates that the development of OFS technology was most active in the period around 1992. This may indicate a decline in investment in research and development in the late 1980s, feeding through in this way.

### 10.3.2 Geographic distribution of OFS patent assignees (1994–1997)

Particular attention has been paid to the period of 4 years from the start of 1994 to the end of 1997. This section and the next (10.3.2 and 10.3.3) cover the detailed analysis which was carried out. There were 486 US patents issued from 1994 to September 1997 in the OFS field, and these are listed in Section 10.6. The geographical distribution of their assignees, who normally are the owners of these intellectual properties, is illustrated in Fig. 10.4 by country of origin.

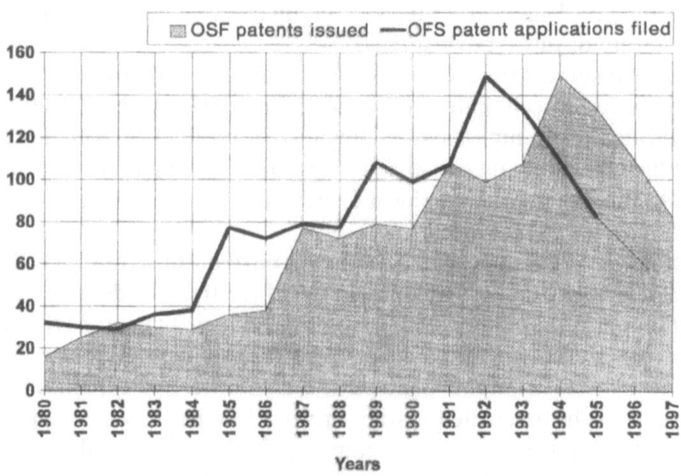


**Fig. 10.3** Numbers of US patents in the OFS field issued from 1980 to 1997.

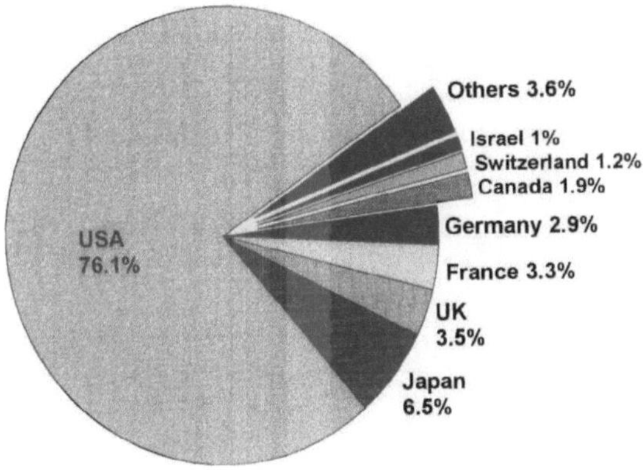


**Fig. 10.4** Geographic distribution of OFS patent assignees by country.

Figure 10.4 gives a more realistic picture of global OFS development activities than Fig. 10.1, which is based mainly upon the OFS product survey carried out by Crossley in 1994. As discussed earlier, the means used for the product survey inevitably limited the response rate from non-English-speaking countries, and consequently a substantial number of manufacturers from those countries could have been overlooked.

Figure 10.4 also shows that the USA is by far the biggest player in OFS technology development, and Japan is second only to the USA, with a share of 6.5%. The UK also has a significant impact, but is contributing a mere 3.5% of these OFS patents. However, this is a slightly higher percentage than that from any other European country, and France and Germany, as expected, are the other major contributors to an EU total of around 12%.

Figure 10.5 illustrates the geographical distribution of OFS patent assignees according to continent. It shows that the activities in OFS development are mostly carried out in the USA, Europe and Asia. In Asia, Japan is the leading country, and in Europe the leaders are the UK, France and Germany, as shown in Fig. 10.4.

### 10.3.3 Measurand distribution of OFS patents (1994–1997)

A further analysis of the data was carried out, and a measurand distribution of those 486 OFS patents issued from 1994 to the end of 1997 is illustrated in Fig. 10.6, where patents are classified according to the measurand categories listed in Table 10.1.

There are several points to note. First, patents which were not associated with a specific measurand and claimed to be 'general purpose', and those falling into the miscellaneous measurand category (referring to Table 10.1), account for a total of 36.4% of those 486 OFS patents. This illustrates the great variety in OFS devices, and is characteristic of the technology itself.

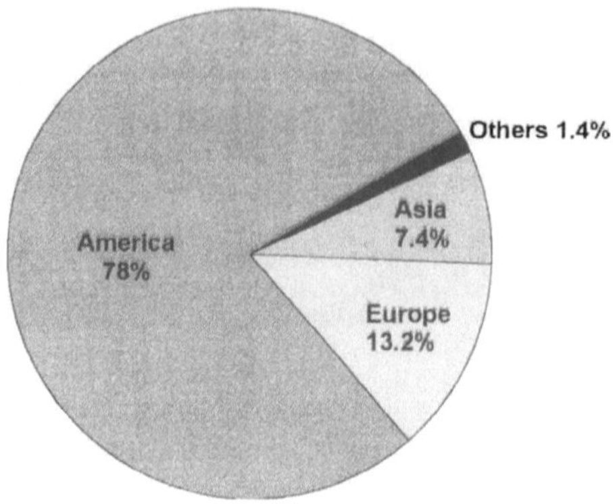


**Fig. 10.5** Geographic distribution of OFS patent assignees by continent.

Second, in contrast to the meager share (1.8%) for fiber optic gyros in the OFS product survey (see Fig. 10.2), fiber optic gyro technology accounts for a very significant proportion (14.4%) of the total of OFS patents surveyed. This is most probably due to the military origin of the early developmental work of this specific technology and an indication of the anticipated early market. With the exploitation of the technology in nonmilitary applications, e.g. the car navigation gyro developed by Hitachi Cable for Nissan Motors, more commercial fiber optic gyros will appear in the market place and at comparatively high volumes.

Third, Fig. 10.7 shows the normalized (to 100%: annual total) measurand distribution of US patents in the OFS field issued over the years from 1994 to 1997, illustrating the comparative level interest of each. Noticeable is the rising trend in the chemical sensor field from 1995, and the significant proportion of strain/stress sensor patents being maintained over these years. Declines in the

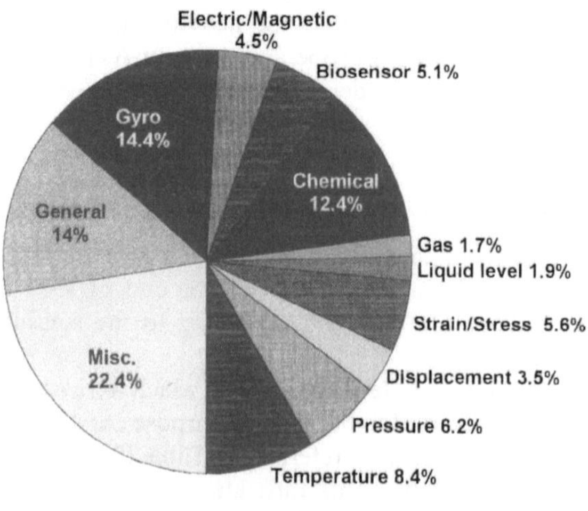


**Fig. 10.6** Measurand distribution of OFS patents issued from 1994 to 1997.

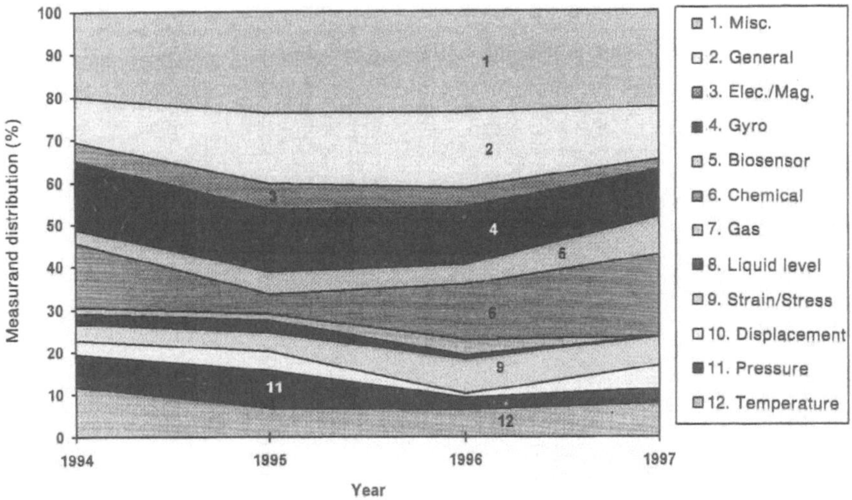


**Fig. 10.7** Measurand distribution of US patents in OFS issued over the years from 1994 to 1997. Z. Y. Zhang, K. T. V. Grattan and A. W. Palmer, Report to the Optical Sensors Collaborative Association (OSCA), 1997 (reproduced with permission).

proportions of other measurands commonly encountered in process control, e.g. liquid level, pressure and temperature, are obvious, although temperature sensor technology has still attracted a relatively high level of interest following a decline from 1994 to 1995, possibly due to the interest in measuring temperature specifically to deconvolve its effect from that of other measurands.

## 10.4 SPECIFICATIONS FOR AVAILABLE SENSORS

Tables 10.2–10.8 show a range of published specifications, categorized according to measurand. These are:

- temperature, Table 10.2;
- pressure, Table 10.3;
- displacement, Table 10.4;
- liquid level, Table 10.5;
- gas, Table 10.6;
- chemical, Table 10.7;
- miscellaneous (including strain, gyro and electric/magnetic), Table 10.8.

## 10.5 SUMMARY AND TRENDS

A study on 'World Fibre Optic Sensor Markets' by Frost and Sullivan [4], published in January 1997, concluded optimistically that the sensor industry

**Table 10.2** Temperature sensors

| Name | Operating principle | Range | Accuracy | Resolution | Application areas | Price guide | Notes |
|---|---|---|---|---|---|---|---|
| ACME Namco | Remote IR detector | 350, 450 and 800°C | n/a | n/a | Hot material detection | $2240 | Switch point |
| Ari Industries Inc. | Blackbody | 300–2000°C | 1% | n/a | Reactive atmospheres, extreme temperatures | | |
| Babcock & Wilcox | Remote pyrometry | 120–180°C | ±1°C | | Temperature of moving web | | Multifiber |
| Celect Electronics Ltd. | Remote pyrometry | 20–800°C | | | | | Low temperature, non-contact (UK Smart award) |
| Conax Buffalo Corp. | Remote pyrometry | 500–1800°C | 1% FS | | Exhaust gas | $5600 | |
| Ericsson AB | Distributed: microbend | 40–90°C | | ±1°C | Overheat | $19 200 | Distributed switch point sensor |
| FIMOD | | | | | | | Development |
| Fuji Electric Co. Ltd. | Pt resistance element E/O converter | | ±0.25% | | | | Fuji system |
| Fuji Electric Co. Ltd. | Thermocouple E/O convertor | | ±0.25% | | | | Fuji system |
| Fujikura | Distributed: Rayleigh | | | 1–3°C | | | Japan ONLY. 3 km range, 3 m resolution |
| G2 Systems Corp. | Distributed: microbend | −200–600°C | ±3°C | | | $20 800 | Multisensor: temperature, pressure, strain (15 km) |
| Kidde Graviner Ltd. | Distributed Raman | | | | Aircraft fire detection | | ex Cossor Thermotrack |
| Luxtron | Fluorescent decay time | −200–450°C | 0.2°C | 0.1°C | | | |
| Luxtron | Blackbody | 150–3000°C | 0.2% | 20 μK | | | ex Accufiber |
| Mikron Instrument Co. | Remote blackbody | | | | | | |
| Nortech Fibronics | Fluorescence | −40–250°C | ±1°C | ±0.1°C | | | 1 km max sensor length |
| Optech Inc. | Microbending | $0.3°C\,min^{-1}$ to $11°C\,min^{-1}$ | ±5% FS | $0.3°C\,min^{-1}$ | Fire detection, overheat | | NB. Temperature rise ONLY |

| | | | | | | | |
|---|---|---|---|---|---|---|---|
| Optrand Inc. | Dual wavelength | 0–700°C | ±1% | | Aircraft etc. engines | | |
| Photonetics SA | FP cavity special modulation | −40–300°C | ±1% FS | n/a | | | |
| Pilkington Security Ltd. | Distributed: intensity | Switch | | | Cryogenic leak | | |
| Pyrometer Instrument Co. | Remote pyrometry | 200–2500°C | ±3°C | | | | |
| Qualimatrix Inc. | Remote pyroelectric | 400–1350°C | | | Hot metal detector | | Fast detection (50 μs) (min distance 6 mm) |
| Rosemount Inc. | Fluorescent decay time | −60–350°C | ±0.5°C | | Aerospace: engine and aircraft | | |
| Sumitomo Electric USA | Distributed Raman backscatter | −100–300°C | ±1°C | ±3°C | Distributed sensing | | Operating range to 4 km (temp. res. at 2 m spatial) |
| Takaoka Electric Ltd. | GaAs fluorescence | 0–200°C | ±1°C | 0.1°C | | $5600 | Ex ASEA sensor |
| Vanzetti Systems Inc. | Remote pyrometry | 50–2800°C | ±1% | ±1°C | | $6400 | |
| York Sensors | Distributed Raman | −140–460°C | ±0.3°C typical | ±1°C | | $80 000 | Range 4 km, sample 10 s for figures quoted. 20 km max, 80 000 points max |
| Pruett Industries | Distributed | To 370°C | | 0.3 m | | | Using stainless tubes |
| Hitachi Cable Ltd. | Distributed | | | | | | e.g. fire alarm system for the Temple of Kinkaku, Kyoto, Japan |
| BICC Thermoheat | Distributed | −140–450°C per 20 km | | 0.02°C per 0.25 m | | | Based on Hitachi Cable's DTS |

**Table 10.3** Pressure sensors

| Name | Operating principle | Range | Accuracy | Resolution | Temperature range | Application areas | Price guide | Notes |
|---|---|---|---|---|---|---|---|---|
| Babcock & Wilcox | Microbend | 0–3000 p.s.i. | ±0.05% | | | Harsh environment | | Temp. co. ±0.00%/°C |
| Dynisco Inc. | Fotonic | 0–20 000 p.s.i. | ±0.5% FS | ±0.01% FS | 400°C max | High-temperature pressure sensing | | Extrusion of plastics |
| Fiber Optic Sensor Tech. | Fotonic | −50–250 mm Hg | | | | Urology | | |
| FIMOD | | | | | | | | Development |
| Fuji Electric Co. Ltd. | Capacitative sensor E/O convertor | 0–500 kg cm$^{-2}$ | ±0.1% | | −30–100°C | | | Fuji system |
| G2 Systems Corp. | Distributed: microbend | 2 g to 3 kg | | | | | $20 800 | Multisensor: temperature, pressure, strain (range to 15 km) |
| Herga Electric Ltd. | Microbending | n/a | n/a | n/a | | | | |
| Lucas Control Systems | Birefringence | | | 0.025% | 10–35°C | | $4800 | |
| Optech Inc. | Microbend | | | | 0–60°C | | | |
| Optelecom | Wavelength modulation | Various | 0.05% FS | 0.003% FS | | | | Development dormant |
| Optrand Inc. | Fotonic (dual diaphragm) | 0–500 p.s.i. | ±0.25% | | 0–650°C | Engines, spark plugs, turbines | | |
| Photonetics SA | FP cavity spectral modulation | 0–1500 p.s.i. | ±1% FS | n/a | 5–50°C | | | Was MetriCor 1450, 4-channel sensor |
| Measurand Inc. | Microbending | 0–100 kPa | ±1.2% FS | | | | | |

has opened up new doors, not just for niche fiber optic applications, but for the whole fiber optic sensing business. According to the study, the fiber optic sensor market was expected to show strong growth in that year. The study noted that revenues for 1995 were at $252.4 million, with an annual growth rate of 9.9% and a compound annual growth rate of 11.8% for the year leading to 2002.

The study undertaken herein may be viewed in terms of the probability that displacement/position/proximity sensors have in the past contributed the majority of the revenue from fiber optic sensors, and it would appear likely that they will still do so, as stated by Frost and Sullivan's report, at least in the short term. According to other recently published data on sensor markets for 1998 from BCC [5], the share of displacement/proximity sensors in terms of the world-wide OFS sensor revenues were 50% in 1993. The next important in revenue terms is the group of temperature sensors, which shared 15% of the revenues in 1993. The image of the OFS technology development activities disclosed in this study shows the contrast between revenue earning and patent activity, where only 3.5% of US patents issued from 1994 are for displacement measurement compared with 8.4% for temperature sensing. BCC predicted that a rise in temperature sensor revenue to 18% would be seen from 1997 onwards, while the share of displacement/proximity sensors would decline to 45%. The most obvious lesson to be drawn is the time from patent activity to revenue stream, and the technology associated with many, if not most, of the commercial displacement sensors is 'tried-and-tested', and the subject of patents of about the early 1980s. The BCC prediction is for the trend to be away from the dominance of displacement sensors, and this is confirmed in the current patent activity, which is expected to be reflected more closely in market share over the next decade.

At the moment, one of the most discussed measurands on the OFS market is the strain sensor. Obviously, this is boosted by the expectation of cheap, reliable and multiplexed devices based on fibre Bragg grating (FBG) sensors, which must be coupled with activity in temperature sensing to deconvolve the two parameters, and this trend has been seen.

An interesting 'snapshot' of activity is seen in the analysis of papers presented at the major international conference in the field at the end of 1997, coinciding with the end point of this survey. Among all the technical papers presented in the 12th International Conference on Optical Fibre Sensors (OFS'12), held in Williamsburg, VA, more than 28% were dedicated or related to FBG technology. Consequently, strain was the most discussed measurand at OFS'12. Figure 10.8 shows each measurand as a percentage of the papers presented at OFS'12, and this conference must be recognized as a broad spectrum of the activity of the most important research groups in the field internationally. The statistics illustrated in Fig. 10.8 compare interestingly with those used in Fig. 10.6 showing patent activity over a recent period. It can be seen that the research activity which they represent correlates more closely to patent activity than do sales and associated revenue generation, based as they are on 'older' technology. This clearly shows the trends for the future: not all patents will be products, but some will,

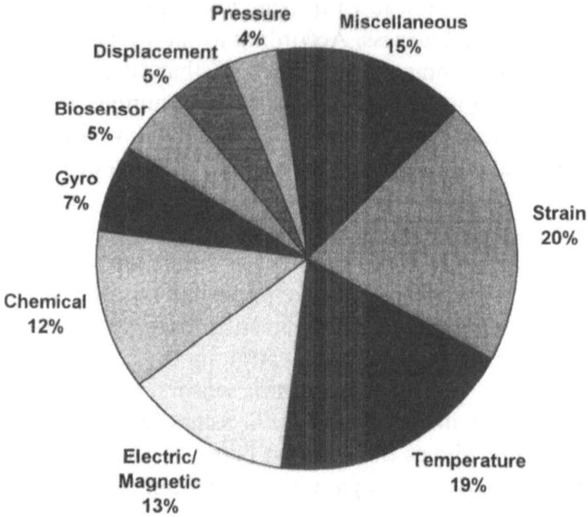


**Fig. 10.8** Distribution of papers presented at OFS'12 by measurand of interest. Z. Y. Zhang, K. T. V. Grattan and A. W. Palmer, Report to the Optical Sensors Collaborative Association (OSCA), 1997 (reproduced with permission).

and if they are a commercial success they will have an impact on revenues for the first decade of the 21st century.

## 10.6 PATENT LISTING (1994–1997)

A full listing of relevant OFS patents during this period is given in Table 10.9. This illustrates their range and diversity, and provides a means for detailed follow-up of specific sensors.

**Table 10.4** Displacement sensors

| Name | Operating principle | Range | Accuracy | Resolution | Temperature range | Application areas | Price guide | Notes |
|---|---|---|---|---|---|---|---|---|
| Anritsu Meter Co. Ltd. | Laser triangulation | ±80 µm to ±40 mm | | 0.0125% | 0–40°C | | | Not FO |
| Carlo Gavazzi Electromatic | Remote photoelectric | 0–10 m max | | | −25–55°C | | | |
| Computer Opt. Products | Angular: code plate | 360° | n/a | To 12 bits | −50–150°C | | | FO-linked optical shaft encoder |
| Computer Opt. Products | Code plate | 25–50 cm | n/a | 8 bits | −50–150°C | | | Linear encoder |
| Dolan Jenner | Remote photoelectric | Switch | | | 600°C max | | | |
| Dr. Johnnes Riegl GmbH | OTDR | 1–500 m | ±0.01 m 0.002% | 0.001 m at 5 s | −10–80°C | Non-contact, high temperature | $12 800 | Remote head is FO coupled |
| Erwin Sick GmbH | n/a | 20 m | 50 mm | | −25–55°C | Safety barrier | $3200 | Safety zone 180°coverage |
| Erwin Sick GmbH | LIDAR | 130 m | ±5 mm | | −25–55°C | | $2400 | |
| Erwin Sick GmbH | Remote photoelectric | To 60 m | | | −25–55°C | Switch; counting etc. | $192 | |
| Haenni | Laser triangulation | 0–100 mm | ±0.5% FS typical | <0.001 | 0–50°C | | | Optical: not FO |
| Herga Electric Ltd. | Distributed microbending | n/a | n/a | n/a | | Safety mats | | |
| IDEC GmbH | Remote photoelectric | Analog o/p | | | | | | |
| Keyence Corp. | Remote photoelectric (laser) | 0.1 mm at 100 mm | | | | | | |
| Koden Industry Co. Ltd. | Remote photoelectric | 10–35 mm | ±5 µm | | 5–40°C | | $192 | |
| Lucas Control Systems | Wavelength modulation – angular rotation | 350° | 0.5% | 1.3° | 0–70°C | | $4800 | |
| MTI Instruments | Fotonic | 50 µm to a few mm | 1–2 µm | | 0–80°C | | | |
| Optech Inc. | Fotonic | 0–400 µm | ±1% FS | <1 µm | | Non-contact displacement/ vibration | | |
| Philtec Inc. | Fotonic | 0.05–250 mm | n/a | 0.5 µm typical | −70–300°C | Precision dynamic measurement | | |
| Pruftechnik AG | Retro alignment | 5 m | <2% | <1 µm | 0–50°C | Shaft alignment | | |
| Pulnix America Inc. | Remote photoelectric | 0–3 m | ±0.1 mm | ±0.001 mm | −50–70°C | | $112 | |

**Table 10.4** (*Continued*)

| Name | Operating principle | Range | Accuracy | Resolution | Temperature range | Application areas | Price guide | Notes |
|---|---|---|---|---|---|---|---|---|
| SCICOM | OTDR | 0.5 mm typical | | 0.1 mm | | Crack width monitoring | | Partial reflectors in quasi-distributed sensor |
| SCICOM | Microbend: spiral overwind | Switch | | 0.001 mm | | Concrete crack monitoring | | Distributed |
| Seeka | Remote photoelectric | Switch | | | | | | |
| Spindler & Hoyer GmbH | Fotonic | | | | | | | MTI sensor, not in UK |
| Volpi AG | Remote photoelectric | | | | | | | |
| Vydas | Remote photoelectric | Switch | | | | | | |
| Wenglor GmbH | Laser photoelectric | 0–50 m | | | −10–50°C | | | |
| Measurand Inc. | Microbending | | ±0.8% | | | | | |

**Table 10.5** Liquid level sensors

| Name | Operating principle | Range | Accuracy | Resolution | Temperature range | Application areas | Price guide | Notes |
|---|---|---|---|---|---|---|---|---|
| Able Instruments & Controls Ltd. | TIR (total internal reflection) | Switch | | | 0–65°C | | $78 | Optical sensor: not FO |
| Carlo Gavazzi Electromatic | TIR | Switch | | | −20–60°C | | | Optical not FO |
| Fuji Electric Co. Ltd. | Pressure E/O convertor | 0–32 m $H_2O$ | ±0.25% | | −30–100°C | | | Fuji system |
| Honeywell Inc. | TIR | ±1 mm | 1 mm | 1 mm | −50–125°C | Set point detection; most liquids | | NB. Not FO |
| IMO Industries | TIR | Switch | | | −20–80°C | | | |
| Keyence Corp. | TIR | Switch | | | | | | |
| Lucas Control Systems | Wavelength modulation – float gauge | 270° | 5% FS | 2° | −20–40°C | Level or rotary position | $4800 | |
| Phoenix Analytec | TIR | Switch | ±0.5 mm | | −30–95°C | | | |
| Vickers Incorporated | TIR | n/a | n/a | n/a | −40–110°C | | | |
| Measurand | ? | | | | | | | |

**Table 10.6** Gas sensors

| Name | Operating principle | Range | Accuracy | Resolution | Temperature range | Application areas | Price guide | Notes |
|---|---|---|---|---|---|---|---|---|
| ADC Ltd. | Absorption – interference filter | To 100% | ±0.1% FS | | 0–50°C | Multi-gas | | |
| AVL Photronics | Fluorescence | | | | | Blood gas analysis – $CO_2$ (off-line) | | |
| Biomedical Sensors | | | | | | Blood gas monitoring | | |
| CDI/3M Healthcare | Fluorescence | 15–75 torr | ±2.5 torr | | | Blood gas monitoring – $CO_2$ | | |
| CDI/3M Healthcare | Fluorescence | 40–220 torr | ±8.5 torr | | | Blood gas monitoring – $O_2$ | | |
| Lightsense Corporation | Fluorescence | n/a | n/a | n/a | | Medical & industrial – $pCO_2$, $pO_2$ | | |
| Neotronics/Solomat | Absorption | 0–25 000 p.p.m. | ±2% | | 0–40°C | $CO_2$ | | |
| Rosemount Analytical Inc. | Resonant spectroscopy – etalon | 0–200 p.p.m. | ±1% FS | | 0–40°C | Spillage/flue gas – $NH_3$ | | |
| Rosemount Analytical Inc. | Gas correlation spectroscopy | 0–10 000 p.p.m. | ±20 p.p.m. ±6% | | −30–65°C | Flue gas – CO | | |
| Scottish Anglo Env Protection | Absorption | 0–100% | ±1% FS typical | | −10–40°C | $CH_4$, $CO_2$ | | |
| Seres | Gas filter correlation – IR | 0–100 p.p.m. | 1% typical | 0.1 p.p.m. | 0–40°C | CO | | |
| Sieger | Absorption – interference filter | 0–100% LEL | | | 5–50°C | $CH_4$ | | |

**Table 10.7** Chemical sensors

| Name | Type | Operating principle | Range | Accuracy | Resolution | Temperature range | Application areas | Price guide | Notes |
|---|---|---|---|---|---|---|---|---|---|
| Synectics Medical | Bilirubin | Dual wavelength absorption | 1–100 μmol | 4 μmol | | | Bile detection | | IROE–CNR development |
| Ciencia Inc. | Fluorescence | AOTF | | | | | Aromatic monitoring | | |
| Fiberchem Inc. | Hydrocarbon | Refractive index change in cladding | | | | | Hydrocarbon detection | | |
| Lightdense Corporation | pH | Fluorescence | n/a | n/a | n/a | | | | |
| CDI/3M Healthcare | pH | Fluorescence | 7.1–7.8 | <0.03 | | | Blood monitoring | | |
| Optical Syst & Sensors | pH | | | | | | | | |
| Chromex Inc. | Raman spectroscopy | FO-linked Raman spectroscopy | | | | | | | |
| Rosemount Analytical Inc. | Spectrometer | IR spectrometry with AOTF | 900–2500 nm | | 0.5% FS | 5–40°C | Multi-component liquids | | |
| Seres | Spectrometer | IR asorbance | 0–30 mg $l^{-1}$ | ±3% FS | 0.1 mg $l^{-1}$ | | THC in water | | Extraction method |
| Seres | Spectrometer | UV-vis spectrometer | 0–200 mg $l^{-1}$ | ±2% FS | 1 mg $l^{-1}$ | | Nitrate monitoring in water | | Holographic grating 200–800 nm |
| Ocean Optics Inc. | Spectrometer | Remote spectrometer | 220–1000 nm | | 1 nm typical | | Remote spectral analysis | $2880 | Low cost; c/w PC interface/multiple ranges/1024 element CCD |
| Photonetics SA | Spectrometer | Filtered photodetectors | 240–1600 nm | ±0.05 Abs units | 0.01 Abs units | 5–40°C | | | OPTOPAC |
| American Holographic | Spectrometer | Holographic grating | 190–1100 nm | | | | | | 4 ranges/256 element CCD |
| Enviro Technology Services plc | Spectrometer | Remote spectrometer | | | | | Environmental monitoring | | Multi-gas analyzer |
| LT Industries Inc. | Spectrometer | Remote spectrometer | 400–2400 nm | ±0.5 nm | 1 nm | −15–45°C | Multiport FO sampling | | |
| Galileo Electro-Optics Corp. | Spectrometer | Remote spectrometer | 900–4500 nm | | | | | | Ruggedized probe |
| Servomex | Spectrophotometer | NIR spectrophotometer | | <1% FS | | To 180°C | CO, $CO_2$, $CH_4$, $NO_x$, $SO_x$ | | |

**Table 10.7** (*Continued*)

| Name | Type | Operating principle | Range | Accuracy | Resolution | Temperature range | Application areas | Price guide | Notes |
|---|---|---|---|---|---|---|---|---|---|
| Acal Auriema | Spectrophotometer | NIR/vis spectrophotometer | | | | | | | |
| Photonetics SA | Spectrophotometer | Remote spectrophotometer head | n/a | n/a | n/a | | | | FISPEC |
| UOP/Guided Wave | Spectrophotometer | NIR spectrophotometer | | | | | Petroleum octane, total aromatics | | |
| Pharmacia Biotek | Surface plasmon resonance | Surface plasmon resonance | | | | | Food industry, biochemical | $80 000 | Not FO |
| AltOptronic AB | Spectrometer | | | | | | | | |

**Table 10.8** Miscellaneous sensors

| Name | Type | Operating principle | Range | Accuracy | Resolution | Temperature range | Application areas | Price guide | Notes |
|---|---|---|---|---|---|---|---|---|---|
| Takaoka Electric MGF Co. Ltd. | Acceleration | Dielectric mirror fluorescence | $0.1$–$700\,m\,s^{-1}\,s^{-1}$ | | $0.1\,m\,s^{-1}\,s^{-1}$ | 0–50°C | | $5600 | ex ASEA sensor/ 10 Hz to 1 kHz |
| Babcock & Wilcox | Acceleration | Microbend | ±100 g | ±0.02 g | | 0–50°C | | | Made to order |
| Optech Inc. | Acceleration | Microbend | 3 μg min | | | 0–60°C | Geophysical use/ general vibration monitoring | | 100 dB dynamic range/100 Hz bandwidth/ 300 Hz resonant frequency/cross-sensitivity – 27 dB |
| IDEC GmbH | Color | 3-wavelength analysis | vis spectrum | | | | | | |
| Pulnix America Inc. | Color | 3-wavelength filters | | | | | | | FO sensor head |
| Keyence Corp. | Color | Reflected intensity | vis spectrum | | | | | | |
| Seeka | Color | Wavelength filters | 3 colors | | | | | | |
| Lucas Control System Products | Current | Faraday effect | 0–2 kA | | ±10 A | −20–70°C | Power distribution | $4800 | |
| Matsushita Electric Industrial Co. Ltd. | Current | Faraday | | | | | | | Optical current transformer |
| Optech Inc. | Current | n/a | mA to >100 kA | ±0.3% | 1 : 35 000 | | HV installations | | 10 Hz to 2.5 kHz. 4 cm × 4 cm × 5 m package |
| Soreq | Current | | | | | | | | |
| Photonetics SA | Current | FP cavity spectral modulation | n/a | n/a | n/a | 5–50°C | EMC testing | | Was MetriCor/ Current transducer is SQUIB mounted temperature sensor |
| Rosemount Analytical Inc. | Dust density | Absorbance | 1% min | ±1% | <0.1% | −40–55°C | Flue gas | | Not FO |
| Flowdata Inc. | Flow | Paddle wheel/ positive displacement FO encoder | Various | ±0.1% | n/a | To 200°C | | $8000 | Figures provided for mid-range instrument |

**Table 10.8** (*Continued*)

| Name | Type | Operating principle | Range | Accuracy | Resolution | Temperature range | Application areas | Price guide | Notes |
|---|---|---|---|---|---|---|---|---|---|
| Fuji Electric Co. Ltd. | Flow | Differential pressure E/O convertor | $0$–$30\ kg\,cm^{-2}$ | ±0.1% | | −30–100°C | | | Fuji system |
| Fuji Electric Co. Ltd. | Flow | Electromagnetic E/O convertor | $0.3$–$10\ m\,s^{-1}$ | ±0.5% | | −30–100°C | | | Fuji system |
| Photonetics SA | Flow | Reflection from flow stream – intensity based | n/a | n/a | n/a | −180–100°C | Heat exchangers, refrigerators, chemical reactors | | Two-phase/ OPTOFLOW |
| Optelecom | Gyro | | | | | | | | Gyro coil only |
| Photonics Systems Lab. | Gyro | Multimode Sagnac interferometer | 50° | | | −40–150°C | Automotive, robotics, toys, security | $112 | Low cost gyro |
| Optech Inc. | Hydrophone | Interferometric | 60 dB re μPa min | n/a | | | Towed/static array | | 10 Hz to 1.2 kHz bandwidth |
| Soreq | Intruder detection | | | | | | | | |
| Ispra Product Research Co. | Oil discharge | Refraction change | n/a | n/a | n/a | −10–60°C | Marine installations, underground storage | | |
| Seres | Oil discharge | Scattering of IR | 0–1000 p.p.m. | ±2% FS typical | | | Bilge/discharge monitoring, oil in condensates | | |
| Sympatec GmbH | Particle size | Laser diffraction | 0.1–3500 μm | | | | | | |
| Boston Advanced Technology | Petrol discharge | Refractive index change | | | 3 mm min | −40–50°C | Petrol pump sumps | | |
| Bertin et Cie | Petroleum octane | Spectrophotometer | 380–750 nm | | 5 nm typical | 5–40°C | On-line petroleum analysis | $56 000 | 128 element CCD/ trichrometeris output |
| LT Industries Inc. | Petroleum octane | Remote spectrometer | | | | −15–45°C | | | |
| Rosemount Inc. | Position | n/a | | | | | | | |
| Photonetics SA | Refractive index | FP cavity spectral modulation | 1.32–1.60 | ±1% FS | n/a | 5–50°C | | $2400 | Was MetriCor, 4-channel sensor |
| Phoenix Analytec | Refractive index | | | | 0.0001 RI | −30–95°C | | | |

| | | | | | | | | |
|---|---|---|---|---|---|---|---|---|
| Protimeter Ltd. | Relative humidity | Remote chilled mirror | 20–98% RH | $< \pm 2\%$ | | −40–100°C | | \$5600 | |
| Firefly AB | Spark detection | Direct optical detection | 400°C min | | 1 mm diameter | 0–60°C | Conveyors, dusts, driers | | Explosion prevention |
| FIMOD | Strain | OTDR | ±10 000 μstrain | ±1% | ±260 μm | −55–350°C | Civil structures, aircraft, composites | | Price approx £60 per segment fiber OTDR |
| Optech Inc. | Strain | Microbending | 2 μm | ±5% FS | 0.001 μm | | Machinery/ structure | | 1 Hz bandwidth |
| G2 Systems Corp. | Strain | Distributed: microbend | 10–1500 μstrain | | | | | \$20 800 | Multisensor: temperature, pressure, strain/ range to 15 km |
| Fiber & Sensor Technologies | Strain | Extrinsic fiber FP cavity | | | 0.03 μstrain | −270–750°C | Embedding: smart structure | | |
| Lucas Control System Products | Strain | Birefringence | | | 0.025% | 10–35°C | Pressure, load cell | \$4800 | |
| Optech Inc. | Tachometer | Remote chopper/ reflector | 25 000 r.p.m. max | <1% reading | n/a | −55–300°C | | | |
| Aurora Optics Inc. | Tachometer | Remote shaft encoder | | 0.001% | | | | | |
| Polytec Gmbh | Thickness | Absorbance | 100 μm to 10 cm | | | | On-line coating | | |
| Seres | Turbidity | Absorbance | 0–10 000 NTU | ±1% FS | 0.1–200 NTU | | Water supply, filter monitoring | | |
| Mettler Toledo AG | Turbidity | Backscatter | 0.1 NTU/FTU to 250 $g\,l^{-1}$ | | | −10–50°C | Filter breakdown, biomass detection typical | | |
| Polytec Gmbh | Velocity | LDV | 30 $mm\,s^{-1}$ to 30 $m\,s^{-1}$ | ±0.1% | | | | | Standoff 500 mm |
| Mindrum Precision Products | Voltage | EO modulator – MM fiber | 10 mV min | | | | Military avionic | | E/O convertor |
| Measurand Inc. | Gas flow | Microbending | | | | | | | |
| Measurand Inc. | Liquid flow | Microbending | | | | | | | |
| Measurand Inc. | Acceleration | Microbending | | | | | | | |
| Measurand Inc. | Curvature | Microbending | | | | | | | |
| Hitachi Cable Ltd. | Gyro | | | | | | High-precision car navigating system | | |

**Table 10.9** US patents in the OFS field from 1994 to 1997

| Patent number: title | Issued (Day/month/year) | Country | Field 1 | Field 2 |
|---|---|---|---|---|
| 5275053: Fiber optic pressure sensor systems | 4/1/94 | USA | Pressure | Diaphragm + fotonic |
| 5276322: Fiber optic accelerometer | 4/1/94 | USA | Miscellaneous | Acceleration |
| 5277496: High-temperature optical probe | 11/1/94 | USA | Temperature | Pyrometer |
| 5277872: Optical fiber pH microsensor and method of manufacture | 11/1/94 | USA | Chemical | pH |
| 5279295: Non-invasive oximeter arrangement | 18/1/94 | Germany | Biosensor | Blood oxygenation |
| 5279793: Optical osmometer for chemical detection | 18/1/94 | USA | Chemical | Osmometer |
| 5280172: Fiber optic sensor for measuring gas | 18/1/94 | France | Gas | Evanescent wave |
| 5280173: Electric and electromagnetic field sensing system including an optical transducer | 18/1/94 | USA | EM | EM field |
| 5280339: Closed loop fiber optic gyroscope with fine angle resolution | 18/1/94 | USA | Gyro | |
| 5280341: Feedback controlled differential fiber interferometer | 18/1/94 | USA | General | Interferometer |
| 5280548: Emission-based fiber optic sensors for pH and carbon dioxide analysis | 18/1/94 | USA | Chemical | pH/$CO_2$ |
| 5280786: Fiberoptic blood pressure and oxygenation sensor | 25/1/94 | USA | Biosensor | Oxygenation/ diaphragm + fotonic |
| 5281820: Radiation detector | 25/1/94 | Germany | Miscellaneous | Radiation |
| 5283429: Fiber optical monitoring system for electrical conductors and the like | 1/2/94 | USA | Temperature/stress | Transmission |
| 5285424: Wide bandwidth fiber optic hydrophone | 8/2/94 | USA | Miscellaneous | Hydrophone |
| 5286777: Preparing a dye-containing polymer | 15/2/94 | USA | Chemical | Dyed polymer |
| 5286980: Product discrimination system and method therefore | 15/2/94 | USA | Miscellaneous | Object discrimination |
| 5289258: Fiber optic gyroscope modulation error reduction | 22/2/94 | USA | Gyro | |
| 5290103: Optical fiber temperature sensor | 1/3/94 | France | Temperature | Fiber amplifier |
| 5290398: Synthesis of tapers for fiber optic sensors | 1/3/94 | USA | Chemical | Fluorescence |
| 5291013: Fiber optical monitor for detecting normal breathing and heartbeat motion based on changes in speckle patterns | 1/3/94 | USA | Displacement | Microbend |
| 5291015: Optical fiber sensor system where variations in temperature of optical source is minimized or equal | 1/3/94 | UK | Auxiliary | Temperature effect |
| 5291032: Fiber optic evanescent wave fuel gauge and leak detector using eccentric core fibers | 1/3/94 | USA | Liquid level | Evanescent wave |
| 5291266: Depolarized light source for fiber optic sensors | 1/3/94 | USA | Auxiliary | Depolarized |

| | | | | |
|---|---|---|---|---|
| 5292196: Optical fiber type temperature distribution measuring apparatus | 8/3/94 | Japan | Temperature | Distributed |
| 5293039: Sandwiched fiber-optic pressure detector | 8/3/94 | France | Pressure | Microbend |
| 5293046: Optical high pressure sensor | 8/3/94 | USA | Pressure | Ruby fluorescence |
| 5293216: Sensor for semiconductor device manufacturing process control | 8/3/94 | USA | Miscellaneous | Film thickness/ roughness/emissivity |
| 5294972: Multi-axis fiber optic rotation rate sensor with parallel sensing coils | 15/3/94 | Germany | Gyro | |
| 5295206: Fiberoptic temperature transducer | 15/3/94 | USA | Temperature | Reflectance |
| 5295207: Optical apparatus for measuring current in a grounded metal-clad installation | 15/3/94 | France | EM | Electric current |
| 5297224: Optically switched sensor array | 22/3/94 | USA | Auxiliary | Multiplexing |
| 5297436: Optical fiber ultrasonic sensor | 29/3/94 | Australia | Miscellaneous | Acoustic |
| 5297597: Container filler indexing counter | 29/3/94 | USA | Miscellaneous | Position |
| 5298741: Thin film fiber optic sensor array and apparatus for concurrent viewing and chemical sensing of a sample | 29/3/94 | USA | Chemical | Dyed thin film |
| 5299141: Automatic fluid monitoring and sampling apparatus and method | 29/3/94 | USA | Chemical | Auxiliary |
| 5299271: System and method of embedding optical fibers in honeycomb panels | 29/3/94 | USA | Auxiliary | Embedding in honeycomb panels |
| 5300769: Method and system of compensating for signal artifacts in a fiber-optic sensing system | 5/4/94 | USA | Auxiliary | Signal artifacts compensation |
| 5301001: Extrinsic fiber optic displacement sensors and displacement sensing systems | 5/4/94 | USA | Displacement | Fizeau cavity |
| 5301558: Testing apparatus for a multilayer piezoelectric actuator | 12/4/94 | USA | Displacement | Fotonic |
| 5302025: Optical systems for sensing temperature and other physical parameters | 12/4/94 | USA | Temperature | Absorption |
| 5303586: Pressure or fluid level sensor | 19/4/94 | USA | Pressure/fluid level | Fotonic |
| 5304809: Luminescent decay time measurements by use of a CCD camera | 19/4/94 | USA | Temperature | Fluorescence |
| 5305075: Magnetostrictive transducer system having three dual-strip fiber optic magnetostrictive transducers bonded into a single fiber Mach–Zehnder interferometer | 19/4/94 | USA | EM | Magnetic field |
| 5306088: Method and apparatus for monitoring the temperature in a turbine component | 26/4/94 | Germany | Temperature | Photoconductive |
| 5307146: Dual-wavelength photometer and fiber optic sensor probe | 26/4/94 | USA | Chemical | Spectrophotometer |

**Table 10.9** (*Continued*)

| Patent number: title | Issued (Day/month/year) | Country | Field 1 | Field 2 |
|---|---|---|---|---|
| 5307437: Optical fibre incorporating a Bragg lattice and its application | 26/4/94 | France | General | In-fiber grating |
| 5308162: Temperature abnormality detecting structure for fluid pipe | 3/5/94 | Japan | Temperature | Distributed |
| 5308973: Method and device for the measurement of force by a fiber optics system by evaluating phase shift of light waves | 3/5/94 | Liechtenstein | Strain/stress/ structure | Strain |
| 5309533: Structure with intrinsic damage control, manufacturing processes and method of use | 3/5/94 | France | Strain/stress/ structure | Smart structure |
| 5309540: Optical fiber sensor and a manufacturing process for making same | 3/5/94 | France | Pressure | Hollow fiber |
| 5311013: Optical fiber distribution system for an optical fiber sensor in a luminescent sensor system | 10/5/94 | USA | Chemical | Fluorescence |
| 5311274: Fiber optic refractometer | 10/5/94 | USA | Chemical | Refractometry |
| 5311283: Fiber optic probe and method for detecting optically active materials | 10/5/94 | USA | Miscellaneous | Fluid constituent |
| 5313065: Fiber optic radiation monitor | 17/5/94 | USA | Miscellaneous | Radiation |
| 5313266: Demodulators for optical fiber interferometers with [3 × 3] outputs | 17/5/94 | USA | General | Interferometer |
| 5313542: Apparatus and method of rapidly measuring hemispherical scattered or radiated light | 17/5/94 | USA | Miscellaneous | Scatterometer |
| 5313957: Guide wire mounted pressure transducer | 24/5/94 | USA | Pressure | Microbend |
| 5315364: Gauge for measuring strong shock waves | 24/5/94 | USA | Pressure | Shock-wave |
| 5315672: Fiber optic chemical sensor | 24/5/94 | USA | Chemical | Evanescent wave |
| 5315673: Optical waveguide vapor sensor | 24/5/94 | USA | Chemical | Spectrophotometer |
| 5315993: Luminescence monitoring with modulation frequency multiplexing | 31/5/94 | USA | Chemical | Luminescence |
| 5315995: Method and apparatus for continuous measurement of central venous oxygen saturation during human cardiopulmonary resuscitation and clinical shock | 31/5/94 | USA | Biosensor | Blood oxygenation |
| 5317147: Method and apparatus for absolute measurement of force by use of polarized, non-coherent light and compensation of strain-induced birefringence effects in a single-mode fiber-optic waveguide | 31/5/94 | Liechtenstein | Strain/stress/ structure | Strain |

| | | | | |
|---|---|---|---|---|
| 5317544: Multiple segment fiber optic hydrophone | 31/5/94 | USA | Miscellaneous | Hydrophone |
| 5317656: Fiber optic network for multi-point emissivity-compensated semiconductor wafer pyrometry | 31/5/94 | USA | Temperature | Pyrometry |
| 5317929: Fiber optic flexural disk accelerometer | 7/6/94 | USA | Miscellaneous | Acceleration |
| 5319434: Laser rangefinder apparatus with fiber optic interface | 7/6/94 | USA | Miscellaneous | Rangefinder |
| 5319440: Fiber optic gyroscopes with depolarized light | 7/6/94 | USA | Gyro | |
| 5319975: Fiber optic moisture sensor | 14/6/94 | USA | Miscellaneous | Moisture |
| 5319978: Optical pressure transducer | 14/6/94 | USA | Pressure | Diaphragm |
| 5320814: Fiber optic array sensors, apparatus, and methods for concurrently visualizing and chemically detecting multiple analytes of interest in a fluid sample | 14/6/94 | USA | Chemical | Spectrophotometer |
| 5321257: Fiber optic bending and positioning sensor including a light emission surface formed on a portion of a light guide | 14/6/94 | Canada | Displacement | Microbend |
| 5321503: Closed loop, depolarized interferometric fiber optic (IFOG) gyroscope with self adjusting serrodyne phase nulling | 14/6/94 | USA | Gyro | |
| 5321778: Method and apparatus for producing a prestressed fiber-optic sensor | 14/6/94 | France | Pressure | Prestressed fiber |
| 5324933: High accuracy and high sensitivity environmental fiber optic sensor with corrugations | 28/6/94 | USA | Chemical | Evanescent wave |
| 5326585: Method and compositions for manufacture of chemical sensors | 5/7/94 | USA | Biosensor | pH/oxygenation/$CO_2$ |
| 5327211: Process for molecular-specific optical fiber probes for Raman detection | 5/7/94 | USA | Chemical | SERS: surface enhanced resonance Raman spectroscopy |
| 5327214: Optical interferometric gyro having reduced return light to the light source | 5/7/94 | Japan | Gyro | |
| 5329349: Method for tuning fiber optic sensor coils | 12/7/94 | USA | Gyro | |
| 5330073: Gasoline dispenser leak detectors and automatic shut-off systems | 19/7/94 | USA | Liquid level | Evanescent wave |
| 5330136: Railway coded track circuit apparatus and method utilizing fiber optic sensing | 19/7/94 | USA | Miscellaneous | Object presence |
| 5330718: Sensor element and method for making the same | 19/7/94 | USA | Chemical | $CO_2$ in sample |
| 5331152: Fiber optic position indicator | 19/7/94 | USA | Miscellaneous | Position |
| 5331272: Optical sensor for detecting an electric field via the Faraday effect | 19/7/94 | Switzerland | EM | Electric current |
| 5331404: Low noise fiber gyroscope system which includes excess noise subtraction | 19/7/94 | USA | Gyro | |

**Table 10.9** (*Continued*)

| Patent number: title | Issued (Day/month/year) | Country | Field 1 | Field 2 |
|---|---|---|---|---|
| 5331961: Scintillation probe and intra-tube cavity insertable radioactive ray detecting apparatus | 26/7/94 | Japan | Miscellaneous | Radiation |
| 5332316: Fiber optic systems for sensing temperature and other physical variables | 26/7/94 | USA | Temperature | Fluorescence |
| 5333047: Optical fiber sensor having digital integrating detection system utilizing electronic phase lock loop | 26/7/94 | USA | Gyro | |
| 5333214: Apparatus for reducing magnetic field-induced bias errors in a fiber optic gyroscope | 26/7/94 | USA | Gyro | |
| 5333487: Spark-excited fluorescence sensor | 26/7/94 | USA | Gas | Exhaust |
| 5334850: Method and device for optically detecting an interface between two fluids and method of setting the parameters for such detection | 2/8/94 | Japan | Miscellaneous | Fluid interface |
| 5335064: Fiber-optic gyroscope for measuring velocity of an object | 2/8/94 | Japan | Gyro | |
| 5335305: Optical sensor for fluid parameters | 2/8/94 | USA | Chemical | Spectrophotometer |
| 5336883: Optical fiber sensor for producing a variable transverse strain in a localized portion of the fiber | 9/8/94 | UK | EM | Electric current |
| 5337142: Optical fiber measuring device using phase difference | 9/8/94 | France | Gyro | |
| 5337376: Chemically sensitive fiber optic cable | 9/8/94 | USA | Chemical | Surface plasmon |
| 5339026: Fiber-optic sensor with first and second bimode fibers | 16/8/94 | Switzerland | EM | Electric field/voltage |
| 5340715: Multiple surface evanescent wave sensor with a reference | 23/8/94 | USA | Chemical | Evanescent wave |
| 5341805: Glucose fluorescence monitor and method | 23/8/94 | USA | Chemical | Glucose |
| 5343035: Strain sensor comprising a strain sensitive, two-mode optical fiber | 30/8/94 | USA | Strain/stress/structure | Strain |
| 5343036: Optical measurement of electrical quantities using inverse piezoelectric effect with dual double-mode fiber | 30/8/94 | Switzerland | EM | Electric field/voltage |
| 5343037: Environmental and physical parameter sensors incorporating polymer-covered fiber field access blocks | 30/8/94 | USA | General | Evanescent wave |
| 5343045: Method and device for measuring moisture content | 30/8/94 | Canada | Miscellaneous | Moisture |
| 5345307: Linear phase ramp-type fiber optic gyro having ramp reset and reset connection signals | 6/9/94 | Japan | Gyro | |

| | | | | |
|---|---|---|---|---|
| 5345519: Temperature-compensated fiber optic external cavity strain sensors and an intensity-based fiber optic sensor system | 6/9/94 | Canada | Strain/stress/ structure | Strain |
| 5345522: Reduced noise fiber optic towed array and method of using same | 6/9/94 | USA | Miscellaneous | Hydrophone |
| 5347354: Sagnac-type optical fiber gyroscope having depolarizers of differing lengths | 13/9/94 | Germany | Gyro | |
| 5348019: Optical fiber pressure sensing catheter | 20/9/94 | USA | Pressure | Fotonic |
| 5348396: Method and apparatus for optical temperature measurement | 20/9/94 | USA | Temperature | Absorption |
| 5349181: Fiber optic chemical sensor having specific channel connecting design | 20/9/94 | USA | Chemical | Sensor configuration |
| 5349182: Fiber optic cable moisture monitoring system | 20/9/94 | Canada | Miscellaneous | Moisture |
| 5349458: Method for detecting external influence on an optical cable | 20/9/94 | Sweden | Miscellaneous | Intrusion/disturbance |
| 5351252: Technique of reducing the Kerr effect and extending the dynamic range in a Brillouin fiber optic gyroscope | 27/9/94 | USA | Gyro | Brillouin |
| 5351268: Modular luminescence-based measuring system using fast digital signal processing | 27/9/94 | USA | Temperature | Fluorescence |
| 5351324: Fiber optic security seal including plural Bragg gratings | 27/9/94 | USA | Miscellaneous | Intrusion/disturbance |
| 5351547: Optical pressure transducer having a fixed reflector and a movable reflector attached to a diaphragm | 4/10/94 | USA | Pressure | Diaphragm + fotonic |
| 5351900: Method of manufacture of quadrupole-wound fiber optic sensing coil | 4/10/94 | USA | Gyro | |
| H001364: Fluorescent fiber diagnostics | 4/10/94 | USA | General | Fluorescence |
| 5352040: Dual neutron flux/temperature measurement sensor | 4/10/94 | USA | Temperature/ radiation | Fluorescence/scintillation |
| 5353110: Method and apparatus for carrying out optical time domain reflectometry using weighing techniques | 4/10/94 | USA | General | OTDR |
| 5354990: In-situ localized tracer injection/fiber optic detection for measurement of flow distributions in fixed bed reactors | 11/10/94 | USA | Miscellaneous | Flow |
| 5355208: Distributed fiber optic sensor for locating and identifying remote disturbances | 11/10/94 | USA | Miscellaneous | Intrusion/disturbance |
| 5355216: Optical fiber sensor having digital integrating detection system utilizing electronic phase lock loop | 11/10/94 | USA | Gyro | |
| 5355423: Optical temperature probe assembly | 11/10/94 | USA | Temperature | Probe fabrication |
| 5356220: Method and apparatus for monitoring temperature of blast furnace and temperature control system using temperature monitoring apparatus | 18/10/94 | Japan | Temperature | Distributed |

**Table 10.9** (*Continued*)

| Patent number: title | Issued (Day/month/year) | Country | Field 1 | Field 2 |
|---|---|---|---|---|
| 5357813: Pressure sensor | 25/10/94 | Israel | Pressure | Microbend |
| 5359405: Hybrid fiber optic sensor including a lead-out optical fiber having a remote reflective end | 25/10/94 | USA | Displacement | FP interferometer |
| 5359445: Fiber optic sensor | 25/10/94 | USA | Miscellaneous | Vibration |
| 5359681: Fiber optic sensor and methods and apparatus relating thereto | 25/10/94 | USA | Chemical | Surface plasmon resonance |
| 5361130: Fiber grating-based sensing system with interferometric wavelength-shift detection | 1/11/94 | USA | General | In-fiber grating |
| 5361313: Two-mode remote fiber optic sensor | 1/11/94 | USA | General | Two-mode fiber |
| 5361314: Micro optical fiber light source and sensor and method of fabrication thereof | 1/11/94 | USA | Auxiliary | Fiber light source |
| 5362971: Fiber optic detection system | 8/11/94 | USA | Liquid level | TIR |
| 5363191: Fiber optic sensor array reading device | 8/11/94 | France | Auxiliary | Interferometer |
| 5363193: Method and arrangement for determining angular velocities | 8/11/94 | Germany | Gyro | |
| 5363195: Automatic gain calibration system for a phase modulator in a fiber optic gyro | 8/11/94 | USA | Gyro | |
| 5363342: High performance extended fiber optic hydrophone | 8/11/94 | USA | Miscellaneous | Hydrophone |
| 5363463: Remote sensing of physical variables with fiber optic systems | 8/11/94 | USA | Temperature/strain | Fluorescence |
| 5365337: Method and apparatus for compensating for the residual birefringence in interferometric fiber-optic gyros | 15/11/94 | USA | Gyro | |
| 5365925: Disposable calibration boot for multi-point calibration in fiber optic sensors | 22/11/94 | USA | Biosensor | Blood oxygenation |
| 5367376: Planar and linear fiber optic acoustic sensors embedded in an elastomer material | 22/11/94 | USA | Miscellaneous | Acoustic |
| 5367583: Fiber optic stress-corrosion sensor and system | 22/11/94 | USA | Strain/stress/structure | Stress/corrosion |
| 5369485: Fiber optic accelerometer with centrally supported flexural disk | 29/11/94 | USA | Miscellaneous | Acceleration |
| 5369487: Device for determining the amplitude of the spectral lines in the output signal from a ring interferometer | 29/11/94 | Germany | Gyro | |

| | | | | |
|---|---|---|---|---|
| 5371593: Sensor coil for low-bias fiber optic gyroscope | 6/12/94 | USA | Gyro | |
| 5371595: Fiber-optic gyroscope using single birefringent material for depolarizer | 6/12/94 | Japan | Gyro | |
| 5371720: Optical fiber pressure sensor for liquid level monitoring | 6/12/94 | USA | Liquid level | Pressure |
| 5372044: Optical sensor, especially pressure sensor, and corresponding optical measurement process | 13/12/94 | France | Pressure | Diaphragm + fotonic |
| 5373361: Fiber optic gyroscope including a 3.times.3 coupler and a contrast adjusting element | 13/12/94 | Germany | Gyro | |
| 5373487: Distributed acoustic sensor | 13/12/94 | USA | Miscellaneous | Acoustic |
| 5374821: Elastomeric optical fiber sensors and method for detecting and measuring events occurring in elastic materials | 20/12/94 | USA | Miscellaneous | Load |
| 5377004: Remote optical measurement probe | 27/12/94 | USA | General | Raman scattering/ fluorescence measurement |
| 5377283: Configuration control of mode coupling errors | 27/12/94 | USA | Gyro | |
| 5378432: Optical fiber pH microsensor and method of manufacture | 3/1/95 | USA | Chemical | pH |
| 5378889: Method and apparatus for detecting hydrocarbon fuels in a vapor state with an absorber–expander member | 3/1/95 | USA | Chemical | Hydrocarbon |
| 5379649: Coriolis effect meter using optical fiber sensors | 10/1/95 | USA | Miscellaneous | Mass flow |
| 5380995: Fiber optic grating sensor systems for sensing environmental effects | 10/1/95 | USA | General | In-fiber grating |
| 5381005: Optical fiber stress detector using amplitude modulation | 10/1/95 | France | Strain/stress/ structure | Stress |
| 5381022: Combined optical waveguide and prismatic liquid-level sensor | 10/1/95 | USA | Liquid level | Prismatic |
| 5381229: Sapphire optical fiber interferometer | 10/1/95 | USA | Temperature | Interferometer |
| 5381237: Multi-purpose optical head probe | 10/1/95 | Israel | Miscellaneous | Refractive index |
| 5381492: Fiber optic vibration sensor | 10/1/95 | USA | Miscellaneous | Vibration |
| 5381493: Optical fiber strain sensor with improved linearity range | 10/1/95 | USA | Strain/stress/ structure | Strain |
| 5384635: Remote sensing in optical fiber networks by synchronously detecting backscattered optical signals | 24/1/95 | USA | Miscellaneous | Intrusion/disturbance |
| 5384637: Fiber optic gyroscope modulation amplitude control | 24/1/95 | USA | Gyro | |
| 5385053: Fiber optic pressure sensor systems | 31/1/95 | USA | Pressure | Diaphragm + fotonic |
| 5385404: Temperature measuring device | 31/1/95 | UK | Temperature | Diamond absorption |
| 5386290: Optical interferometric angular rate meter with a self-diagnostic capability | 31/1/95 | Japan | Gyro | |

**Table 10.9** (*Continued*)

| Patent number: title | Issued (Day/month/year) | Country | Field 1 | Field 2 |
|---|---|---|---|---|
| 5386729: Temperature compensated microbend fiber optic differential pressure transducer | 7/2/95 | USA | Pressure | Microbend |
| 5387791: Impurity-doped optical shock, detonation and damage location sensor | 7/2/95 | USA | Miscellaneous | Intrusion/disturbance |
| 5388110: Superluminescent light source for an interferometric fiber optic gyroscope | 7/2/95 | USA | Gyro | |
| 5390546: Fiber optic diaphragm sensors for engine knock and misfire detection | 21/2/95 | USA | Pressure | Diaphragm + fotonic |
| 5391878: Multiplexed fiber readout of scintillator arrays | 21/2/95 | USA | Miscellaneous | Radiation |
| 5392117: Fabry–Perot optical sensing device for measuring a physical parameter | 21/2/95 | Canada | General | FP interferometer |
| 5392655: Optical pickup and signal analyzing system for mass flowmeters | 28/2/95 | USA | Miscellaneous | Mass flow |
| 5394242: Fiber optic resonant ring sensor and source | 28/2/95 | USA | Gyro | |
| 5394377: Polarization-insensitive hydrophone | 28/2/95 | USA | Miscellaneous | Hydrophone |
| 5394378: Hydrophone transduction mechanism | 28/2/95 | USA | Miscellaneous | Hydrophone |
| 5394488: Optical fiber grating-based sensor | 28/2/95 | USA | Miscellaneous | Speed |
| 5396079: Fiber optic detector and depth sensor and method for doing same | 7/3/95 | USA | Liquid level | TIR |
| 5396166: Fiber optic interferometric electric field and voltage sensor utilizing an electrostrictive transducer | 7/3/95 | USA | EM | Electric field/voltage |
| 5396568: Temperature compensation for fiber optic sensors | 7/3/95 | USA | Pressure | Microbend |
| 5397411: Method for making optical probe | 14/3/95 | USA | Auxiliary | Fiber bundle |
| 5397891: Sensor systems employing optical fiber gratings | 14/3/95 | USA | General | In-fiber grating |
| 5399854: Embedded optical sensor capable of strain and temperature measurement using a single diffraction grating | 21/3/95 | USA | Temperature/strain | In-fiber grating |
| 5399866: Optical system for detection of signal in fluorescent immunoassay | 21/3/95 | USA | Biosensor | Fluorescence immunoassay |
| 5400142: Fiber optic angular rate sensor including digital phase modulation | 21/3/95 | USA | Gyro | |
| 5400653: Coriolis effect meter using optical fiber sensors | 28/3/95 | USA | Miscellaneous | Mass flow |
| 5401954: Product ripeness discrimination system and method therefore with area measurement | 28/3/95 | USA | Miscellaneous | Object discrimination |

| | | | | |
|---|---|---|---|---|
| 5401956: Diagnostic system for fiber grating sensors | 28/3/95 | USA | Auxiliary | In-fiber grating |
| 5402230: Heterodyne interferometric optical fiber displacement sensor for measuring displacement of an object | 28/3/95 | China/Japan | Displacement | Interferometer |
| 5402231: Distributed Sagnac sensor systems | 28/3/95 | USA | General | Interferometer |
| 5402508: Fiber optic probe having fibers with endfaces formed for improved coupling efficiency and method using the same | 28/3/95 | USA | Auxiliary | Probe configuration |
| 5404218: Fiber optic probe for light scattering measurements | 4/4/95 | USA | Chemical | Raman spectroscopy |
| 5406370: Winding technique for decreasing the pump power requirement of a brillouin fiber optic gyroscope | 11/4/95 | USA | Gyro | |
| 5408326: Dual-wavelength absorption detector adapted for continuous-flow detection | 8/4/95 | USA | Miscellaneous | Flow continuity |
| 5408999: Fiber-optic probe for the measurement of fluid parameters | 25/4/95 | USA | Chemical | pH/O/$CO_2$ |
| 5410404: Fiber grating-based detection system for wavelength encoded fiber sensors | 25/4/95 | USA | General | In-fiber grating |
| H001436: Interferometric fiber optic sensor configuration with pump-induced phase carrier | 2/5/95 | USA | General | Interferometer |
| 5414266: Measuring system employing a luminescent sensor and methods of designing the system | 9/5/95 | USA | Temperature | Fluorescence |
| 5414507: Fiber optics pressure sensor transducer having a temperature compensator | 9/5/95 | USA | Pressure | Interferometer |
| 5416467: Security system utilizing loosely contained optical fiber | 16/5/95 | Japan | Miscellaneous | Intrusion/disturbance |
| 5416579: Method for determining concentration in a solution using attenuated total reflectance spectrometry | 16/5/95 | Netherlands | Chemical | Spectrometry |
| 5416585: Fiber optic gyro drift rate compensation based on temperature | 16/5/95 | USA | Gyro | Temperature effect compensation |
| 5416860: Method and apparatus for optically measuring electric current and/or magnetic field with temperature compensation | 16/5/95 | China (Taiwan) | EM | Electric current/magnetic field |
| 5418361: Optical displacement sensor employing reflected light of four wavelengths to determine displacement and the refractive index of the medium | 23/5/95 | UK | Displacement | Fotonic |
| 5419312: Multi-function endoscope apparatus | 30/5/95 | USA | Biosensor | Endoscopic temperature/pressure |
| 5419636: Microbend fiber-optic temperature sensor | 30/5/95 | USA | Temperature | Microbend |
| 5420684: Resonant interferometer fiber optic gyroscope | 30/5/95 | USA | Gyro | |
| 5420688: Interferometric fiber optic displacement sensor | 30/5/95 | USA | Displacement | Interferometer |
| 5421195: Fiber optic microbend sensor for engine knock and misfire detection | 6/6/95 | USA | Pressure | Microbend |

**Table 10.9** (*Continued*)

| Patent number: title | Issued (Day/month/year) | Country | Field 1 | Field 2 |
|---|---|---|---|---|
| 5422478: Fiberoptic pressure sensor having drift correction means for in situ calibration | 6/6/95 | USA | Pressure | Diaphragm + fotonic |
| 5422495: Optical sensor having a flotation means for detecting fluids through refractive index measurement | 6/6/95 | USA | Liquid level | TIR |
| 5424546: Scintillation counter having two scintillation fibers provided in series and which emit at different wavelengths | 13/6/95 | Japan | Miscellaneous | Radiation |
| 5425371: Fiberoptic pressure transducer | 20/6/95 | USA | Pressure | Diaphragm + fotonic |
| 5425624: Optical fluid-level switch and controls for bilge pump apparatus | 20/6/95 | USA | Fluid level | Transmission |
| 5426297: Multiplexed Bragg grating sensors | 20/6/95 | USA | General | In-fiber grating |
| 5426981: Vibrating sensor | 27/6/95 | UK | Miscellaneous | Vibration |
| 5427114: Dual pressure sensing catheter | 27/6/95 | USA | Pressure | Diaphragm + fotonic |
| 5430544: Process and an apparatus for correcting the measuring signals of a fiber optic gyro | 4/7/95 | Germany | Gyro | |
| 5430545: Fiber optic angular rate sensor including arrangement for reducing output signal distortion | 4/7/95 | USA | Gyro | Angular rate |
| 5430813: Mode-matched, combination taper fiber optic probe | 4/7/95 | USA | Biosensor | Fluorescence immunoassay |
| 5430814: Temperature transient detection system for superconducting magnets | 4/7/95 | USA | Temperature | Interferometer |
| 5430815: Optical fiber water sensor | 4/7/95 | USA | Miscellaneous | Water presence |
| 5432637: Fiber optic depolarizer for optical time domain reflectometer and fiber optical communication systems | 11/7/95 | USA | General | OTDR |
| 5434415: Radiation-detecting light-transmission apparatus | 18/7/95 | Japan | Miscellaneous | Radiation |
| 5434501: Polarization insensitive current and magnetic field optic sensor | 18/7/95 | USA | EM | Electric current/magnetic field |
| 5434557: Intrusion detecting apparatus | 18/7/95 | Israel | Miscellaneous | Intrusion/disturbance |
| 5435308: Multi-purpose multi-parameter cardiac catheter | 25/7/95 | USA | Biosensor | Blood oxygenation |
| 5436167: Fiber optics gas sensor | 25/7/95 | USA | Gas | Surface plasmon resonance |
| 5436444: Multimode optical fiber motion monitor with audible output | 25/7/95 | USA | Displacement | Modal noise |

| | | | | |
|---|---|---|---|---|
| 5436454: Optical probe for remote attenuated total reflectance measurements | 25/7/95 | USA | Chemical | Spectrometer |
| 5437186: Integrated optical acceleration sensor | 1/8/95 | Germany | Miscellaneous | Acceleration |
| 5438411: Electronic phase-tracking open-loop fiber optic gyroscope | 1/8/95 | USA | Gyro | |
| 5438873: Fiberoptic sensor using tapered and bundled fibers | 8/8/95 | USA | Pressure | Tapered fiber |
| 5439000: Method of diagnosing tissue with guidewire | 8/8/95 | USA | Biosensor | Tissue |
| 5440414: Adaptive polarization diversity detection scheme for coherent communications and interferometric fiber sensors | 8/8/95 | USA | General | Interferometer |
| 5440927: Fiber optic moisture sensor | 15/8/95 | USA | Miscellaneous | Moisture |
| 5441530: Photochemotherapy dosimeter | 15/8/95 | USA | Miscellaneous | Irradiation dosage |
| 5442720: Switched array | 15/8/95 | USA | Auxiliary | Sensor array multiplexing |
| 5444533: Fiber-optic gyroscope obtaining a signal from a perturbed light source | 22/8/95 | Japan | Gyro | |
| 5444534: Coil mounting arrangement for fiber optic gyroscope | 22/8/95 | Japan | Gyro | Fiber coiling |
| 5444803: Fiber-optic devices and sensors using fiber grating | 22/8/95 | Korea | General | Fiber grating |
| 5446278: Fiber optic sensor employing successively destroyed coupled points or reflectors for detecting shock wave speed and damage location | 29/8/95 | USA | Miscellaneous | Intrusion/disturbance |
| 5446279: Fiber optic sensor sensing curvature of a diaphragm | 29/8/95 | USA | Pressure | Diaphragm + fotonic |
| 5446280: Split-spectrum self-referenced fiber optic sensor | 29/8/95 | USA | Auxiliary | Light source referencing |
| 5446533: Fiber optic measuring apparatus and method | 29/8/95 | USA | General | Distributed measurement |
| 5446808: Photomechanical positioning and stabilization method and devices using optical fibers and feedback | 29/8/95 | USA | Displacement | Fotonic |
| 5446811: Thermally actuated optical fiber switch | 29/8/95 | USA | Temperature | Thermal switch |
| 5448055: Direct-contact-type image sensor using optical fiber array with light-absorbing cladding | 5/9/95 | Japan | Miscellaneous | Image sensor array |
| 5448657: Polarimetric fiber laser sensors | 5/9/95 | Korea | General | Polarimetric |
| 5449233: Temperature distribution measuring apparatus using an optical fiber | 12/9/95 | Japan | Temperature | Distributed |
| 5450203: Method and apparatus for determining an objects position, topography and for imaging | 12/9/95 | USA | Miscellaneous | Position |
| 5451772: Distributed fiber optic sensor | 19/9/95 | USA | Strain/stress/structure | Strain |
| 5452087: Method and apparatus for measuring pressure with embedded non-intrusive fiber optics | 19/9/95 | USA | Pressure | FP interferometer |

**Table 10.9** (*Continued*)

| Patent number: title | Issued (Day/month/year) | Country | Field 1 | Field 2 |
|---|---|---|---|---|
| 5453248: Cross-linked gas permeable membrane of a cured perfluorinated urethane polymer, and optical gas sensors fabricated therewith | 26/9/95 | USA | Gas | $O_2/CO_2$ in fluid |
| 5453827: Fiberoptic in-line filter and technique for measuring the transmission quality of an optical fiber through the use of a fiberoptic in-line filter | 26/9/95 | USA | Miscellaneous | Fault detection |
| 5453836: Fiber optic laser rotation sensor utilizing a fiber loop reflector | 26/9/95 | Korea | Gyro | |
| 5453838: Sensing system with a multi-channel fiber optic bundle sensitive probe | 26/9/95 | USA | General | Multi-mechanical parameters |
| 5456252: Induced fluorescence spectroscopy blood perfusion and pH monitor and method | 10/10/95 | USA | Biosensor | Blood perfusion |
| 5457532: Harmonic phase modulation error reducer | 10/10/95 | USA | Gyro | |
| 5459575: Laser fiber optic gyroscope having phase modulation | 17/10/95 | UK | Gyro | |
| 5460048: Ultrasonic sensor using a polarization maintaining fiber | 24/10/95 | Australia | Miscellaneous | Ultrasonic sensor |
| 5460182: Tissue penetrating apparatus and methods | 24/10/95 | USA | Biosensor | Tissue penetration monitoring |
| 5461914: Optical fiber gravity meter | 31/10/95 | USA | Miscellaneous | Gravity |
| 5461926: Single-ended optical fiber strain sensor for measuring maximum strain | 31/10/95 | USA | Strain/stress/ structure | Strain |
| 5461927: Optical fiber strain sensor for measuring maximum strain | 31/10/95 | USA | Strain/stress/ structure | Strain |
| 5462244: System for detecting trains | 31/10/95 | Netherlands | Miscellaneous | Object presence |
| 5463311: Apparatus and method for measuring operating voltage electrical signals in a high voltage environment | 31/10/95 | USA | EM | Electric voltage |
| 5463312: Faraday-effect sensing coil with stable birefringence | 31/10/95 | USA | EM | Electric current/magnetic field |
| 5463458: Optical sensor for duct fan discharge Mach numbers | 31/10/95 | USA | Miscellaneous | Mach number |
| 5465150: Optical fiber gyroscope sensing coil having a reduced sensitivity to temperature variations occurring therein | 7/11/95 | UK | Gyro | |
| 5469257: Fiber optic gyroscope output noise reducer | 21/11/95 | USA | Gyro | |

| | | | | |
|---|---|---|---|---|
| 5470155: Apparatus and method for measuring temperatures at a plurality of locations using luminescent-type temperature sensors which are excited in a time sequence | 28/11/95 | USA | Temperature | Fluorescence |
| 5471139: Fiber optic line magnetometer having thermal compensation | 28/11/95 | USA | Miscellaneous | Object presence |
| 5471197: Tamper-proof bracelet for home arrest system | 28/11/95 | USA | Miscellaneous | Intrusion/disturbance |
| 5471301: Optical fiber gyro | 28/11/95 | Japan | Gyro | |
| 5471546: Fiber-optic transmission sensor with modulator | 28/11/95 | Switzerland | General | Interferometer |
| 5473156: Optical displacement sensor using dual reference reflectors | 5/12/95 | UK | Displacement | Dual wavelength |
| 5473714: Optical fiber system using tapered fiber devices | 5/12/95 | USA | Auxiliary | Tapered fiber |
| 5475216: Fiber optic sensor having mandrel wound reference and sensing arms | 12/12/95 | USA | General | Interferometer |
| 5475298: Method and apparatus for measurements dependent on the Faraday effect using polarized counter-propagating light | 12/12/95 | UK | EM | Electric current |
| 5475489: Determination of induced change of polarization state of light | 12/12/95 | Denmark | General | Polarimetry |
| 5475773: Fiber-optic quartz voltage sensor | 12/12/95 | Switzerland | EM | Electric voltage |
| 5475774: Optical fiber sensor coil using reverse quadrupoles and method of winding same | 12/12/95 | USA | Gyro | |
| 5477323: Fiber optic strain sensor and read-out system | 19/12/95 | USA | Strain/stress/ structure | Strain |
| 5479094: Polarization-insensitive current and magnetic field optic sensor | 26/12/95 | USA | EM | Electric current/magnetic field |
| 5479828: Structure with intrinsic damage control, manufacturing processes and method of use | 2/1/96 | France | Strain/stress/ structure | Stress |
| 5481358: Coil mounting arrangement for fiber optic gyroscope using a gel loaded with particles | 2/1/96 | USA | Gyro | Coil mounting |
| 5481922: Elastic transducer designs incorporating finite length measurement paths | 9/1/96 | USA | Displacement | Interferometer |
| 5483546: Sensor system for remote spectroscopy | 9/1/96 | USA | General | Spectroscopy |
| 5483607: Two-mode remote fiber optic sensor | 9/1/96 | USA | General | Two-mode fiber |
| 5483958: Fluorescent-tipped dosimeter probe | 16/1/96 | USA | Miscellaneous | Irradiation dosage |
| 5485007: Optical fiber quadrant detector having a sensor head with intermixed receiving and transmitting optical fibers | 16/1/96 | USA | Miscellaneous | Quadrant |
| 5485269: System for minimizing native fiber fluorescence variations in fiber optic sensors | 16/1/96 | USA | Chemical | Fluorescence |

**Table 10.9** *(Continued)*

| Patent number: title | Issued (Day/month/year) | Country | Field 1 | Field 2 |
|---|---|---|---|---|
| 5485741: Vacuum calibration method for an optical fiber pressure transducer | 23/1/96 | USA | Pressure | |
| 5486921: Optimum coupler configuration for fiber optic rate gyroscope using [3 × 3] coupler | 23/1/96 | USA | Gyro | |
| 5486922: Sensor coil with thermomechanically matched spool for fiber optic gyroscope | 23/1/96 | USA | Gyro | |
| 5488224: System for characterizing flow pattern, pressure and movement of a fluid | 30/1/96 | USA | Pressure | Microbend |
| 5488475: Active fiber cavity strain sensor with temperature independence | 30/1/96 | USA | Strain/stress/ structure | Strain |
| 5489536: Detection of chlorinated aromatic compounds | 6/2/96 | USA | Chemical | Fluorescence |
| 5490490: On-board gas composition sensor for internal combustion engine exhaust gases | 13/2/96 | USA | Gas | Fluorescence |
| 5491335: Fiber optic transducer with fluid coupling means between optical fiber and transduction | 13/2/96 | USA | EM | Electro-magnetic field |
| 5491470: Vehicle security apparatus and method | 13/2/96 | USA | Miscellaneous | Intrusion/disturbance |
| 5492552: Holder for annealing fiber optic coils | 20/2/96 | USA | EM | Electric current |
| 5492674: Evanescent wave immunoassay system | 20/2/96 | USA | Biosensor | Evanescent wave immunoassay |
| 5493113: Highly sensitive optical fiber cavity coating removal detection | 20/2/96 | USA | Strain/stress/ structure | Corrosion detection |
| 5493390: Integrated optical instrumentation for the diagnostics of parts by embedded or surface attached optical sensors | 20/2/96 | Italy | Strain/stress/ structure | Smart structure |
| 5493623: PZT fiber optic modulator having a robust mounting and method of making same | 20/2/96 | USA | General | Fiber modulator |
| 5493629: Liquid core heat exchangers for fiber optic sensing and method using same | 20/2/96 | USA | Temperature | Liquid core fiber |
| 5494798: Fiber optic evanescent wave sensor for immunoassay | 27/2/96 | USA | Biosensor | Evanescent wave immunoassay |
| 5495850: Method for the measurement of oxygen concentration levels by the steady-state determination of fluorescence lifetime | 5/3/96 | USA | Gas | Fluorescence |
| 5496700: Optical immunoassay for microbial analytes using non-specific dyes | 5/3/96 | USA | Biosensor | Evanescent wave immunoassay |

| | | | | |
|---|---|---|---|---|
| 5496979: Accelerometer with optical switch | 5/3/96 | USA | Miscellaneous | Acceleration |
| 5496997: Sensor incorporating an optical fiber and a solid porous inorganic microsphere | 5/3/96 | USA | General | Fluorescence |
| 5499313: Distributed and spatially averaged fiber optic temperature sensors and method using same | 12/3/96 | USA | Temperature | Fluorescence (luminescence) |
| 5500730: Method and apparatus for determining the distance to a reflective event | 19/3/96 | USA | Miscellaneous | Reflective fault detection |
| 5500733: Interferometric system for the detection and location of reflecting faults of light-guiding structures | 19/3/96 | France | Miscellaneous | Reflective fault detection |
| 5500909: Sensor head for a fiber-optic current measuring device | 19/3/96 | Switzerland | EM | Electric current |
| 5501226: Short coherence length, Doppler velocimetry system | 26/3/96 | USA | Miscellaneous | Doppler velocimetry |
| 5502301: Fiber optic sensors and control systems for automobile occupant protection apparatus | 26/3/96 | USA | Miscellaneous | Force/load |
| 5502781: Integrated optical devices utilizing magnetostrictively, electrostrictively or photostrictively induced stress | 26/3/96 | USA | General | Fiber modulator |
| 5503559: Fiber-optic endodontic apparatus and method | 2/4/96 | USA | Biosensor | Dental anatomical structure |
| 5504580: Tuned integrated optic modulator on a fiber optic gyroscope | 2/4/96 | USA | Gyro | |
| 5504617: Optical time domain reflectometry | 2/4/96 | UK | General | OTDR |
| 5504720: Fiber optic planar hydrophone | 2/4/96 | USA | Miscellaneous | Hydrophone |
| 5506923: Apparatus and method for trimming of fiber optic winding | 9/4/96 | USA | Gyro | |
| 5509039: Automatic pellet stack length recording switch | 16/4/96 | USA | Miscellaneous | Position |
| 5510895: Probe for monitoring a fluid medium | 23/4/96 | USA | Miscellaneous | Fluid monitoring |
| 5513002: Optical corona monitoring system | 30/4/96 | USA | Miscellaneous | Corona |
| 5513913: Active multipoint fiber laser sensor | 7/5/96 | USA | General | Fiber laser sensor |
| 5515461: Polarization-independent optical wavelength filter for channel dropping applications | 7/5/96 | USA | Auxiliary | Fiber WDM |
| 5517022: Apparatus for measuring an ambient isotropic parameter applied to a highly birefringent sensing fiber using interference pattern detection | 14/5/96 | Canada | General | Special fiber |
| 5517303: Coherence selective sensor system | 14/5/96 | USA | General | Interferometer |
| 5517998: Closed loop pressure determination system and method for fiber optic pressure transducer system | 21/5/96 | USA | Pressure | Microbend |
| 5519211: Method and apparatus for monitoring resin crystallization and shrinkage during polymer molding | 21/5/96 | USA | Miscellaneous | Polymer molding monitoring |

**Table 10.9** (*Continued*)

| Patent number: title | Issued (Day/month/year) | Country | Field 1 | Field 2 |
|---|---|---|---|---|
| 5525466: Multiple output referencing system for evanescent wave sensor | 11/6/96 | USA | Chemical | Evanescent wave assay |
| 5525796: Fiber optic sensing apparatus for detecting a fracture in a metallic workpiece and an associated method of attaching a fiber optic sensing element to the metallic workpiece | 11/6/96 | USA | Strain/stress/structure | Structure fracture |
| 5525800: Selective multi-chemical fiber optic sensor | 11/6/96 | USA | Chemical | Evanescent wave spectroscopy |
| 5526112: Probe for monitoring a fluid medium | 11/6/96 | USA | Miscellaneous | Fluid monitoring |
| 5526115: Light source-detection fiber-optic gyroscope using depolarized beam propagation | 11/6/96 | Japan | Gyro | |
| 5527349: Photochemotherapy dosimeter | 18/6/96 | USA | Miscellaneous | Irradiation dosage |
| 5528367: In-line fiber etalon strain sensor | 18/6/96 | USA | Strain/stress/structure | Strain |
| 5528409: Fiber-optic interface system | 18/6/96 | USA | Auxiliary | Sensor interface |
| 5528715: Trimming of fiber optic winding and method of achieving same | 18/6/96 | USA | Gyro | |
| 5529267: Railway structure hazard predictor | 25/6/96 | USA | Strain/stress/structure | Structure hazard |
| 5530430: Vibration responsive barbed tape security system | 25/6/96 | USA | Miscellaneous | Intrusion/disturbance |
| 5530545: Method for reducing random walk in fiber optic gyroscopes | 25/6/96 | USA | Gyro | |
| 5530779: Method and apparatus for the immobilization of vitreous supports at the ends of plastic or glass optical fibers for the construction of fiber-optic chemical sensor | 25/6/96 | Italy | Chemical | Chromophone |
| 5531343: Cylindrical fiber probe devices and methods of making them | 2/7/96 | USA | Miscellaneous | Microscope |
| 5536783: Fluorescent polymers useful in conjunction with optical PH sensors | 16/7/96 | USA | Chemical | pH |
| 5537671: Technique of reducing the Kerr effect and extending the dynamic range in a Brillouin fiber optic gyroscope | 16/7/96 | USA | Gyro | |
| 5541405: Method and device for continuous pattern sensing using fiber optics | 30/7/96 | USA | Miscellaneous | Position/patterns |

| | | | | |
|---|---|---|---|---|
| 5545892: Gyro sensor coil with low-friction hub interface | 13/8/96 | USA | Gyro | |
| 5546180: Apparatus for inserting light signals into, or receiving light signals from, one of a series of optical paths and usable as optical time domain reflectometry apparatus | 13/8/96 | Canada | General | OTDR |
| 5546482: Potted fiber optic gyro sensor coil for stringent vibration and thermal environments | 13/8/96 | USA | Gyro | |
| 5547545: Digester temperature distribution control system | 20/8/96 | Japan | Temperature | Distributed |
| 5551484: Pipe liner and monitoring system | 3/9/96 | USA | Strain/stress/ structure | Stress |
| 5552880: Optical radiation probe | 3/9/96 | USA | Miscellaneous | Radiation |
| 5552887: Fiber optic rotation sensor or gyroscope with improved sensing coil | 3/9/96 | USA | Gyro | |
| 5553173: Faraday-effect sensing coil with stable birefringence and method of making same | 3/9/96 | USA | EM | Electric current |
| 5553613: Non-invasive blood analyte sensor | 10/9/96 | USA | Biosensor | Blood analyte |
| 5555086: Coherence multiplexed interferometric signal processing system and method | 10/9/96 | USA | General | Interferometer |
| 5555235: Optical fiber gyro with memory storing data measured under application of angular velocity | 10/9/96 | Japan | Gyro | |
| 5556204: Method and apparatus for detecting the temperature of a sample | 17/9/96 | Japan | Temperature | Fluorescence |
| 5557099: Signal device for turbo-engines using an optical probe with an elliptical shaped light beam | 17/9/96 | Germany | Miscellaneous | Vibration/position |
| 5557100: Flatness-measuring roller for a continuously manufactured strip by analyzing the birefringence caused by the application of lateral force to an optical fiber wrapped around the roller | 17/9/96 | France | Miscellaneous | Flatness |
| 5557406: Signal conditioning unit for fiber optic sensors | 17/9/96 | USA | General | Interferometer |
| 5557415: Apparatus and method for inducing and detecting fluorescence | 17/9/96 | Denmark | Chemical | Fluorescence |
| 5559908: Method for damping optical fiber on a bias optical phase modulator | 24/9/96 | USA | Auxiliary | Phase modulator damping |
| 5560712: Optical systems for sensing temperature and thermal infrared radiation | 1/10/96 | USA | Temperature | Pyrometry |
| 5561522: Integrated birefringent-biased pressure and temperature sensor system | 1/10/96 | USA | Temperature/ pressure | Birefringence |

**Table 10.9** *(Continued)*

| Patent number: title | Issued (Day/month/year) | Country | Field 1 | Field 2 |
|---|---|---|---|---|
| 5563705: Optical power balancing in interferometric fiber optic gyroscopes | 8/10/96 | USA | Gyro | |
| 5563967: Fiber optic sensor having a multicore optical fiber and an associated sensing method | 8/10/96 | USA | General | Special fiber |
| 5564832: Birefringent active fiber laser sensor | 15/10/96 | USA | General | In-fiber grating and fiber laser sensor |
| 5565976: Method and apparatus for detecting and compensating for a kink in an optic fiber | 15/10/96 | USA | Miscellaneous | Fiber kink detection |
| 5567622: Sensor for detection of nitrogen dioxide and nitrogen tetroxide | 22/10/96 | USA | Gas | $NO_2$, $NO_3$ |
| 5567624: Carbazine dyes and derivatives for pH measurement | 22/10/96 | USA | Chemical | pH |
| 5567932: Geomembrane barriers using integral fiber optics to monitor barrier integrity | 22/10/96 | USA | Strain/stress/structure | Structure integrity |
| 5567933: Optical fiber detection system with disturbance and positive cut-loop detection capabilities | 22/10/96 | USA | Miscellaneous | Intrusion/disturbance |
| 5569911: Fiber optic system for remote fluorescent sensing using excitation and return fibers | 29/10/96 | USA | Chemical | Fluorescence |
| 5569923: Fiber optic reflectance probe | 29/10/96 | Israel | Chemical | Spectroscopy |
| 5572609: Optical fiber vibration modal filter for flexible structures produced by the photorefractive effect | 5/11/96 | USA | Miscellaneous | Vibration |
| 5576534: Error reduction by quasi non-multiplexed signal processing in a multiplexed fiber optic rotation sensor loop | 19/11/96 | USA | Gyro | |
| 5577137: Optical chemical sensor and method using same employing a multiplicity of fluorophores contained in the free volume of a polymeric optical waveguide or in pores of a ceramic waveguide | 19/11/96 | USA | Chemical | Fluorescence |
| 5580527: Polymeric luminophores for sensing of oxygen | 3/12/96 | USA | Gas | Luminescence |
| 5581641: Optical fiber multiport apparatus for detecting phase shifts | 3/12/96 | UK | General | Coherent detection |
| 5582170: Fiber optic sensor for in vivo measurement of nitric oxide | 10/12/96 | USA | Chemical | Nitric oxide |
| 5583051: Use of a fiber optic probe for organic species determination | 10/12/96 | USA | Chemical | Fluorescence |

| | | | | |
|---|---|---|---|---|
| 5585634: Attenuated total reflectance sensing | 17/12/96 | USA | Chemical | Spectroscopy |
| 5585914: Apparatus and method for measuring a temperature of a high temperature liquid contained in a furnace | 17/12/96 | Japan | Temperature | Pyrometry |
| 5585915: Light detecting device | 17/12/96 | Japan | Miscellaneous | Corona/spark detection |
| 5587791: Optical interferometric current sensor and method using a single mode birefringent waveguide and a pseudo-depolarizer for measuring electrical current | 24/12/96 | Canada | EM | Electrical current |
| 5588084: Scintillating optical fibers containing non-uniformly distributed dopants and process for preparing same | 24/12/96 | USA | Miscellaneous | Radiation |
| 5588558: Liquid beverage dispenser | 31/12/96 | USA | Liquid level | TIR |
| 5589937: Fiber optic self-multiplexing amplified ring transducer and force transfer sensor with pressure compensation | 31/12/96 | USA | General | Fiber amplifier sensor |
| 5590457: Terminal insertion guiding apparatus | 7/1/97 | Japan | Miscellaneous | Position |
| 5591965: Multiparameter sensor system using a multiple grating fiber optic birefringent fiber | 7/1/97 | USA | General | In-fiber grating |
| 5592149: Security fence | 7/1/97 | Israel | Miscellaneous | Intrusion/disturbance |
| 5594239: Measuring system for monitoring buildings, terrain sections or the like | 14/1/97 | Germany | Displacement | Microbend |
| 5594253: Hybrid luminescent device for imaging of ionizing and penetrating radiation | 14/1/97 | USA | Miscellaneous | Radiation |
| 5594819: Field-mountable fiber optic sensors for long term strain monitoring in hostile environments | 14/1/97 | USA | Strain/stress/ structure | Strain |
| 5596988: Multi-parameter sensor apparatus | 28/1/97 | UK | Chemical | pH, $pCO_2$ |
| 5598489: Depolarized fiber optic rotation sensor with low Faraday effect drift | 28/1/97 | USA | Gyro | |
| 5600070: Fiber optic combustion pressure sensor with improved long-term reliability | 4/2/97 | USA | Pressure | Diaphragm + fotonic |
| 5600125: Compensation and status monitoring devices for fiber optic intensity-modulated sensors | 4/2/97 | USA | Auxiliary | Signal referencing |
| 5600133: Structural analyzer using acousto-optic sensor | 4/2/97 | USA | Miscellaneous | Acoustic sensing |
| 5600147: Temperature measuring system having improved signal processing and multiple optical sensors | 4/2/97 | USA | Temperature | Fluorescence |
| 5600433: Optical fiber waist refractometer | 4/2/97 | USA | Miscellaneous | Refractive index |
| 5601087: System for diagnosing tissue with guidewire | 11/2/97 | USA | Biosensor | Tissue diagnosis |
| 5601611: Optical blood flow measurement apparatus and method and implantable defibrillator incorporating same | 11/2/97 | USA | Biosensor | Blood flow |
| 5602387: Method of protecting an RF receiver in a hostile electromagnetic environment | 11/2/97 | USA | EM | RF field |

**Table 10.9** (*Continued*)

| Patent number: title | Issued (Day/month/year) | Country | Field 1 | Field 2 |
|---|---|---|---|---|
| 5602642: Magnetically insensitive fiber optic rotation sensor | 11/2/97 | USA | Gyro | |
| 5602946: Fiber optic sensor system for detecting movement or position of a rotating wheel bearing | 11/2/97 | USA | Displacement | Magneto-optical material |
| 5604318: Optical pressure detector | 18/2/97 | Germany | Pressure | Bending |
| 5605152: Optical glucose sensor | 25/2/97 | USA | Chemical | Fluorescence |
| 5606170: Multifunctional sensor system | 25/2/97 | USA | Auxiliary | Intrinsic sensing |
| 5606633: Chemical detector employing surface plasmon resonance excited using an optical waveguide configured as an asymmetric waveguide coupler | 25/2/97 | USA | Chemical | Surface plasmon resonance |
| 5608524: Coherence collapsed multimode source for a fiber optic gyro | 4/3/97 | USA | Gyro | |
| 5608525: Phase adjustable optical fiber gyro using both even-order harmonics and an orthogonal component of an even-order harmonic | 4/3/97 | Japan | Gyro | |
| 5610393: Diode laser interrogated fiber optic reversible hydrazine-fuel sensor system and method | 11/3/97 | USA | Chemical | Colorimetric |
| 5611338: Multi-purpose multi-parameter cardiac catheter | 18/3/97 | USA | Biosensor | Blood oxygenation |
| 5612778: Fiber optic sensor for multiple variables | 18/3/97 | USA | Temperature/pressure | Fotonic |
| 5617022: Fiberoptic velocity transducer including dielectric coating for filtering and velocity determination | 1/4/97 | USA | Miscellaneous | Velocity |
| 5618108: Temperature distribution measuring apparatus using an optical fiber | 8/4/97 | Japan | Temperature | Distributed |
| 5621522: Fiber optic probe for determination of trace levels of organic pollutants using Raman spectroscopy | 15/4/97 | USA | Chemical | Raman spectroscopy |
| 5625196: Method and apparatus for monitoring/calibrating a process measuring system | 29/4/97 | USA | Miscellaneous | Sheet material monitoring |
| 5625459: Diffuse reflectance probe | 29/4/97 | USA | Miscellaneous | Diffuse reflectance |
| 5625605: Optic bundle towed array | 29/4/97 | USA | Miscellaneous | Hydrophone |
| 5625724: Fiber optic hydrophone having rigid mandrel | 29/4/97 | USA | Miscellaneous | Hydrophone |
| 5627637: Fully distributed optical fiber strain sensor | 6/5/97 | USA | Strain/stress/structure | Distributed strain |

| | | | | |
|---|---|---|---|---|
| 5627644: Kerr effect compensation by optical power balancing in interferometric fiber optic gyroscopes | 6/5/97 | USA | Gyro | |
| 5627921: Optical fiber for sensors including holes in cladding | 6/5/97 | Sweden | General | Special fiber |
| 5627922: Micro-optical fiber light source and sensor and method of fabrication thereof | 6/5/97 | USA | Auxiliary | Sensor light source |
| 5627927: Fiber with multiple overlapping gratings | 6/5/97 | USA | General | In-fiber grating |
| 5627934: Concentric core optical fiber with multiple-mode signal transmission | 6/5/97 | USA | General | Special fiber |
| 5628311: Chemical sensor with variable volume sensor cell and method | 13/5/97 | USA | Chemical | Probe design |
| 5629774: Optical sensor | 13/5/97 | UK | Chemical | Surface plasmon |
| 5630987: Method and apparatus for the measurement of pollutants in liquids | 20/5/97 | UK | Chemical | Absorption spectromer |
| 5631170: Method for improving measurement precision in evanescent wave optical biosensor assays | 20/5/97 | Netherlands | Biosensor | Ligands |
| 5633494: Fiber optic bending and positioning sensor with selected curved light emission surfaces | 27/5/97 | Canada | Displacement | Bending |
| 5633748: Fiber optic Bragg grating demodulator and sensor incorporating same | 27/5/97 | USA | General | In-fiber grating |
| 5633960: Spatially averaging fiber optic accelerometer sensors | 27/5/97 | USA | Miscellaneous | Acceleration |
| 5633975: Security system responsive to optical fiber having Bragg grating | 27/5/97 | USA | Miscellaneous | Security key |
| 5634720: Multi-purpose multi-parameter cardiac catheter | 3/6/97 | USA | Biosensor | Blood oxygenation |
| 5636021: Sagnac/Michelson distributed sensing systems | 3/6/97 | USA | General | Interferometer/fiber grating |
| 5636022: Closed loop unmodulated fiber optic rate gyroscope with 3.times.3 coupler and method | 3/6/97 | USA | Gyro | |
| 5637797: Optical fiber gravity meter | 10/6/97 | USA | Miscellaneous | Gravity |
| 5637865: Fiber optic self-multiplexing amplified ring transducer and force transfer sensor with pressure compensation | 10/6/97 | USA | Miscellaneous | Acoustic perturbation |
| 5638165: Crack detection system | 10/6/97 | UK | Strain/stress/structure | Structure crack |
| 5639162: Temperature distribution measuring apparatus using an optical fiber | 17/6/97 | Japan | Temperature | Distributed |
| 5639668: Optical apparatus for performing an immunoassay | 17/6/97 | USA | Biosensor | Fluorescence immunoassay |
| 5639968: Optical fiber strain-to-failure sensor | 17/6/97 | USA | Strain/stress/structure | Strain-to-failure |

**Table 10.9** (*Continued*)

| Patent number: title | Issued (Day/month/year) | Country | Field 1 | Field 2 |
|---|---|---|---|---|
| 5640017: Remote radiation detection device with inorganic scintillating detecting crystal and fiber optic | 17/6/97 | France | Miscellaneous | Radiation |
| 5640234: Optical sensor for detection of chemical species | 17/6/97 | Germany | Chemical | Thin film |
| 5640240: Probe for surface measurement | 17/6/97 | UK | Miscellaneous | Surface measurement |
| 5640246: Apparatus for measuring reflected light utilizing spherically arranged optical fibers | 17/6/97 | USA | Miscellaneous | |
| 5640470: Fiber-optic detectors with terpolymeric analyte-permeable matrix coating | 17/6/97 | USA | Biosensor | Blood analyte |
| 5640472: Fiber optic sensor for magnetic bearings | 17/6/97 | USA | Displacement | Fotonic |
| 5641955: Reconfigurable birefringent fiber-optic sensor with shape-memory alloy elements | 24/6/97 | USA | Temperature/strain | Birefringence |
| 5641956: Optical waveguide sensor arrangement having guided modes–non-guided modes grating coupler | 24/6/97 | USA | General | In-fiber grating |
| 5646401: Fiber optic grating and etalon sensor systems | 8/7/97 | USA | Temperature/strain | Fiber grating |
| 5647030: Fiber optic sensor and methods and apparatus relating thereto | 8/7/97 | USA | Chemical | Surface plasmon resonance |
| 5647038: Narrow bandwidth Bragg grating reflector for use in an optical waveguide | 8/7/97 | USA | General | Fiber grating/amplifier |
| 5647359: Measuring apparatus | 15/7/97 | Japan | Biosensor | Calorimetric analysis |
| 5649035: Fiber optic strain gauge patch | 15/7/97 | USA | Strain/stress/ structure | Strain |
| 5650331: Optical high acidity sensor | 22/7/97 | USA | Chemical | Acidity |
| 5650846: Microcolumnar analytical system with optical fiber sensor | 22/7/97 | USA | Chemical | Fluorescence |
| 5650849: Optical rate sensor having modulated clockwise and counterclockwise beams | 22/7/97 | UK | Gyro | |
| 5650856: Fiber laser intra-cavity spectroscope | 22/7/97 | USA | Chemical | Fiberlaser spectroscope |
| 5651047: Maneuverable and locateable catheters | 22/7/97 | USA | Miscellaneous | Radiation |
| 5652810: Fiber optic sensor for site monitoring | 29/7/97 | USA | Chemical | Fluorescence |
| 5652810: Fiber optic sensor for site monitoring | 29/7/97 | USA | Chemical | Fluorescence |
| 5653537: Non-contacting infrared temperature thermometer detector apparatus | 5/8/97 | USA | Temperature | Infrared radiation |

| | | | | |
|---|---|---|---|---|
| 5654539: Laser Doppler optical sensor for use on a monitoring probe | 5/8/97 | USA | Miscellaneous | Flow |
| 5655035: Differential fiber optic angular rate sensor for minimizing common mode noise in the sensor output | 5/8/97 | USA | Gyro | |
| 5656241: Method for manufacturing fiber optic sensors | 12/8/97 | USA | Chemical | pH, $O_2$, $CO_2$ in fluid |
| 5656934: Optical method of measuring an alternating electrical current, including temperature compensation, and a device for carrying out the method | 12/8/97 | Japan | EM | Electric current |
| 5657405: Optical fiber sensor for measuring pressure or displacement | 12/8/97 | Japan | Pressure/ displacement | Diaphragm + fotonic |
| 5657411: Negative trimming of fiber optic winding | 12/8/97 | USA | Gyro | |
| 5661246: Fiber optic displacement sensor for high temperature environment | 26/8/97 | USA | Displacement | Microbend |
| 5668324: Strain sensors having ultra-high dynamic range | 16/9/97 | USA | Strain/stress/ structure | Strain |
| 5668908: Fiber optic gyro sensor coil with improved temperature stability | 16/9/97 | USA | Gyro | |
| 5671191: Variable coupler fiber optic sensor hydrophone | 23/9/97 | USA | Miscellaneous | Hydrophone |
| 5672515: Simultaneous dual excitation/single emission fluorescent sensing method for pH and $pCO_2$ | 30/9/97 | USA | Chemical | pH, $pCO_2$ |
| 5674751: Diode laser interrogated fiber optic hydrazine-fuel sensor system and method | 7/10/97 | USA | Chemical | Colorimetric |

## REFERENCES

1. ERA Technology Ltd., Report to the Optical Sensors Collaborative Association, Chislehurst (1988).
2. Crossley, Report to the Optical Sensors Collaborative Association, Chislehurst (1994).
3. News Column, *Opto & Laser Europe* 42, p. 5 (1997).
4. Available from Frost and Sullivan, 2525 Charieston Road, Mountain View, CA 94043, USA (1997).
5. Business Communications Co., Inc., 25 Van Zant Street, Norwalk, CT 06855, USA (1996).

# Index

Page numbers in *italics* refer to figures and tables

MIX
Papier aus verantwortungsvollen Quellen
Paper from responsible sources
FSC® C105338

If you have any concerns about our products,
you can contact us on
**ProductSafety@springernature.com**

In case Publisher is established outside the EU,
the EU authorized representative is:
**Springer Nature Customer Service Center GmbH**
**Europaplatz 3, 69115 Heidelberg, Germany**

Printed by Libri Plureos GmbH
in Hamburg, Germany